Building Enterprise IoT Applications

Chandrasekar Vuppalapati

Senior Vice President–Products & Programs
Hanumayamma Innovations and Technologies, Inc.
Fremont, California, USA

and

College of Engineering
San Jose State University
San Jose, California, USA

CRC Press
Taylor & Francis Group
Boca Raton   London   New York

CRC Press is an imprint of the
Taylor & Francis Group, an **informa** business

A SCIENCE PUBLISHERS BOOK

CRC Press
Taylor & Francis Group
6000 Broken Sound Parkway NW, Suite 300
Boca Raton, FL 33487-2742

© 2020 by Taylor & Francis Group, LLC
CRC Press is an imprint of Taylor & Francis Group, an Informa business

No claim to original U.S. Government works

Printed and bound by CPI Group (UK) Ltd, Croydon, CR0 4YY

Version Date: 20190923

International Standard Book Number-13: 978-0-367-17385-2 (Hardback)

Library of Congress Cataloging-in-Publication Data

Names: Vuppalapati, Chandrasekar, 1972- author.
Title: Building enterprise IoT applications / Chandrasekar Vuppalapati,
    Senior Vice President, Products & Programs, Hanumayamma Innovations and
    Technologies, Inc., Fremont, California, USA and College of Engineering,
    San Jose State University, San Jose, California, USA.
Description: First edition. | Boca Raton : CRC Press/Taylor & Francis
    Group, [2020] | Includes bibliographical references and index. |
    Summary: "The book deals with the complete spectrum IoT software
    development, from startup, industry and academic perspectives, with the
    objective of teaching readers to develop state of art IoT applications.
    The cases presented in the book will encourage readers to create
    innovative blue ocean software products with the goal to serve the
    humanity and create value for future generations"-- Provided by
    publisher.
Identifiers: LCCN 2019040297 | ISBN 9780367173852 (hardcover ; acid-free
    paper)
Subjects: LCSH: Internet of things. | Business enterprises--Computer
    networks. | Application software--Development.
Classification: LCC TK5105.8857 .V87 2020 | DDC 005.3--dc23
LC record available at https://lccn.loc.gov/2019040297

Visit the Taylor & Francis Web site at
http://www.taylorandfrancis.com

and the CRC Press Web site at
http://www.crcpress.com

Preface

The Internet of Things (IoT) is a game changer, transforming many enterprises into digital businesses and facilitating new business models, blue-ocean product strategies, Data Monetization opportunities, just in time (JIT) product service offerings, service contracts with proactive product maintenance, Enablement of citizenry services, Smart meters, Smart Cities, Smart Buildings, Smart Manufacturing and many other next-generation products & services. IoT architectures and deployments can be found in various industries including: Chemical, Healthcare, Industrial, Pharmaceutical, Construction, Agriculture, Smart Buildings, Smart cities, E-Governance, Connected Stadiums, Retail, Contact Center, Gaming, Chip Fabrication, Transportation, IT Services (Managed) and many more. It is clear from market adoption that the Internet of Things (IoT) is the platform of the future.

IoT based architectures are projected to inter-connect 20 billion new devices with a potential of $11.1 trillion per year in economic value by 2025. At the time of writing this book, IoT has already connected 39% of the world's population and has generated more data in 1 year than in the previous 5000 years put together.

In order to harvest the benefits of the IoT revolution, traditional software development paradigms must be upgraded. The purpose of this book is to prepare current and future software engineering teams with the skills and tools that fully utilize IoT capabilities. The book introduces essential IoT concepts from the perspectives of full-scale software development with the emphasis on creating niche blue ocean products. It also:

- Outlines a fundamental full stack architecture for IoT
- Describes various development technologies in each IoT layer
- Explains deployment of Machine Learning Models at the Edge Level
- Describes Machine Learning Models and Artificial Intelligence
- Presents IoT Use cases from Agriculture, Healthcare, Mobile industry verticals
- Explains IoT solution development from Product management perspective
- Extensively covers security and applicable threat models as part of IoT stack

The book provides details of several IoT reference architectures with emphasis on data integration, edge analytics, cluster architectures and closed loop responses.

Dedication

The author dedicates his works, book, and efforts of this book to his late father Murahari Rao Vuppalapati, who had inspired him to write the book and guided the author's pursuits of life, and to his loving mother Hanumayamma Vuppalapati, who was the inspiration to develop company and dairy products for the betterment of humanity and future generations.

Acknowledgement

The author is deeply indebted to the love, support, and encouragement of his wife Anitha Ilapakurti, and the active support of his daughters, Sriya Vuppalapati and Shruti Vuppalapati who helped draft the first version of the manuscript.

The author is deeply thankful to the support of his elder brother, Rajasekar Vuppalapati, and the joyful support of his younger brother, Jaya Shankar Vuppalapati. Additionally, the author is also very thankful to his elder sisters Padmavati Vuppalapati and Sridevi Vuppalapati.

The author sincerely thanks Santosh Kedari and his team in India for conducting field studies in India for Dairy and Sanjeevani Healthcare analytics products and bringing the market feedback for the betterment of services.

The author is deeply indebted to Sharat Kedari, who helped to draft the book structure, code examples of the book and helped in setting Cloud and machine learning algorithms. In addition, the author is thankful for the support, hard work and dedication of Vanaja Satish who helped to solve embedded systems and build of iDispenser and SQLite.

The author sincerely thanks the support of fellow researchers and engineers including Sudha Rama Chandran, Shruti Agrawal, MeenuYadav, his students Renato Silveira Cordeiro, Sadab Qureshi, and Suchishree Jena and his graphic illustrator Rohit Ranjan.

Contents

## CHAPTER 1

Introduction

This Chapter Covers:
- The Internet of Things Addressable Market (TAM) potential
- Augmented Analytics
- Industry 4.0
- IoT & New Technology Stack
- Porter Five Forces that Shape Competition & IoT Influence
- The Cross-Industry Standard Process for Data Mining (CRISP-DM)

The Internet of Things (IoT) has become the economic and social disruption of the century. In its report on the most disruptive technologies—advances that will transform life, business, and the global economy, the McKinsey global institute has listed Internet of Things as the third major force that will transform our lives and global economy and commerce (please see Figure 1). Due to the diffusion of IoT, there has been 300% increase in connected machine-to-machine (M2M) devices in the past 5 years. The IoT has resulted in 80%–90% decline in MEMS (Microelectromechanical systems) Sensors costs in the past 5 years. The IoT could connect more than 1 trillion things to the Internet across industries such as manufacturing, healthcare, and mining. Finally, IoT could generate cost synergies of magnitude $36 Trillion costs of key connected industries (manufacturing, health care, and mining) [1].

Speed, scope, and economic value at stake of 12 potentially economically disruptive technologies[1]

The Internet of Things—embedding sensors and actuators in machines and other physical objects to bring them into the connected world—is spreading rapidly. From monitoring the flow of products through a factory, to measuring the moisture in a field of crops and tracking the flow of water through utility pipes, the Internet of Things allows businesses and public-sector organizations to manage assets, optimize performance, and create new business models. With remote monitoring, the Internet of Things also has great potential to improve the health of patients with chronic illnesses and attack a major cause of rising health-care costs [2].

The IoT has connected 39% of the world population and is generating more data in one year than in the previous 5000 years put together. To support the promise and growth of IoT (50 billion Smart Objects

---

[1] McKinsey 12 most disruptive technologies - https://www.mckinsey.com/~/media/McKinsey/Business%20Functions/ McKinsey%20Digital/Our%20Insights/Disruptive%20technologies/MGI_Disruptive_technologies_Full_report_ May2013.ashx

		Illustrative rates of technology improvement and diffusion	Illustrative groups, products, and resources that could be impacted[1]	Illustrative pools of economic value that could be impacted[1]
	Mobile Internet	**$5 million vs. $400[2]** Price of the fastest supercomputer in 1975 vs. that of an iPhone 4 today, equal in performance (MFLOPS) **6x** Growth in sales of smartphones and tablets since launch of iPhone in 2007	**4.3 billion** People remaining to be connected to the Internet, potentially through mobile Internet **1 billion** Transaction and interaction workers, nearly 40% of global workforce	**$1.7 trillion** GDP related to the Internet **$25 trillion** Interaction and transaction worker employment costs, 70% of global employment costs
	Automation of knowledge work	**100x** Increase in computing power from IBM's Deep Blue (chess champion in 1997) to Watson (Jeopardy winner in 2011) **400+ million** Increase in number of users of intelligent digital assistants like Siri and Google Now in past 5 years	**230+ million** Knowledge workers, 9% of global workforce **1.1 billion** Smartphone users, with potential to use automated digital assistance apps	**$9+ trillion** Knowledge worker employment costs, 27% of global employment costs
	The Internet of Things	**300%** Increase in connected machine-to-machine devices over past 5 years **80–90%** Price decline in MEMS (microelectromechanical systems) sensors in past 5 years	**1 trillion** Things that could be connected to the Internet across industries such as manufacturing, health care, and mining **100 million** Global machine to machine (M2M) device connections across sectors like transportation, security, health care, and utilities	**$36 trillion** Operating costs of key affected industries (manufacturing, health care, and mining)
	Cloud technology	**18 months** Time to double server performance per dollar **3x** Monthly cost of owning a server vs. renting in the cloud	**2 billion** Global users of cloud-based email services like Gmail, Yahoo, and Hotmail **80%** North American institutions hosting or planning to host critical applications on the cloud	**$1.7 trillion** GDP related to the Internet **$3 trillion** Enterprise IT spend
	Advanced robotics	**75–85%** Lower price for Baxter[3] than a typical industrial robot **170%** Growth in sales of industrial robots, 2009–11	**320 million** Manufacturing workers, 12% of global workforce **250 million** Annual major surgeries	**$6 trillion** Manufacturing worker employment costs, 19% of global employment costs **$2–3 trillion** Cost of major surgeries

Figure 1: McKinsey Global Major Influencer [1]

[3]—adoption rate 5X faster than electricity and telephony),[3] the enterprise compute platforms need to scale to the needs of data.

Case Study: IoT Devices play pivotal role in driving Augmented Analytics[4]

Augmented Intelligence [4] focuses on Artificial Intelligence (AI)'s assistive role [5], that is, enhancing human intelligence rather than replacing it. The choice of the word augmented, which means "to improve", reinforces the role that human intelligence plays when using machine learning and deep learning algorithms to discover relationships and solve problems.[5]

Augmented analytics[6] [5] helps enterprises perform KDD in order to discover new insights. It helps, importantly, to citizen data science[7] to the enabler of data analytics [6].

Mobile is increasingly ubiquitous. With 6.8 billion mobile subscriptions worldwide, access anytime, anywhere through smart gadgets is now putting cheap and connected, mobile computing power in the hands of millions of consumers and healthcare practitioners. Mobile Sensors—accelerometers, location detection, wireless connectivity and cameras—offer another big step towards closing the feedback loop in personalized medicine [7]. There is no more personal data than on the-body or in-the-body sensors. With so many connected devices generating various vital health related data, integrating these data points to Electronic Health Records not only provide more accurate picture of the patient but also helps to connect with the doctor. Mobile device being universally classic IoT device, through IoT device and application of augmented intelligence helping humanity.

---

2   Samsung - https://www.samsung.com/global/business/networks/solutions/iot-solutions/

3   IoT Platform - https://www.slideshare.net/Cisco/enabling-the-internet-of-everything-ciscos-iot-architecture?from_action=save

4   Gartner Identifies Augmented Analytics as a Top Technology Trend for 2019 - https://solutionsreview.com/business-intelligence/gartner-identifies-augmented-analytics-as-a-top-technology-trend-for-2019/

5   Augmented Intelligence - https://whatis.techtarget.com/definition/augmented-intelligence?track=NL-1823&ad=922091&src=922091&asrc=EM_NLN_103914963&utm_medium=EM&utm_source=NLN&utm_campaign=20181122_Word%20of%20the%20Day:%20augmented%20intelligence

6   Augmented Analytics: The Next Disruption in Analytics and BI - https://www.gartner.com/webinar/3846165

7   Gartner Identifies the Top 10 Strategic Technology Trends for 2019 - https://www.gartner.com/en/newsroom/press-releases/2018-10-15-gartner-identifies-the-top-10-strategic-technology-trends-for-2019

Industry 4.0

We are in the midst of a digital transformation powered and fueled by Data and Analytics. The Industry 3.0 has introduced Computer and Information Technology. However, the Industry 4.0 is due to availability of sensors, connecting via smart and high-speed networking, and application of data science and machine learning to automate business processes with intelligence. Industry 4.0[8] is here to stay [8].

The Internet of Things and Smart Connected Objects are the main catalysts for the Industry 4.0.[9]

Machine Learning and Data Science are the backbone for deriving insights from the Industry 4.0 connected objects (please see Figure 2 and Figure 3) [9].

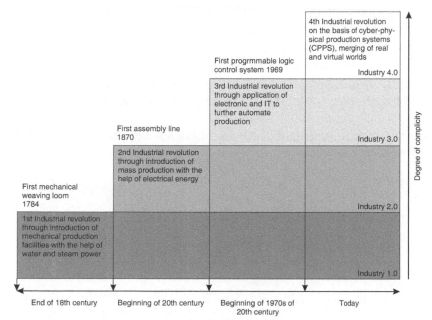

Figure 2: Industry 4.0

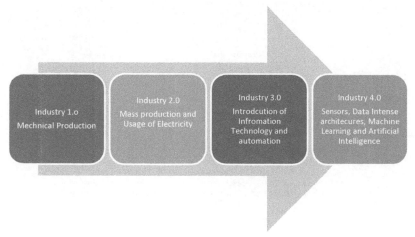

Figure 3: Industry 4.0 Changes

8 Industry   4.0   -   https://www.forbes.com/sites/bernardmarr/2018/09/02/what-is-industry-4-0-heres-a-super-easy-explanation-for-anyone/#4b13dbe39788

9 Deloitte   Industry   4.0   -   https://www2.deloitte.com/content/dam/Deloitte/ch/Documents/manufacturing/ch-en-manufacturing-industry-4-0-24102014.pdf

The New Technology Stack

The Smart connected products have a new technology stack[10] that integrates hardware, software, services, analytics, and recommendations (please see Figure 4) [10].

Product Hardware: The Product Hardware comprises of Embedded Sensors, Actuators, Processors, Intelligent Microcontrollers, Micro-Electro-Mechanical-Systems (MEMS), and Embedded Systems.

Product Software: Product Software consists of Edge-based systems that process the sensor data close to the source. This area also consists of Filters, Transformers, Data Aggregators, and Machine learning modules.

Connectivity: Connectivity relays the data from the Edge level to the Cloud. The connectivity protocols include: Wi-Fi, Cellular, Bluetooth, Zigbee, HTTPS, REST, CoAP, MQTT and other protocols.

Product Database: allows aggregation, normalization, and the management of real-time & historical data.

Rules/Analytics Engine: Rules and Analytics engines enable the application of machine learning and artificial intelligence algorithms to predict the future events.

Smart Product Recommendation Systems: The recommendation systems connect the Users with actionable insights. The item based, and collaborative filtering recommendations are applied as part of the system.

Security Systems: Manage authentication, authorization of resources. The Role Based Access (RBAC) enables guarding of the information.

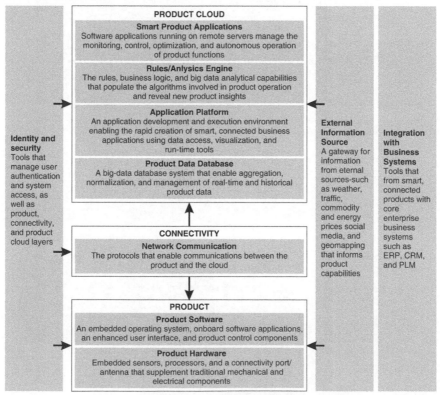

Figure 4: Product Stack [10]

---

10   How Smart, Connected Products Are transforming Competition - https://hbr.org/2014/11/how-smart-connected-products-are-transforming-competition

External Systems: Integrating with external systems, such as social media data, twitter data, and weather data enables context aware analytics and geolocation driven insights.

Business Databases: Integrating IoT Sensor data with traditional enterprise relational databases (ERP, CRM) provides 360-degree view of the customers and enables the formation of blue ocean strategies.

The Man and the Machine—Robots may guide collaboration with Humans[11]

The sensor makes it possible to track the movements and actions of workers inside a factory or warehouse [11].

Existing sensors, including cameras and depth sensors, are often affected by lighting and offer only a rough idea of a person's position in three-dimensional space. Emerging safety systems allow people to work in closer proximity to powerful robots, but they shut these systems down completely if a person moves too close. By tracking a person's motions very precisely, the new system could make it possible for a powerful robot to work in concert with a person, perhaps handing something to a human coworker [12]. "You could get rid of the cages around big robots", Mindell says.[12]

Flow Information

Smart Connected Objects enable Flow of Information for the greater good:

The Five Forces that shape Industry Competition and Smart Connected Objects

Smart, connected products will have transformative effect on Industry structure. The Five Forces that shape competition[13] provide a framework to understand the influence of Smart connected objects on the competition (please see Figure 5).

Bargaining Power of Buyer: Traditional Segmentation of the market is based on market research, voice customer details, and unmet needs & value perception. The Smart Connected Objects make the bargaining power of the buyer more exclusive and driven by digital and traditional strategies. For instance, mobile recommendations to the shopping loyalists are derived from historical shopping behaviors, Market basket Just in Time buying behavior, Location awareness, weather cognitive and population sentiment or mood.

Threat of new entrants: New entrants in a smart, connected world face significant new obstacles, starting with the high fixed costs of more-complex product design, embedded technology, and multiple layers of new IT infrastructure.[14]

It is prohibitively expensive to manufacture the products with limited or new resources. Without adequate resources, it's very difficult to establish the business.

Though the demand for the Cow Necklace is huge, the new entrants, a Hanumayamma Innovations and Technologies Inc.,[15] a Delaware based company, must face huge challenges in setting up the production and running the sale and business operations.

---

[11] Intelligent Machines - https://www.technologyreview.com/s/608863/a-radar-for-industrial-robots-may-guide-collaboration-with-humans/

[12] Humantics - https://www.technologyreview.com/s/608863/a-radar-for-industrial-robots-may-guide-collaboration-with-humans/

[13] The Five Forces that Shape Competition - https://hbr.org/2008/01/the-five-competitive-forces-that-shape-strategy

[14] https://hbr.org/2014/11/how-smart-connected-products-are-transforming-competition

[15] Hanumayamma Innovations and Technologies, Inc – http://www.hanuinnotech.com

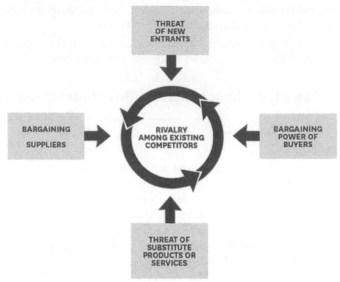

Figure 5: Porter Five Forces [14]

Threat of Substitutes: Smart, connected products can offer superior performance, customization, and customer value relative to traditional substitute products, reducing substitution threats and improving industry growth and profitability. Data provides competitive advantages and creates new monetization models.

Bargaining Power of Suppliers: Unlike traditional Market forces model, the Suppliers in the Smart Connected Objects are connected by the incentives that the data provides. For product manufacturers the Smart Connected Object is a way to get the data to develop exclusive, augmentative, network, and actionable insights to monetize the data. For suppliers, the data provide a holistic improvement of product design. For instance, small diagnostic digital equipment sensor producers[16] have incorporated free sharing of their digital IoT sensors to Farmers for capturing data from the various geographical locations and for refining and fine-tuning the analytics. For Farmers, the sensors are a new way of approaching agriculture, with more focus towards addressing the issues that are not addressable without adequate data.

Digital Twin

A digital twin is a virtual model of a process, product or service. This pairing of the virtual and physical worlds allows analysis of data and monitoring of systems to head off problems before they even occur, prevent downtime, develop new opportunities and even plan for the future by using simulations. The concept of Digital Twin[17] has existed since 2002. However, the Internet of Things (IoT) has made it possible to implement Digital Twin [16].

How does a digital twin work?

Digital Twin acts as a bridge between a Physical system and the virtual world.

First, smart components that use sensors to gather data about real-time status, working condition, or a position are integrated with a physical system. The components are connected to a cloud-based system that receives and processes all the data that the sensors monitor. This input is analyzed against business and other contextual data.

Lessons are learned, and opportunities are uncovered within the virtual environment that can be applied to the physical world—ultimately, to transform your business.

IoT Sensors, Cloud Connected Data Analytics Platforms, and real-time analytics enable the "Digital Twins".

---

[16]   Please see Dairy IoT Sensor Manufacturing Company – www.hanuinnotech.com

[17]   Digital Twin - https://www.forbes.com/sites/bernardmarr/2017/03/06/what-is-digital-twin-technology-and-why-is-it-so-important/#4a3dfca82e2a

Enterprise IoT Platforms

The IoT plays a key role in enterprise digital transformation. For instance, integration mobile and sensor data into Electronic Health Records (EHR) or integration of Wearables, Mobile Data, and Electronic Data Interchange (EDI) processing for better insurance processing. With the right level of integration, IoT acts as a crucial bridge between the physical assets, legacy systems, business processes, newer competitive models and the IT infrastructure.

IoT enables enterprises to create a better connected physical world[18] by enabling (a) flexible deployment of architectures, (b) multiple connections, (c) Intelligent management, (d) data security, and (e) an Open Ecosystem [16]. To successfully monetize the capabilities, for any sizeable company, a state-of-the-art data and analytics platform is necessary.[19] The Enterprise grade of the IoT platforms is weighed on Data Platform and requires following Key Performance Indicator (please see Figure 6) in order to fulfill customer and business needs. What this entails is the development of IoT data analytics platform that include: Three major value drivers to deliver value for customers and businesses [17].

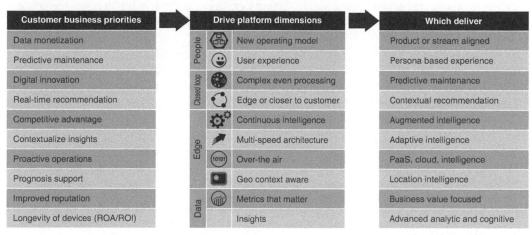

Figure 6: KPIs

1. **Customer business priorities:** includes integration of IoT data sources with enterprise data assets to deliver digital innovation and predictive maintenance to provide contextual intelligence of the devices and to provide adaptive prognosis to transpose from reactive to proactive operations for device operational longevity and data monetization through service/blue-ocean models.

2. **Platform dimensions:** include data insights and metrics that have positive impact through the implementation of adaptive edge processing with closed loop event handling and customer excellence.

3. **Delivers:** advanced analytics and cognitive intelligence that deliver predictive maintenance and proactive exclusive recommendations.

There are many public Cloud platforms that offer enterprise IoT platforms. Here is the current trend of those platforms:[20]

1. Data is the biggest driver of the IoT Platform

2. IoT Solutions that need integration of several sensors are complex.

---

18 https://e.huawei.com/us/solutions/technical/iot

19 Building a great data analytics platform - https://www.mckinsey.com/industries/electric-power-and-natural-gas/our-insights/building-a-great-data-platform

20 Current IoT Platforms - https://www.forbes.com/sites/janakirammsv/2018/09/10/the-current-state-of-public-cloud-enterprise-iot-platforms/#12f6779256b2

3. IoT sensor integration requires steep data processing, connectivity, cloud and analytics knowledge. Given the complexity, many CIOs and companies are planning to Use Cloud platforms [18].

4. Contextual real-time actionable insights require Complex Event Processing (CEP) closer to the Data Source. Edge Processing bridging the gap between heavy data intense platforms and real-time insights.

5. Lack of standards is hurting the IoT.

6. Security is playing an important role in choosing an enterprise platform.

Human Touch—artificial intelligence infused Mobile Companion

In Britain and the United States, roughly one in three people older than 65 live alone, and in the United States, half of those older than 85 live alone. Studies in both countries show that the prevalence of loneliness among people older than 60 ranges from 10 percent to 46 percent. Elderly people who are socially isolated and lonely may be at greater risk of early death. Loneliness increases the likelihood of mortality by 26 percent.[21] Loneliness is comparable to other mortality risks factors such as obesity, substance abuse, and access to healthcare.[22] Senior Loneliness is a public health issue. Importantly, Senior Loneliness is one of the risk factors for suicide. The rate of suicide[23] for those over 65 is almost 15 of every 100,000 suicides [20,21].

Mobile technologies with artificial intelligence, importantly, can detect activity and engagement levels of senior citizens. Interweaving such activity levels with nostalgic and location or time specific, historical, valuable data leads to phycological and neural simulation. The combined effects of these two aspects will deter loneliness and facilitate positive engagement ques; The overall effect is a better outcome to the societies and carrying for senior citizens.

The Sharing History through Active Reminiscence and Photo-imagery (SHARP) Program, run by the Oregon Health and Science University (OHSU), brings together small groups of older African Americans for mile-long GPS-guided walks. Historic photos of local interest, like the one above, appear on the group's digital device at selected locations. At each of these Memory Markers, the group pauses to discuss the photograph and flex their memories.

The CRISP-PM Process

The Cross-Industry Standard Process for Data Mining (CRISP-DM) is the model [22] (please see Figure 7) that is designed as a general model and can be applied to a wide variety of industries and business problems. The following diagram provides the several techniques in CRISP-DM:[24]

Business Understanding

In my experience, this is one of the most important phases in an ML process. The success of initiative and expectations to stakeholders and project outcomes are derived from clear understanding of the business requirements. This phase also helps to uncover current as-is process and the proposed machine learning objectives.

---

[21]   One is the Loneliest Number: Combating Senior Isolation - https://stonegatesl.com/one-is-the-loneliest-number-combating-senior-isolation/

[22]   Senior Loneliness and Public Health Response - https://www.publichealthpost.org/research/senior-loneliness-public-health-response/

[23]   Elderly Suicide Rate - https://www.sageminder.com/SeniorHealth/MentalHealth/Suicide.aspx

[24]   Introduction to IBM SPSS Modeler and Data Mining, ftp://ftp.software.ibm.com/software/analytics/spss/support/Modeler/Documentation/14/UserManual/CRISP-DM.pdf

Phases and Tasks

Business understanding	Data understanding	Data preparation	Modelling	Evaluation	Deployment
Determine business objective *Background* *Business objective* *Business success Criteria* **Assess situation** *Inventory of resource* *Requirements, assumption, and constraints* *Risk and contingencies* *Terminology* *Costs and benefits* **Determine data mining goals** *Data mining goals* *Data mining success criteria* **Produce project plan** *Project plan* *Initial assessment of tools and techniques*	**Collect initial data** *Initial data collection report* **Describe data** *Data description report* **Explore data** *Data exploration report* **Verify data quality** *Data quality report*	**Data set** *Data et description* **Select data** *Rationale for inclusion exclusion* **Clean data** *Data cleaning report* **Construct data** *Derived attributes* *Generated records* **Integrate data** *Merged data* **Format data** *Reformatted data*	**Select modelling Technique** *Modelling technique* *Modelling assumption* **Generate test design** *Test design* **Build model** *Parameter settings* *models* *Model description* **Assess model** *Model assessment* *Revised parameter settings*	**Evaluate results** *Assessment of data mining results w.r.t. business success criteria* *Approved model* **Review process** *Review of process* **Determine next steps** *List of possible actions* *Decision*	**Plan deployment** *Deployment plan* **Plan monitoring and maintenance** *Monitoring and maintenance plan* **Produce final report** *Final report* *Final presentation* **Review project** *Experience documentation*

Figure 7: CRISP-DM

Some of the steps that will help in this phase include:

- Establish Good understanding with Data and Machine Learning champion.
- Understand the business process—especially, accumulate as much tacit knowledge as possible from experts and champions involved.
- Make sure the Machine Learning process and outcome are directly connected to strategic drivers and the future roadmap of the organization.
- Assess the ML process is for cost synergies vs. monetization & long-term competitive advantage.
- Finally, understand the objectives are driven from Cost vs. Revenue Business lines.

Data Understanding

The preamble or pre-requisite for the phase is the business understanding phase. During the data understanding phase, involve data scientists and data engineers, as they will be creating data models and ml algorithms. Generally, this phase accounts for 30 to 40 percent of the project time.

Life Cycle - Industry Standards:[25] please see below the table for ML Life cycle standards [23]:

Data Preparation Phase

Data preparation phase involves working with several origination units in order to collect the required data for developing Machine Learning algorithms. During this phase, we need to work with biasness and product groups in order to develop data processing and transformation rules. These rules will subsequently be deployed at the Edge and at the Cloud level.

---

[25] Preparing and Architecting for Machine Learning - https://www.gartner.com/binaries/content/assets/events/keywords/catalyst/catus8/preparing_and_architecting_for_machine_learning.pdf

Table 1: Life Cycle – Industry Standards

Task (Proportion of Effort)	Subtasks	Business	Data Scientist	IT/Operations
1. Problem Understanding (5% to 10%)	a) Determine Objective	×	×	
	b) Define Success Criteria	×	×	
	c) Assess Constraints	×	×	×
2. Data Understanding (10% to 25%)	a) Assess Data Situation	×	×	×
	b) Obtain Data (Access)		×	×
	c) Explore Data	×	×	×
3. Data Preparation (20% to 40%)	a) Filter Data		×	×
	b) Clean Data		×	×
	c) Feature Engineering	×	×	
4. Modeling (20% to 30%)	a) Select Model Approach		×	
	b) Build Models		×	
5. Evaluation of Results (5% to 10%)	a) Select Model		×	
	b) Validate Model		×	
	c) Explain Model	×	×	
6. Deployment (5% to 15%)	a) Deploy Model		×	×
	b) Monitor and Maintain	×	×	×
	c) Terminate	×	×	×

Major work areas under this phase include:

- Select Data
  - Based on business objectives, identify key data attributes—these attributes, later, will be input to ML models.
  - Not all attributes are intrinsically available within the data—need to develop synthetic variables that explains the models' outcome.
- Clean Data
  - Data is gold and no data are bad data; all we need to develop is a process to cleanse the data to be utilized under modeling and understanding purposes.
- Construct Data
  - Based on the business objectives, construct the data to fully describe the ML charter.
- Integrate Data
  - Integrate data involves working with several business units within the organization or working with 3rd party providers to collect necessary data.
- Format Data
  - Format data to be pluggable usable for model development.

Data Modelling

This phase is also known as machine learning phase. In this phase, the actual mission learning algorithm is built, trained, validated and tested in order to meet the biasness objectives. During this phase, both data scientists and business Subject Matter Experts (SMEs) are closely involved and worked with Users. The successful development of algorithms is more of a collaborative prosses than siloed approaches.

This phase comprises of following major tasks:

- Select Modeling Technique
  - Supervised:
    - Classification

- – Decision Trees, Naïve Bayesians, Support Vector Machines, K-Nearest Neighbor
  - ▪ Regression
    - – Linear Regression, Non-linear Regression
  - ○ Unsupervised
    - ▪ Clustering
      - – Hierarchical (Agglomerative, divisive)

Difference between Supervised and unsupervised algorithms:

	Supervised Learning	Unsupervised Learning
Input Data	Use Known and Labeled Data	Uses unknown data
Computational Complexity	Very computational intense	Less computational
Real Time	Use off-line analysis	Use Real-time analysis
Number of Classes	Known	Unknown
Accuracy of results	Accurate and reliable	Reliable

- • Generate Test Design
  - ○ Develop algorithm (on paper or with small datasets)
- • Build Model
  - ○ Model description
  - ○ Codification of Model
- • Assess Model
  - ○ Train Model
  - ○ Validate Model
  - ○ And Test Model

Best to Divide data into Training and Validation:

Assuming you have enough data to do proper held-out test data (rather than cross-validation), the following is an instructive way to get a handle on variances:

- • Split your data into training and testing (80/20 is indeed a good starting point).
- • Split the training data into training and validation (again, 80/20 is a fair split).
- • Subsample random selections of your training data, train the classifier with this, and record the performance on the validation set.
- • Try a series of runs with different amounts of training data: Randomly sample 20% of it, say, 10 times and observe performance on the validation data, then do the same with 40%, 60%, 80%. You should see both greater performance with more data, but also lower variance across the different random samples.
- • To get a handle on variance due to the size of test data, perform the same procedure in reverse. Train on all of your training data, then randomly sample a percentage of your validation data a number of times and observe performance. You should now find that the mean performance on small samples of your validation data is roughly the same as the performance on all the validation data, but the variance is much higher with smaller numbers of test samples.

Data Holdout

If a lot of data are available, simply take two independent samples and use one for training and one for testing. The more training, the better the model. The more test data, the more accurate the error estimate.

Problem: Obtaining data is often expensive and time consuming. Example: Corpus annotated with word senses, Dairy Sensor data, experimental data on sub-cat preferences.

Solution: Obtain a limited data set and use a holdout procedure. Most straightforward: Random split into test and training set. Typically, between 1/3 and 1/10 held out for testing.

Data Stratification

Problem: The split into training and test set might be unrepresentative, e.g., a certain class is not represented in the training set, thus, the model will not learn to classify it.

Solution: Use stratified holdout, i.e., sample in such a way that each class is represented in both sets.

Example: Data set with two classes A and B. Aim: Construct a 10% test set. Take a 10% sample of all instances of class A plus a 10% sample of all instances of class B.

However, this procedure doesn't work well on small data sets.

Deployment

The successful ML algorithms are deployed for business and user purposes. The deployment phase consists of working with HPC, data engineers and IT groups.

Chapter Summary

- You should see the potential positive economic impact of IoT.
- The New Stack that is required to process IoT Data feeds.
- You should see how Industry 4.0 evolved.
- You would learn the importance of augmented analytics.
- You would see the role of analytics and mobile platform as a human companion.
- The Key Performance Indicators (KPIs) of Analytics platform.
- You should understand the phases in CRISP-DM.

References

1. James Manyika, Michael Chui, Jacques Bughin et al. Disruptive technologies: Advances that will transform life, business, and the global economy. May 2013, https://www.mckinsey.com/~/media/McKinsey/Business%20Functions/McKinsey%20Digital/Our%20Insights/Disruptive%20technologies/MGI_Disruptive_technologies_Full_report_May2013.ashx , Access date: 09/18/2018
2. Samsung Software Development. IoT Solutions. 2017, https://www.samsung.com/global/business/networks/solutions/iot-solutions/, Access date: 09/27/2018
3. Kip Compton and Vikas Butaney. Enabling the Internet of Everything's, Cisco's IoT Architecture, https://www.slideshare.net/Cisco/enabling-the-internet-of-everything-ciscos-iot-architecture?from_action=save , 02/05/2015, Access date: 09/18/2018
4. Timothy King. Gartner Identifies Augmented Analytics as a Top Technology Trend for 2019. November 9, 2018, https://solutionsreview.com/business-intelligence/gartner-identifies-augmented-analytics-as-a-top-technology-trend-for-2019/, Access Date: 12/25/2018
5. Margaret Rouse. Augmented Intelligence. July 2017, https://whatis.techtarget.com/definition/augmented-intelligence?track=NL-1823&ad=922091&src=922091&asrc=EM_NLN_103914963&utm_medium=EM&utm_source=NLN&utm_campaign=20181122_Word%20of%20the%20Day:%20augmented%20intelligence

6. Rita L. Sallam. June 2018. Augmented Analytics: The Next Disruption in Analytics and BI Register now for this free 60 minute Disruption and Innovation webinar, https://www.gartner.com/webinar/3846165, Access date: 12/20/2018

7. Jennifer Garfinkel. Gartner Identifies the Top 10 Strategic Technology Trends for 2019. October 15, 2018, https://www.gartner.com/en/newsroom/press-releases/2018-10-15-gartner-identifies-the-top-10-strategic-technology-trends-for-2019, Access Date: December 29, 2018

8. Bernard Marr. What is Industry 4.0? Here's A Super Easy Explanation For Anyone. Sep 2, 2018, 11: 59 pm, https://www.forbes.com/sites/bernardmarr/2018/09/02/what-is-industry-4-0-heres-a-super-easy-explanation-for-anyone/#75023f039788

9. Dr. Ralf C. Schlaepfer, Markus Koch, Dr. Philipp Merkofer, October 2014. Industry 4.0 - Challenges and Solutions for the digital transformation and use of exponential technologies, https://www2.deloitte.com/content/dam/Deloitte/ch/Documents/manufacturing/ch-en-manufacturing-industry-4-0-24102014.pdf, Access Date: December 10, 2018

10. Michael E. Porter,James E. Heppelmann How Smart, Connected Products Are Transforming Competition. NOVEMBER 2014 ISSUE, https://hbr.org/2014/11/how-smart-connected-products-are-transforming-competition

11. Will Knight. A Radar for Industrial Robots May Guide Collaboration with Humans. September 20, 2017, https://www.technologyreview.com/s/608863/a-radar-for-industrial-robots-may-guide-collaboration-with-humans/

12. Will Knight. A Radar for Industrial Robots May Guide Collaboration with Humans The sensor makes it possible to track the movements and actions of workers inside a factory or warehouse. September 20, 2017, https://www.technologyreview.com/s/608863/a-radar-for-industrial-robots-may-guide-collaboration-with-humans/, Access Date: September 18, 2018

13. Michael E. Porter. The Five Competitive Forces That Shape Strategy. THE JANUARY 2008 ISSUE, https://hbr.org/2008/01/the-five-competitive-forces-that-shape-strategy

14. Bernard Marr, What Is Digital Twin Technology - And Why Is It So Important?, March 06, 2017, https://www.forbes.com/sites/bernardmarr/2017/03/06/what-is-digital-twin-technology-and-why-is-it-so-important/#2db6e6432e2a

15. Sam Lucero. Enterprise IoT Leading IoT, Driving Industry Digital Transformation. March 2017, https://e.huawei.com/us/solutions/technical/iot, Access date: 12/14/2018

16. Adrian Booth, Jeff Hart and Stuart Sim. Building a great data platform. August 2018, https://www.mckinsey.com/industries/electric-power-and-natural-gas/our-insights/building-a-great-data-platform

17. Janakiram MSV. The Current State of Public Cloud Enterprise IoT Platforms. Sep 10, 2018, 10:17 pm, https://www.forbes.com/sites/janakirammsv/2018/09/10/the-current-state-of-public-cloud-enterprise-iot-platforms/#6f28f4c056b2

18. Stonegate. One is the Loneliest Number: Combating Senior Isolation. https://stonegatesl.com/one-is-the-loneliest-number-combating-senior-isolation/, Access date: 12/14/2018

19. Gilbert Benavidez. Senior Loneliness and the Public Health Response. September 13, 2017, https://www.publichealthpost.org/research/senior-loneliness-public-health-response/

20. D. Bowes. Elderly Suicide Rate. May 15, 2015, https://www.sageminder.com/SeniorHealth/MentalHealth/Suicide.aspx, Access Date: May 27, 2019

21. Pete Chapman, Julian Clinton, Randy Kerber et al. CRISP-DM 1.0. March 2000, ftp://ftp.software.ibm.com/software/analytics/spss/support/Modeler/Documentation/14/UserManual/CRISP-DM.pdf , Access Date: 11/22/2018

22. Carlton E. Sapp. Preparing and Architecting for Machine Learning. 17 January 2017, https://www.gartner.com/binaries/content/assets/events/keywords/catalyst/catus8/preparing_and_architecting_for_machine_learning.pdf, Access Date: 11/22/2018

CHAPTER **2**

Foundation Architectures

This Chapter Covers:
- IoT Foundations Architectures
- Data Mining Reference architecture
- Intel IoT Architecture
- IoT Case Study Water Leak management
- IoT Platform Review: Artik
- Sensors for Safety – Angle of Attack (AoA)
- Data at Rest vs. Data in Motion Differences
- CAP Theorem for IoT
- Edge and Cloud Architecture

The focus of the chapter is to provide foundation all architecture to process the IoT Data. IoT Data exhibit 5Vs (Volume, Velocity, Variability, Veracity, and Value) and to process such high-density data the architectures should scale to provide actionable insights to Users.

IoT Platform

IoT Platform (see Figure 1) provides an end-to-end connectivity to bring it online and brings intelligence to "Things" [6].

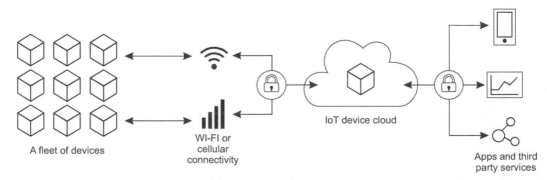

A fleet of devices   WI-FI or cellular connectivity   IoT device cloud   Apps and third party services

Figure 1: IoT Platform

At the source level, you have several types of devices that generate data at several different frequencies and velocities. The Connectivity layer is the one that collects data from the Sensors and uploads to either edge level or to the Cloud level. On the Cloud, the IoT data is processed through Analytics frameworks and Machine learning engines to derive useful insights. Finally, the end insights are stored in data base or looped back to the Users.

Here is the end to end platform view from Intel (see Figure 2): The goal of the System Architecture Specification is very straight—to connect any device any type to the Cloud. The specification is also a guideline for third party developers, Original Equipment Manufacturers (OEM), and other developer community to develop products for IoT [7].

The tenets for the application developers to deploy IoT solutions are as follows:

- Services to Monetize the IoT Infrastructure.

- Seamless data ingestion.

- Intelligence deployment to any-form-factor devices by running analytics on the Cloud or on to the device.

In the grand scheme of IoT, the philosophy is very straightforward: ***Deployment of distributed data collection, contextual intelligence, centralized data platform, and holistic integration of devices to the Enterprise Data architecture systems.***

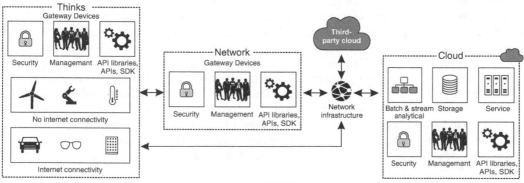

Figure 2: Intel IoT Platform

IoT Platform Types

IoT Platform Types can be classified under End-to-End, Connectivity, Cloud, and Data Platforms.

End to End Platform

End to End platform, generally, is used by businesses or service providers as a completely vertically integrated service offering or product architecture. For instance, Samsung, offers IoT platform to integrate all data sources.

End to End platform offers all the parts of architecture: (a) Hardware or Sensor Product, (b) Gateway or Edge Collector, (c) Mobile App as an Edge device, (d) Middleware component, (e) Message Bus or Message Queue injection, (f) Compute or Analytics platform, and (g) UI or Insights recommendation.

Case Study: IoT for Water Leak management

The cities of Doha, São Paulo, and Beijing all use sensors on pipes, pumps, and other water infrastructure to monitor conditions and manage water loss, identify and repair leaks or change pressure as necessary. On average, these cities have reduced leaks by 40 to 50 percent. Smart meters at the user end allow real-time monitoring of demand and leak detection by residents and property managers, reducing costs. Dubuque and Indianapolis in the United States, as well as Malta, New Delhi, and Barrie (Ontario), have seen, on average, a 5 to 10 percent reduction in water usage via the use of smart water meters.

The total potential economic impact from traffic applications, smart waste handling, and smart water systems in urban areas could total **$100 billion to $300 billion** per year by 2025. This assumes that 80 to 100 percent of cities in advanced economies and **25 to 50** percent of cities in the developing world could have access to this technology by that time.

The Reference architecture for End to End IoT is as follows:

The reference architecture for IoT is very much like that of the architecture of data processing systems except that the IoT architectures consume the data before the platform can massage or compact the data.

On the left side (see Figure 3), the data is generated by Sensors, Human Activity, Motors or Sensors/Actuators. The sensors could be of active, passive or hybrid of the **data generators**. The data from the generators are collected by the main processing units Data Controllers (microcontroller units). The Data Controllers could be in-memory memories or Sensor Units. The Sensor Units comprise silicon circuit board with RTOS (real-time operating system) that collects data either by polling a particular sensor hardware address or by triggered by interrupt routine. The Data is then processed at the edge level or Sensor Unit level for immediate action or Closed loop behavior. This is achieved by combing events from several data sources, Complex Event Processing. The processed data is stored or relayed to traditional cloud platforms.

The End to End architecture contains Sensors or Data generators that feed the data into Edge Data Nodes processors. The purpose of the Edge node is to evaluate rules on incoming high velocity multi-sensor data. The data is validated against any anomalies for alerting the User. In addition, the context-based analytics can be run on the in-coming data in order to derive contextual intelligence. Once data is analyzed, it can be compressed in order to reduce the data footprint and data transmissions costs.

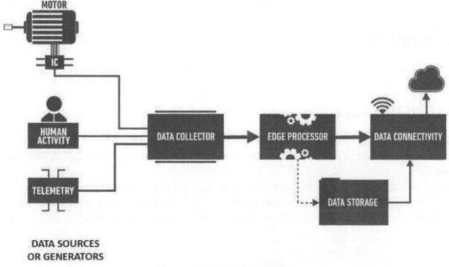

Figure 3: IoT End to End View

As shown in the above diagram (see Figure 4):

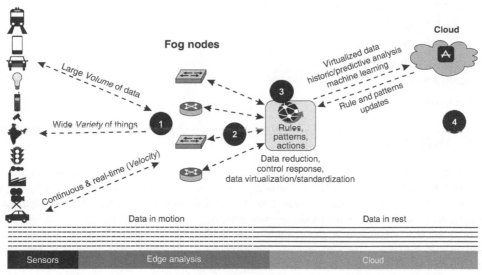

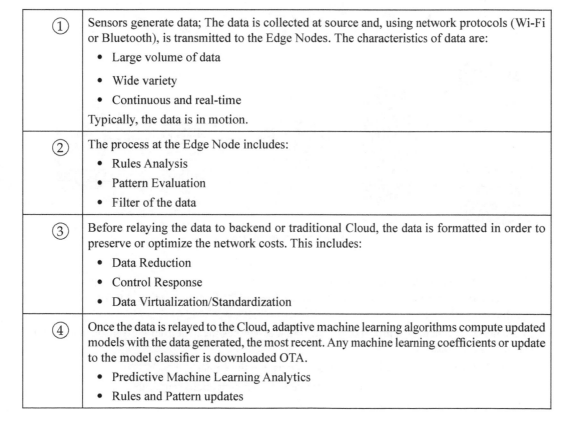

Figure 4: Fog Nodes

①	Sensors generate data; The data is collected at source and, using network protocols (Wi-Fi or Bluetooth), is transmitted to the Edge Nodes. The characteristics of data are: • Large volume of data • Wide variety • Continuous and real-time Typically, the data is in motion.
②	The process at the Edge Node includes: • Rules Analysis • Pattern Evaluation • Filter of the data
③	Before relaying the data to backend or traditional Cloud, the data is formatted in order to preserve or optimize the network costs. This includes: • Data Reduction • Control Response • Data Virtualization/Standardization
④	Once the data is relayed to the Cloud, adaptive machine learning algorithms compute updated models with the data generated, the most recent. Any machine learning coefficients or update to the model classifier is downloaded OTA. • Predictive Machine Learning Analytics • Rules and Pattern updates

The end to end platform provides several benefits:
- Over the air Firmware (OTA) Upgrade
- Device Management
- Cloud Connection
- Cellular or Connectivity
- Analytics and
- User Interface

Connectivity Platform

Connectivity IoT platforms enable direct integration of Sensors via edge device or gateway to the Cloud. The Sensors upload the collected data, periodically, to the on-site collector device. The Collector on the availability of network connection, uploads to the Cloud. Connectivity Platform enforces tighter API or Data integration between the Sensor and Gateway devices.

The connectivity platform, as the name implies, is ideal for the Study Cases that solve or address network convexity issues. The Connectivity Platform potentially allows IoT devices to run for years on small batteries, occasionally sending out small packets of data, waiting for a short time for response messages, and then closing the connection until more data needs to be sent.

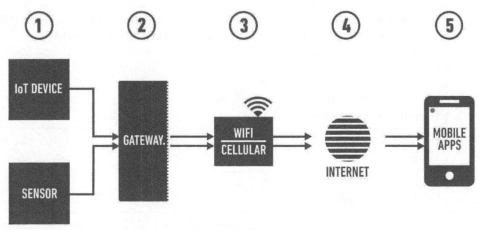

Figure 5: Integrated view

In the above diagram (see Figure 5):
1. The sensors communicate to the gateway either Near Field Communication, MQTT over UDP, or LoRa.[1]
2. The Gateway, generally, a processing system that assembles the data from the Sensors and perform basic logic operations.
3. Using Cellular or Wi-Fi, the gateway relays the data to the Cloud.
4. The data is processes at the Cloud and actionable recommendations sent to the User.
5. Mobile recommendations are delivered.

---

[1]   LoRa - https://www.loraserver.io/overview/

LoRa

LoRaWAN (please see Figure 6) is a long range, low power, wireless protocol that is intended for use in building IoT networks. IoT devices ("nodes") send small data packets to any number of "gateways" that may be in the several-kilometer range of a node via the LoRaWAN wireless protocol. The gateways then use more traditional communications, such as wired Internet connections, to forward the messages to a network-server, which validates the packets and forwards the application payload to an application-server [8].

LoRa Architecture[2] [9]

- LoRaWAN devices: IoT Devices that send data to LoRa network (through LoRa gateways).
- LoRa Gateway: LoRa Gateways are central in the network. They receive data from the Sensor devices and implement packet forward software in order to send data to LoRa Servers.
- LoRa Gateway Bridge: The LoRa gateway bridge takes the Packet forward message (UDP) and translate into MQTT. The advantage of using MQTT is:
  - It makes debugging easier.
  - Sending downlink data only requires knowledge about the corresponding MQTT topic of the gateway, the MQTT broker will route it to the LoRa Gateway Bridge instance responsible for the gateway.
  - It enables a secure connection between your gateways and the network (using MQTT over TLS—see Figure LoRa).
- LoRa Server [9]: LoRa Server manages the WAN and provides the network. It also maintains registry of active devices on the network. LoRa Server performs one more additional functionality: When data is received by multiple gateways, LoRa Server will de-duplicate this data and forward it once to the LoRaWAN application-server. When an application-server needs to send data back to a device, LoRa Server will keep these items in queue, until it is able to send to one of the gateways.
- LoRa Geo Server: It is an optional component. providing geo resolution of Gateways.
- LoRa App Server: The LoRa App Server component implements a LoRaWAN application-server compatible with the LoRa Server component. To receive the application payload sent by one of your devices, you can use one of the integrations that LoRa App Server provides (e.g., MQTT, HTTP or directly write to an InfluxDB database).

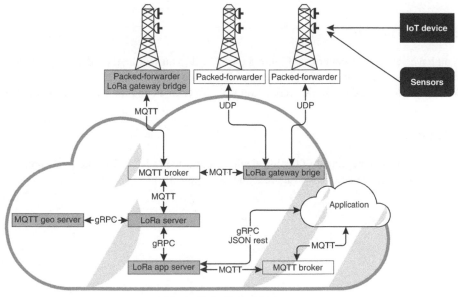

Figure 6: LoRa Architecture

---

[2] LoRa Architecture - https://www.loraserver.io/overview/architecture/

Case Study: Artik Connectivity Platform

Artik platform (see Figure 7) provides connectivity platform for the IoT devices [10]. The platform provides connecting IoT devices to the platform and enable to create custom rules processing engine to evaluate conditions and integrate action—email or alert.

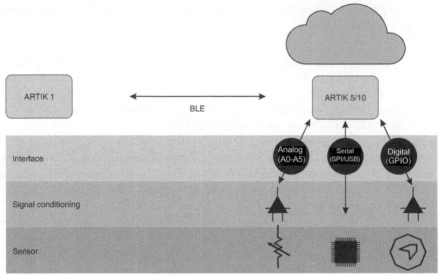

Figure 7: ARTIK Architecture

Artik Platform Architecture

Artik platform architecture ties the sensors to the Cloud.[3] The Sensors data (see Figure 7)—analog (voice data), digital (temperature) and serial—collected at the Artik edge; the custom rules process the data to provide immediate alert or notify to the Users [11].

Artik Cloud Setup

Once you have set up an account (see Figure 8) with Artik, you will be provided a signup account; The login dashboard will provide a way to add sensor data to the Artik cloud.

With Artik, we can add devices, scenes, and Rules to enable communication of IoT devices.

Add Device

Add devices enables management of the IoT devices connected to ARTIK Cloud (see Figure 9).

Device Added Successfully (see Figure 9 and see Figure 10).

---

[3]   Artik Architecture - https://www.artik.io/blog/2015/10/iot-101-sensors/

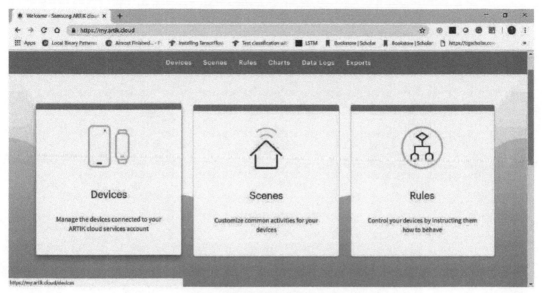

Figure 8: Artik Setup

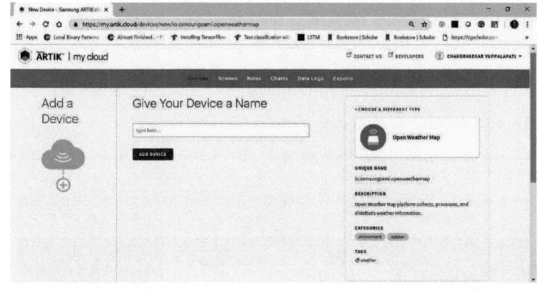

Figure 9: Add device

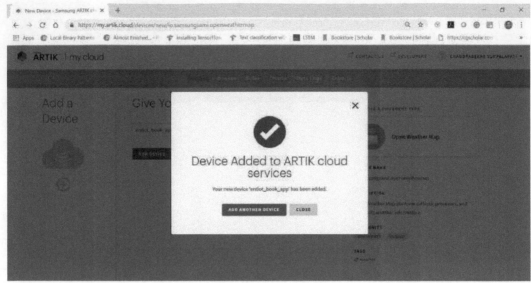

Figure 10: Device Add

Once added, we can customize the device;

Data Simulation

Devices are activated and enabled simulation by running simulation data points.

Create a simulation data run

Create data simulation run provides various data generation run facilities to feed data into IoT architectures (see Figure 11).

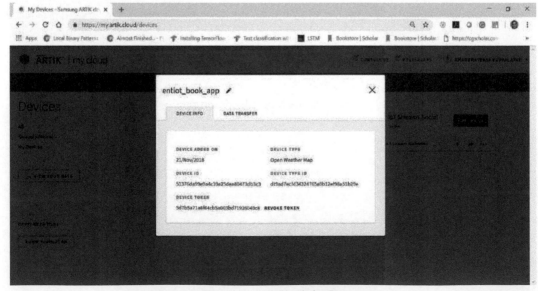

Figure 11: Data Simulation Run

Data Interval enables the data generation as per mentioned Interval (simulation interval—see Figure 12);

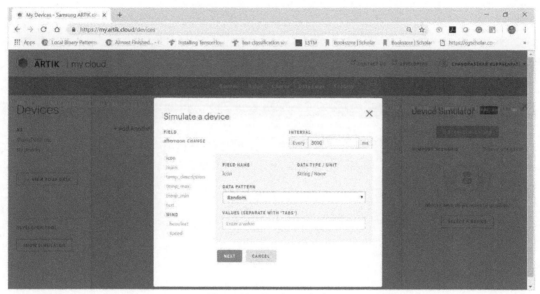

Figure 12: Simulate a device

Data Pattern

The data can be generated Randomly, in increments, at a constant value, or you can cycle the data (see Figure 13).

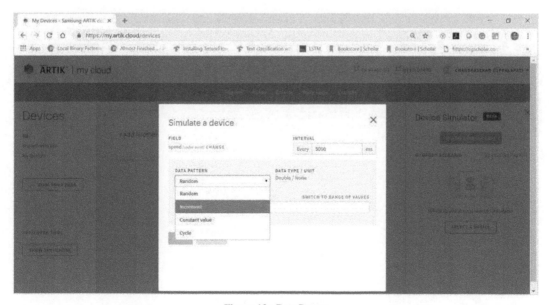

Figure 13: Data Pattern

Data Charts

The data is fed to the IoT Device and application of rules provides edge level simulation of the data (see Figure 14).

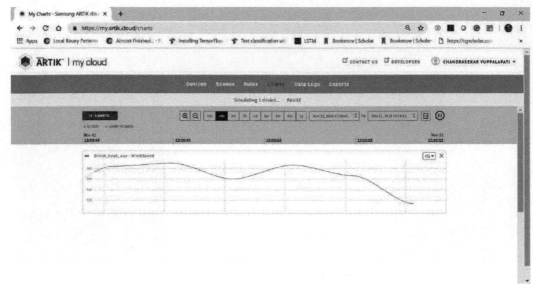

Figure 14: Data charts

Data Fields or Attributes

Simulate the data on a wide varieity of Edeg attributes (see Figure 15).

Figure 15: Edge Parameters

Main: Temperature, Humidity, Temp

Morning: Temp_Min, Text, Wind Speed, Wind Text

Wind, Sys, Weather, and other parameters.

Device monitor: Choose device activity to monitor (please see Figure 16)

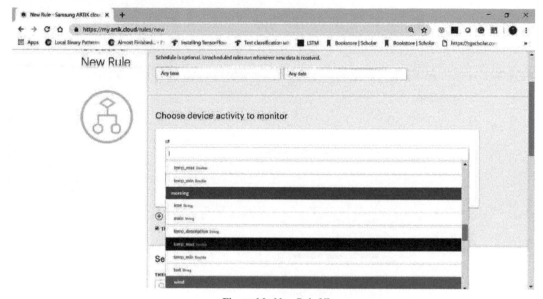

Figure 16: New Rule UI

Multiple Charts

Once configured, the device data can be viewed (see Figure 17)

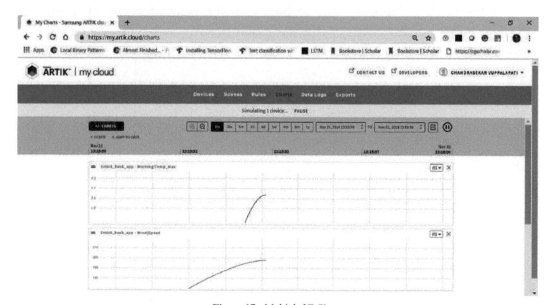

Figure 17: Multiple UI Charts

Edge Rules

Create Edge Rules to process the data before relayed to the Cloud. Graphically we can provide rules (see figure Create Rule UI).

Anomaly detection

Specified Anomaly detection for the data points (see Figure 18); Can be provided data edge conditions.

Figure 18: Anomaly detection

Rule Created

TEMP > 40—the rule[4] is fully qualified. In our case, if morning   temperature is more than 40 degrees Celsius, send email to User email—cvuppalapati@hanuinnotech.com

Rule meta[5] data:

```
1       {
2         "if": {
3           "and": [ {
4             "sdid": ... the device ...,
5             "field": ... the field ...,
6             "operator": ... the operator ...,
7             "operand": { "value": ... the value ... },
8             "duration": {
9               "value": ... duration in seconds ...,
10              "acceptUpdate": ... true/false ...
11            }
12          } ]
13        }
14      }
```

4   Rule - https://developer.artik.cloud/documentation/rules-engine.html
5   Rules engine - https://developer.artik.cloud/documentation/rules-engine.html

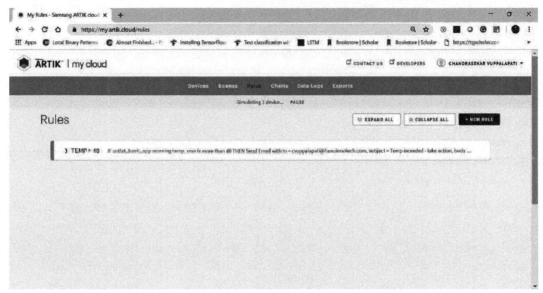

Figure 19: Rules

Email Notification

Once data violates the rule, email notification[6] is sent to the User (see Figure 20).

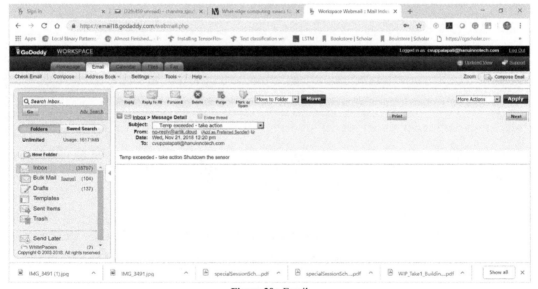

Figure 20: Email

---

6   Cloud Integration - https://developer.artik.cloud/documentation/data-management/push-to-amazon-kinesis.html

Technical Case Study: Using IoT to detect water leakages with Powel[7]

Water leakage is a costly issue for the water governing bodies across the world [12]. Given the aging infrastructure across the many municipalities, the water leakage is one of the on-going expenses that governments and tax paying citizen unwillingly share. In the following Study Case, the municipalities in Norway used IoT to detect water losses due to leakages (than 30% due to leakages in their water distribution network and this represents a significant cost that ends up in the hands of consumers).

Key Technologies used:

- Telemetry provided by SCADA Sensors, measuring waterflow and pressure
- Azure Function
- Azure Cloud & Storage

Process: understanding from RAW data to creation of model is needed to see the solution (see Figure 21).

Figure 21: Process

The Data from Sensors are collected by Edge Gateways [12] (see Figure 22) and relayed to IoT Hub for the data ingestion purposes. The data is retrieved using Sensor Analytics to identify leakage issues by applying machine learning algorithms. The API and app services enable generation reports for the management consumption; finally, the data is repented on UI using HTML 5 and Power BI application.

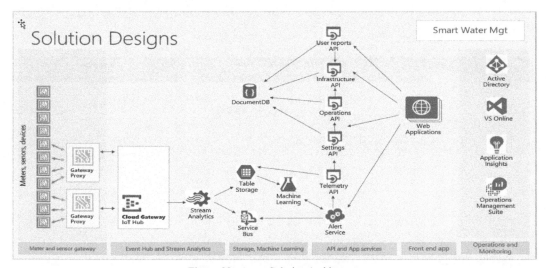

Figure 22: Azure Solution Architecture

Overlying historical data with machine learning prediction (see Figure 23) results in identification of water flow outliers, a good indicator of leakage, at an interval of time. In general [12].

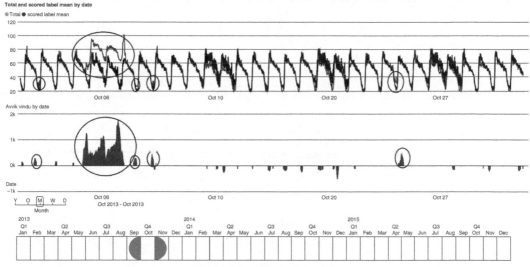

Total and scored label mean by date

⊕ Total ● scored label mean

Figure 23: Anomaly Detection

Sensors – Safety & Efficacy

Sensors play an important role in our day to day life. From enabling personal assistant mobile applications to commerce critical air industry. For instance, consider one of the most important air flight sensors: Angle of Attack (AOA) Sensor (see figure AOA Schematic) [13].

Angle of Attack (AOA) Sensor

Angle of Attack (see Figure 24) is a fluid dynamics chief principle for floating an object. The angle of Attack (AOA) is the angle between a reference line on a body and the vector representing the relative motion between the body and the fluid through which it is moving.

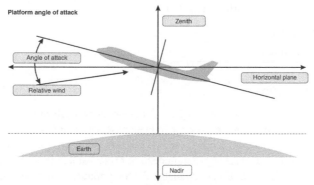

Figure 24: Angle of Attack (AOA) Schematic

In the figure above, the angle of attack is measures by taking horizontal plan reference and relative wind positions that flow through the wing mounted AOA sensor.

The Angle of Attack (AoA) Probe provides AoA or Sideslip (SS) by sensing the direction of local airflow. It is mounted on the fuselage with the sensing probe extending through the aircraft fuselage[8] [13].

At any specific AOA, the airflow over some percentage of the wing surface (please see figure above) will generate lift as well as some amount of drag. Maximum lift is usually obtained at a relatively high angle (see Figure 25).

---

[8]   Angle of Attack - https://www.aerosonic.com/products/angle-of-attack/

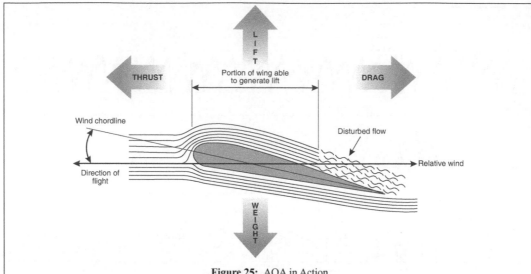

Figure 25: AOA in Action

Stall is defined as the condition which arises when the angle grows so large that the flow is completely disrupted, and not enough lift is generated to overcome the weight of the aircraft. This AOA, the stall angle, is constant for a particular aircraft, although various wing designs stall at differing angles. The amount of useful lift and drag generated by any wing at some specific AOA will depend upon the influence of variables as the wing geometry, density altitude, aircraft gross weight and velocity.

Knowledge Discovery in Databases (KDD)

The Knowledge Discovery in Databases are traditionally applied in order to unravel information in large databases. KDD has following important steps [14,15]:

- Cleaning and Integration
- Selection and Transformation
- Data Mining
- Evaluation and Presentation

At the base of the KDD [14] (see Figure 26), we have data sources or databases. The data from data sources are cleansed and integrated in order to be stored in at Data bases or Data Warehouses. For instance, we're developing foundation architecture for a real-time data platform, this step represents cleansing of data for data validation rules. For instance, checking Longitude or Latitude values are set for location-based services or input temperature or humidity values are available for temperature or humidity-based case study. The next step, Selection and Transformation is to enable data store for running machine learning or data mining algorithms. Generally, the selection represents key attributes that are derived from or required to run a machine learning algorithm. The transformation purpose is to translate the input data into intermediator form in order to run the algorithm. The transformation, for instance, is to extract feature engineering data for audio or video-based input signals. In the case of text or document-based systems, for instance Natural Language Processing, it is to convert key word indexes into matrix form. The transformed input is either persisted on database or loaded into memory for performing machine learning processes. The machine learning algorithms vary based on the case study or based on the type of the data. The processed output of the machine learning process is stored into data base for visualization and reporting purposes.

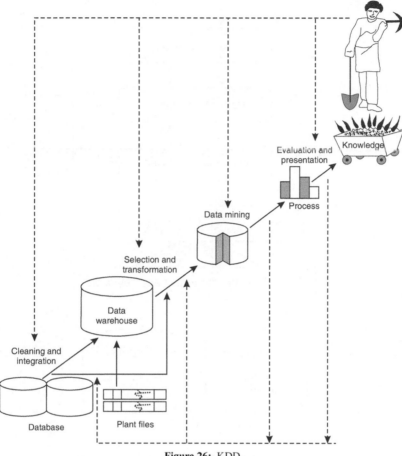

Figure 26: KDD

The Machine Learning reference architecture (see Figure 27) [14]:

IoT & KDD

The Knowledge Discovery in Databases is applied to data in rest (see Table 1). That is, the data is stored in databases and there is no variation of the data. The data is a source of truth. On the other hand, Data from IoT devices are in motion. The data is continuously changing. In short, the data has velocity and is in constant flux [16].

The IoT application constructs are different from conventional applications from algorithms and processing sides. Some of the differences that make IoT applications unique include:

- Data Filtering
- Aggregation and Correlation
- Data Enrichment (e.g., Spectrogram or histograms)
- Location (Geo-context)
- Time Windows
- Temporal Platforms

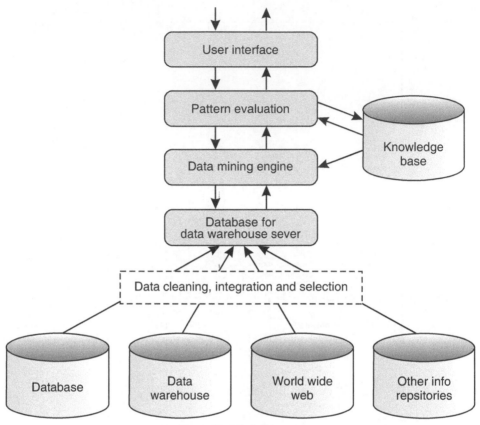

Figure 27: ML Architecture

Table 1: Streams vs. Lakes

Analyzing data as streams and lakes[9]	
Lakes	**Streams**
• Ingested and Stored in Data Warehouse	• Data Collected real-time
• Several Sources of Data (Traditional Structured and Unstructured Data Sources)	• Multiple Sensors or Sources of Data
• Analysis or Analytics run weekly, daily, or hourly	• Analytics run continuously, second and sub-second responses
• Insights used to modify future actions	• Insights used to proactively adjust immediate and future actions

5Vs and IoT

Unlike analytics for the traditional platforms, the analytics for the IoT platforms has two parts: Data in Motion and Data in Rest. Let's look from 5V's of Data on IoT[10] [17]. Before going details onto 5V's in IoT, let's look 5Vs (see Figure 29):

---

9   The Next Big Thing in the Internet of Things: Real-time Big Data Analytics - http://www.vitria.com/pdf/Forrester-Vitria-IOT-Dale-Slides.pdf

10   Why only one of the 5 Vs of big data really matters - https://www.ibmbigdatahub.com/blog/why-only-one-5-vs-big-data-really-matters

Volume

Volume refers to the vast amount of data generated every second. The data could be from traditional sources such real-time transaction system generating huge amounts of transaction data. For instance, holiday sales at a popular store.

In IoT's context, the Volume refers to the number of sensors generating the data at a manufacturing plan level, or the data generated by oil rigs. For instance, A large refinery generates 1TB of raw data per day.

Velocity

Velocity refers to the speed at which new data is generated and the speed at which data moves around. The data generated by Sensors is in the denominator of microsecond or millisecond level. Let's say a sensor generates 256 bytes data at every microsecond level and all the data points are required to process the business case study, then in a one day the sensor generates:

24 X 60 X 60 X 1000 X 256 bytes

$$22118,400,000 \text{ bytes} \tag{1}$$

22.1 GB per day

If thousands of sensors collect data, we can see that the Volume and Velocity of data are huge—Big Data level (see Figure 5Vs) [18].

Variety

Variety refers to the different types of data we can now use. In the past, we focused on structured data that neatly fits into tables or relational databases, such as financial data (for example, sales by product or region). In fact, 80 percent of the world's data is now unstructured and therefore can't easily be put into tables or relational databases—think of photos, video sequences or social media updates. In IoT's context, the data generated by Sensors falls into variety data (see Table Data Formats). For instance, security camera data is streaming video data. The voice assistant audio sensors capture the data in the audio format and the environmental or activity sensors collect data on a time-series basis. Please see the following table for audio, video and environmental sensor data formats.

Veracity

Veracity refers to the messiness or trustworthiness of the data. Veracity plays an important role in the IoT—especially the sensors that generate data with out of balance calibration factors.

To achieve the best possible accuracy, a sensor should be calibrated (see Figure 28) in the system where it will be used.[11] This because:

- No Sensor is perfect
  - Sensors subject to heat, cold, shock, humidity, etc., during storage, shipment and/or assembly may show a change in response.
  - Some sensor technologies 'age' and their response will naturally change over time—requiring periodic re-calibration [19].
- The Sensor is only one component in the measurement system [20]. For example:
  - With analog sensors, your ADC is part of the measurement system and subject to variability as well.

---

[11] Why Calibrate? - https://learn.adafruit.com/calibrating-sensors/why-calibrate

Table 2: Data Formats

Source	Format
Audio Data	 Spectrogram
Image Data	 Histogram
Text Data	\<START\> this a rip roaring western and i have watched it many times and it entertains on every level however if your after the true facts about such legends as \<UNK\> cody and \<UNK\> jane then look elsewhere as john ford suggested this is the west when the truth becomes legend print the legend the story moves with a cracking pace and there is some great dialogue between gary cooper and jean arthur two very watchable stars who help to make this movie the sharp eyed amongst you might just spot \<UNK\> hayes as an indian \<UNK\> also there is a very young anthony quinn making his debut as \<UNK\> warrior he actually married one of \<UNK\> daughters in real life indeed its \<UNK\> character who informs cooper of the massacre of \<UNK\> told in flash back the finale is well done and when the credits roll it \<UNK\> the american west with american history so please take time out to watch this classic western \<PAD\> \<PAD\> \<PAD\> \<PAD\> \<PAD\> \<PAD\> \<PAD\> \<PAD\> \<PAD\> \<PAD\> \<PAD\> \<PAD\> \<PAD\> \<PAD\> \<PAD\> \<PAD\> \<PAD\> \<PAD\> \<PAD\> \<PAD\> \<PAD\> \<PAD\> \<PAD\> \<PAD\> \<PAD\> \<PAD\> \<PAD\> \<PAD\> \<PAD\> \<PAD\> \<PAD\> \<PAD\> \<PAD\> \<PAD\> \<PAD\> \<PAD\> \<PAD\> \<PAD\> \<PAD\> \<PAD\> \<PAD\> \<PAD\> \<PAD\> \<PAD\> \<PAD\> \<PAD\> \<PAD\> \<PAD\> \<PAD\> \<PAD\> \<PAD\> \<PAD\> \<PAD\> \<PAD\> \<PAD\> \<PAD\> \<PAD\> \<PAD\> \<PAD\> \<PAD\> \<PAD\> \<PAD\> \<PAD\> \<PAD\> \<PAD\> \<PAD\> \<PAD\> \<PAD\> \<PAD\> \<PAD\> \<PAD\> \<PAD\> \<PAD\> \<PAD\> \<PAD\> \<PAD\> \<PAD\> \<PAD\> \<PAD\> \<PAD\> \<PAD\> \<PAD\> \<PAD\> \<PAD\> \<PAD\> \<PAD\> \<PAD\> Document Vector

    ○ Temperature measurements are subject to thermal gradients between the sensor and the measurement point.

Sensor said to be accurate if it follows ship settings expected behavior:

- Precision
  - ○ The ideal sensor will always produce the same output for the same input.
- Resolution
  - ○ A good sensor will be able to reliably detect small changes in the measured parameter.

The precision is affected by Noise and Hysteresis.

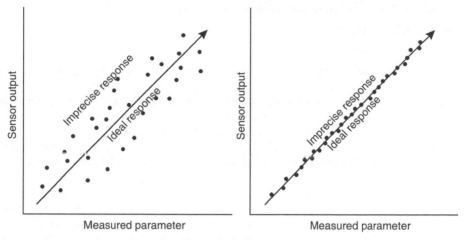

Figure 28: Calibration

Sensor quality issues are a common issue in data analytics and automated operations, with customers citing sensor-level issues as the cause of 40% of issues they experience.[12]

4 Steps to data readiness to Industrial Analytics Establish Sensor Data Trust

Data quality issues related to sensor noise, null/static values, and calibration issues are a leading contributor of poor outcomes when doing industrial analytics. Why? Sensor data are inherently noisy, often contain large gaps in the data, and can experience calibration drift over time. When dealing with millions of sensors, these quality issues become significant. Some customers have stated that these sensor level quality issues are the root cause of 40% of the issues they experience in their highly automated operations. Establishing sensor trust requires being able to effectively identify bad actor sensors through fleet-wide reporting on null values, data gaps, flat lined sensors, calibration and drift issues, as well as noise.

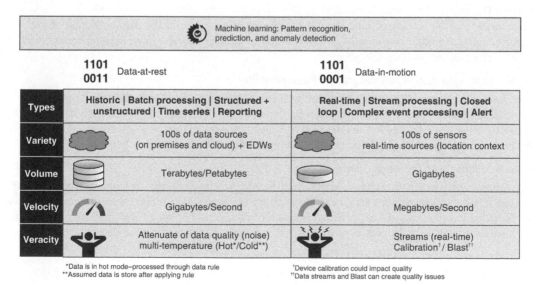

Figure 29: Data at Rest vs. Data in motion

---

[12] The Importance of Sensor Calibration in real-time data analytics - https://www.rtinsights.com/importance-sensor-calibration-real-time-data-analytics/

The proposed framework consists of two sub-frameworks [16]:

1. Offline Predictive Analysis
2. Online Decision Making

Offline Predictive Analysis can be divided into three parts, Data Integration, Pattern detection, and Process Detection.

During the Data Integration phase first, the remote sensors, which can be wired and wireless, are incorporated into the sensor networks. Next, the numerical physics-based computer models the outputs. These outputs provide Ancillary information and encoded domain knowledge.

The next phase is Pattern Detection. During this phase, offline data mining is done from sensor observations and models. This checks the computational efficiency and scalability to massive data, here the anomalies, extremes and nonlinear processes in space and time are evaluated.

The last phase within the offline predictive analysis is process detection. Here, the numerical models are evaluated with sensor data assimilation schemes. Extraction of dynamics from massive sensor observations is carried out, as well as extraction of dynamics from incomplete and noisy information.

The next phase links the offline predictive analysis with online decision making. Decision Support consists of online (real-time) analysis from models and observations, algorithmic efficiency for dynamic and distributed processing of sensor observations, resiliency, vulnerability, and impacts of observations, and lastly visualization and decision or policy aids models [21,22].

Step-1: Data Generation: Sensors and Tags, Hardware and OS, Power

IoT devices have sensors which are their major source of data, especially spatial sensors, geo-spatial sensors, wearables, health sensors and environment sensors. Tags can be like QR code patterns and generate the data that usually require a scanner device in place. The active type RFID tags require the power source, for broadcasting the signals and can be either Low Frequency type or ultra-high frequency type.

Step-2: Data Collection: Discovery, Management, Transmission, Context and Fog

For the data collection phase, most of the work is to be done for the development of the middleware-software layer, which is responsible for connecting components such as device and network together. This layer has the responsibilities of resource management, code management and event management. For the discovery of devices, there are a number of technologies that are supported: Micro Plug and Play, DNS Service Discovery, DNS, and Simple Service Discovery Protocol, etc.

Transmission of the generated data is another very important stage in the data collection phase. The transmission technologies are usually divided into different categories, like those used for communication within sensor networks such as Zigbee.

Step-3: Data Aggregation and Integration: Interoperability

McKinsey [4] estimate that the interoperability will unlock approximately an extra 40 to 60 percent of the total estimated future IoT market value. Berrios et al. [5] describes how the various cross-industry consortia concerned with the IoT are merging and converging on the semantic interoperability inside the application layer which they split across into interoperability of business semantics.

Step-4: Architectures for Storage and Compute

Architectures help us to define and guide us to build an infrastructure and lays out method on how to handle the big IOT data in storage for analytics. One such architecture which is recommended here is lambda architecture, which comprises of speed, serving and batch layers. The speed layer here compensates for the high latency issue by looking at recent data and providing fast incremental updates.

Step-5: Compute Technologies and IoT Analytics Applications

The computation on the cloud is distributed into-server less computing, container technologies and orchestration.

Morevover, the container technologies are also becoming popular for portability and reduction for dependencies. They provide lower overhead, faster launch times and also the orchestration of the containers. Kubernetes, docker and mesosphere are such container orchestration technologies.

Intel IoT Data Flow Diagram

The following diagram (see figure Intel IoT Data Flow [7]) provides the Data Flow of typical Sensors that are not connected directly to the internet—Intel IoT Data Flow Architecture.[13] The Sensors and Actuators collect the data relays to the Gateways. The Gateways perform one or all of the operations: Edge Analytics, Secure the data, and Manage the data. The Data is posted the to Cloud using Message Queue Telemetry Transport (MQTT), Hyper Text Transport Secure (HTTPS), Constrained Application Access Protocol (CoAP), Representational State Transfer (REST) Protocol and others. The Data is processed and computed using Cloud technologies.

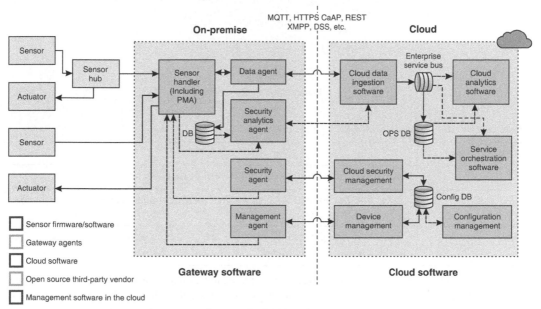

Figure 30: Intel IoT Data Flow

CAP Theorem

CAP Theorem (see Figure 31) states that it is impossible for any distributed computing system to simultaneously address consistency, availability and partition tolerance. A system can have two out of three. For instance, a system with higher partition tolerance and availability (like Kafka and Cassandra) will sacrifice some consistency [23].

Consistency: A guarantee that every distributed cluster returns to the same, most recent, successful Write state. There are various types of consistency and the CAP refers to Linearizability, the most complex form of consistency.

Availability: Every non-failing node returns a response for all read and write requests in a reasonable amount of time.

Partition Tolerance: The system continues to function and upholds its consistency guarantees despite network partitions. Distributed systems guaranteeing partition tolerance can gracefully recover from partitions once the partition heals.

---

[13] Intel IoT Reference Architecture - https://www.intel.com/content/www/us/en/internet-of-things/white-papers/iot-platform-reference-architecture-paper.html

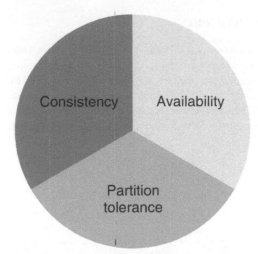

Figure 31: CAP Theorem

Three categorizations based on CAP Theorem [24]:

Categorization	
CP	Availability is sacrificed only in the case of network partition.
CA	Under this categorization, systems are Consistent and Available in the absence of Network Partition.
AP	Under this categorization, systems are available and partition tolerant but cannot guarantee Consistency.

Kind of Applications with AP[14] data store: Social Applications, such as Facebook, LinkedIn, YouTube, and others, will benefit from AP, as these applications are concerned about having the data, old data is fine.

Kind of Applications with CP data store: Banking and e-commerce applications fall into this category. These applications need consistent data; stale data is more harmful than not having the data.

CAP Theorem and IoT[15]
Smart City implementations heavily rely in CA side of CAP. For instance, a city government may choose to enable IoT based SMART City lights (see figure Smart Traffic Lights). The desired result is functioning of traffic lights (Availability) with consistent data despite network failures or other signals or lights failure [25]. The system will likely employ several nodes throughout the traffic grid in order to collect the data and make it available to the applications. If one node fails, however, the data it collects and processes must still be available to the rest of the system and possibly to other central applications

---

IoT Streams and Reference Architecture

The Data processing architecture for IoT has two phases (see Figure 32): First, processing data at the edge level and then processing IoT Streams data at the Cloud level. The processing of streams

---

14 No, You Can't Cheat the CAP Theorem. But Pivotal Cloud Cache on PCF Comes Close. Here's How - https://content. pivotal.io/blog/acid-cap-and-pcc

15 https://internetofthingsagenda.techtarget.com/blog/IoT-Agenda/How-IoT-is-making-distributed-computing-cool-again

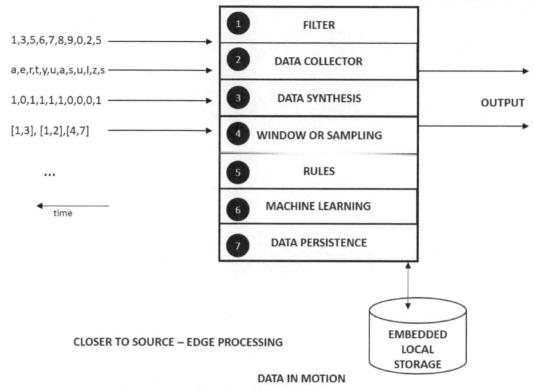

Figure 32: Edge - Data In Motion

at the edge level has the following steps: (1) Data Filter, (2) Data Collector, (3) Data Synthesizer, (4) Sampling, (5) Rules, (6) Machine Learning, and (7) Data Persistence.

Data Filter performs a crucial step: It identifies outliers in the data. As the data is processed through the edge, first data filters such as Kalman filters identify the abnormalities of the data. Abnormal increase in temperature, for example, can easily be identified through Filter process.

The Next step, Data Collector, is in-memory representation of the data. The Data at the Edge is continuous stream data that is moving rapidly. The goal of the data collector is to represent the data on time interval level to be processed. Data Synthesis is merge of data from two or more sources for analytical purposes. Sampling or Sliding window is to perform the evaluation of data in the streams. The sliding window (see Figure 33) provides a data comparison between the current event and previous events. The length of previous window size could be adjusted.

The Machine learning processes performs supervised or unsupervised machine learning algorithms in order to enable contextual awareness and enable adaptive device behavior intelligence.

Operational View

The Operational View of IoT Processing at the Edge Level includes Kalman Filter at the Filter processing, In-Memory Data Cubes or Data Collectors, Data Transformation and Synthesis from various data sources, Data Windowing algorithms (see Figure 34) such as sliding window, Rules application, Machine Learning module that provides contextual and geolocation intelligence, and data persistence using local file system and/or external drives.

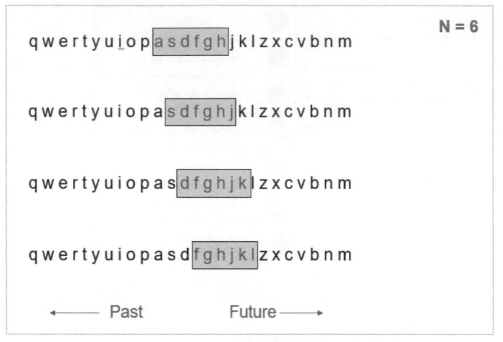

Figure 33: Sliding Window

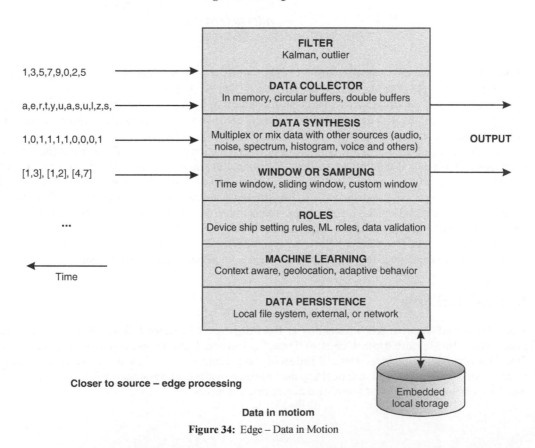

Figure 34: Edge – Data in Motion

Data at Rest

At the Edge level, data is in motion (see Figure 35 and Figure 36). It's fast and needs to be processed in a millisecond or less time (sub-millisecond). Once the data is relayed to traditional Cloud backend, data is at rest and is combined with other sources in order to generate quality reporting or insights. The processing of the data follows pretty much KDD steps, as shown in the figure below.

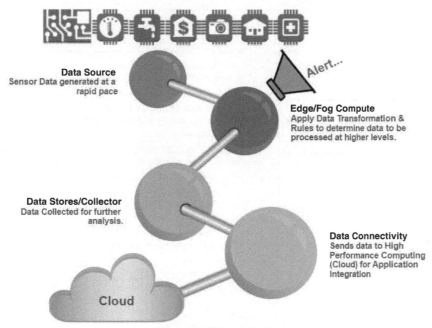

Figure 35: Edge on the Cloud

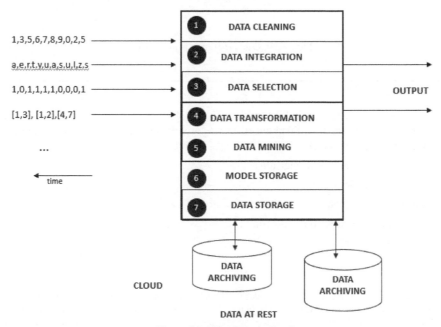

Figure 36: Edge Data at Cloud

Operational View

Data is Cleansed for simple data rules, such as missing dates and Sensor values. The Cleansed output (see Figure 37) data is integrated with Enterprise or 3rd party data sources in order to get a complete view of the data. The next step is the iterative process for picking important data attributes. The attributes selection is based on the business case study or algorithm under the case.

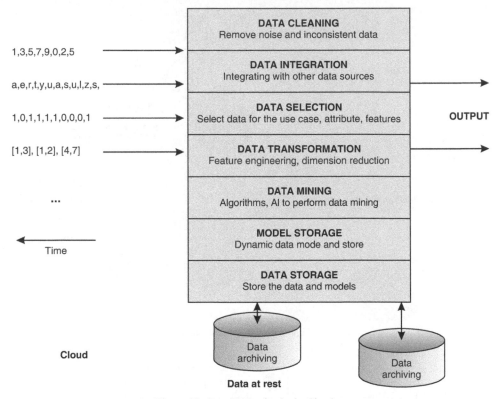

Figure 37: Data Processing in the Cloud

Digital Transformation:[16]

In McKinsey's survey of top executives making a transformation at traditional companies with digital technology [26] it was found that the enterprise architecture teams can play an integral role in transforming companies with technologies [** R1]. The five key factors include:

1. Engage Top Executives in key decisions
2. Emphasize in strategic planning
3. Focus on Business outcomes
4. Use Capabilities to connect business and IT
5. Develop and retain high caliber talent

---

[16] Five enterprise-architecture practices that add value to digital transformations - https://www.mckinsey.com/~/media/McKinsey/Business%20Functions/McKinsey%20Digital/Our%20Insights/Five%20enterprise%20architecture%20practices%20that%20add%20value%20to%20digital%20transformations/Five-enterprise-architecture-practices-that-add-value-to-digital-transformations-vF.ashx

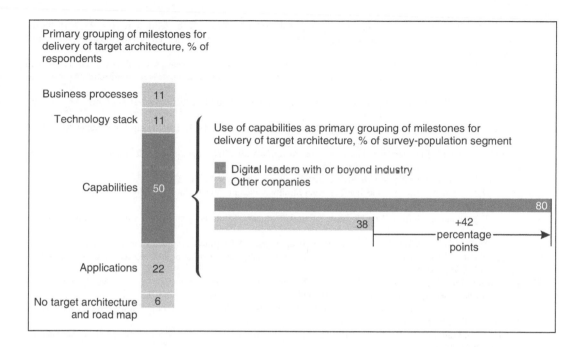

Primary grouping of milestones for delivery of target architecture, % of respondents

Business processes — 11
Technology stack — 11

Use of capabilities as primary grouping of milestones for delivery of target architecture, % of survey-population segment

■ Digital leaders with or beyond industry
□ Other conpanies

Capabilities — 50

80
38   +42 percentage points

Applications — 22

No target architecture and road map — 6

Chapter Summary:

After reading the chapter:

- Should able to blue-print for Knowledge Discovery in Data bases
- Should able to develop IoT devices reference architecture
- Customize a third-party Cloud & IoT platform, specifically, Artik
- Should able to understand Sensors in action for Water leak detection and Angle of Attack
- Should able to list 5Vs for Data at rest vs. Data in motion
- Should able to apply CAP theorem for IoT Devices
- Able to apply IoT Streams architecture

References

1. James Manyika, Michael Chui, Jacues Bughin et al. Disruptive technologies: Advances that will transform life, business, and the global economy. May 2013, https://www.mckinsey.com/~/media/McKinsey/Business%20Functions/McKinsey%20Digital/Our%20Insights/Disruptive%20technologies/MGI_Disruptive_technologies_Full_report_May2013.ashx, Access Date: April 30 2018

2. Kip Compton and Vikas Butaney. Enabling the Internet of Everything's, Cisco's IoT Architecture. 02/05/2015, https://www.slideshare.net/Cisco/enabling-the-internet-of-everything-ciscos-iot-architecture?from_action=save, Access date: 09/18/2018

3. Timothy King. Gartner Identifies Augmented Analytics as a Top Technology Trend for 2019. November 9, 2018, https://solutionsreview.com/business-intelligence/gartner-identifies-augmented-analytics-as-a-top-technology-trend-for-2019/, Access Date: 12/25/2018

4. Margaret Rouse. Augmented Intelligence. July 2017, https://whatis.techtarget.com/definition/augmented-intelligence?track=NL-1823&ad=922091&src=922091&asrc=EM_NLN_103914963&utm_medium=EM&utm_source=NLN&utm_campaign=20181122_Word%20of%20the%20Day:%20augmented%20intelligence

5. Rita L. Sallam. June 2018. Augmented Analytics: The Next Disruption in Analytics and BI Register now for this free 60 minute Disruption and Innovation webinar. https://www.gartner.com/webinar/3846165, Access date: 12/20/2018

6. Jeffrey Lee, How to Choose the Right IoT Platform: The Ultimate Checklist, April 25, https://hackernoon.com/how-to-choose-the-right-iot-platform-the-ultimate-checklist-47b5575d4e20 , Access Date: August 06 2018

7. David McKinney. Intel Champions Internet of Things and The Intel IoT Reference Architecture. April 23, 2015, https://www.intel.com/content/www/us/en/internet-of-things/white-papers/iot-platform-reference-architecture-paper.html, Access Date: August 06 2018

8. Donna Moore. The LoRa Server project. March 2016, https://www.loraserver.io/overview/ Access Date: 11/22/2018

9. Alper Yegin. The LoRa Architecture. December 2016, https://www.loraserver.io/overview/architecture/Access Date: 11/22/2018

10. Kevin Sharp. IoT 101: Sensors. October 7, 2015 https://www.artik.io/blog/2015/10/iot-101-sensors/ Access Date: 10/30/2018

11. Wei Xiao. Artik Cloud Developer, Develop Rules for devices. February 7, 2017 , https://developer.artik.cloud/documentation/rules-engine.html, Access Date: 10/30/2018

12. Pedro Dias, Using IoT to detect water leakages with Powel, Nov 29, 2016, https://microsoft.github.io/techcasestudies/iot/2016/11/29/Powel.html, Access Date: 11/08/2018

13. Rick Snyder. Angle of Attack. March 2018, https://www.aerosonic.com/angle-of-attack, Access Date: 11/08/2018

14. Jiawei Han, Micheline Kamber and Jian Pei, Data Mining: Concepts and Techniques, Morgan Kaufmann; 3 edition (July 6, 2011)

15. Anand Rajaraman and Jeffrey David Ullman, Mining of Massive Datasets, Cambridge University Press (December 30, 2011)

16. Nilamadhab Mishra, Chung-Chih Lin and Hsien-Tsung Chang. A Cognitive Adopted Framework for IoT Big-Data Management and Knowledge Discovery Prospective. First Published October 5, 2015 Research Article https://doi.org/10.1155/2015/718390

17. Dale Skeen. The Next Big Thing in the Internet of Things: Real-time Big Data Analytics. March 2017, http://www.vitria.com/pdf/Forrester-Vitria-IOT-Dale-Slides.pdf , Access Date: 11/08/2018

18. Bernard Marr. Why only one of the 5 Vs of big data really matters. March 19 2015, https://www.ibmbigdatahub.com/blog/why-only-one-5-vs-big-data-really-matters, Access Date: Nov 22 2018

19. Bill Earl. Calibrating Sensors. July 23 2018, https://learn.adafruit.com/calibrating-sensors/why-calibrate, Access Date: November 22 2018

20. Kayla Matthews. The Importance of Sensor Calibration in Real-Time Data Analytics. April 19, 2018, https://www.rtinsights.com/importance-sensor-calibration-real-time-data-analytics/, Nov 14 2018

21. Andy Bane. 4 Steps to Data Readiness for Industrial Analytics. July 21 2016, https://www.elementanalytics.com/blog/4-steps-to-data-readiness-for-industrial-analytics, Access Date: November 14, 2018

22. Bernard Marr. Why only one of the 5 Vs of big data really matters. MARCH 19, 2015, https://www.ibmbigdatahub.com/blog/why-only-one-5-vs-big-data-really-matters, Access Date: November 14, 2018

23. Dr. Eric A. Brewer. Towards Robust Towards Robust Distributed Systems Distributed Systems. July 19 2000, http://pld.cs.luc.edu/courses/353/spr11/notes/brewer_keynote.pdf, Access Date: November 14, 2018

24. Mike Stolz. No, You Can't Cheat the CAP Theorem. But Pivotal Cloud Cache on PCF Comes Close. Here's How. June 4 2018, https://content.pivotal.io/blog/acid-cap-and-pcc, Access Date: Sep 18 2018

25. Adam Wray. How IoT is making distributed computing cool again. 04 OCT 2016, https://internetofthingsagenda.techtarget.com/blog/IoT-Agenda/How-IoT-is-making-distributed-computing-cool-again, Access Date: Sep 18 2018

26. Sven Blumberg, Oliver Bossert and Jan Sokalski, Five enterprise-architecture practices that add value to digital transformations, November 2018, https://www.mckinsey.com/business-functions/digital-mckinsey/our-insights/five-enterprise-architecture-practices-that-add-value-to-digital-transformations, Access Date: Dec 10, 2018.

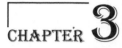

CHAPTER 3

Hardware Design

This Chapter Covers:

- Arduino IDE Installation
- Sketch Code
- IoT Sensor Code Module
- Temperature and Humidity Sensors
- Microcontroller code
- Integration of BLE & Sensor Code

Arduino IDE Installation[1]

The open-source Arduino Software (IDE) (see Figure 1) makes it easy to write code and upload it to the board. It runs on Windows, Mac OS X, and Linux. The environment is written in Java and based on Processing and another open-source software [1].

Download: You can download Arduino IDE from https://www.arduino.cc/en/Main/Software Once downloaded, you can start using Sketch IDE (see Figure 2).

Arduino Uno Pinout Diagram and Guide

The Arduino Uno Microcontroller is one of the most versatile boards on the market today and many of Case study hardware (Dairy IoT Sensor and iDispenser) used in this book are based on Arduino Uno boards.

The Arduino[2] Uno is based on the **ATmega328** by Atmel. The Arduino Uno pinout consists of 14 digital pins, 6 analog inputs, a power jack, USB connection and ICSP header [2]. The versatility of the pinout provides many different options such as driving motors, LEDs, reading sensors and more.

[1]   Arduino - https://www.arduino.cc/en/Main/Software
[2]   Arduino Pin Diagram - https://www.circuito.io/blog/arduino-uno-pinout/

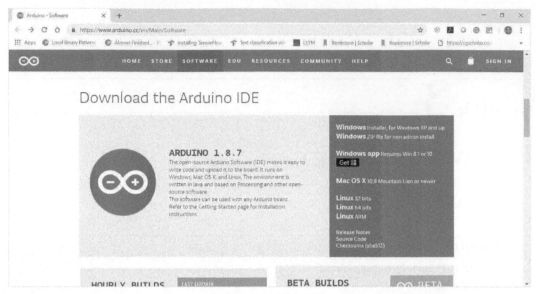

Figure 1: Arduino IDE

Figure 2: Sketch

Capabilities of the Arduino Uno pinout

Let's see Arduino pin layout and its capabilities (see Figure 3)

Arduino Uno Pinout – Power Supply

There are 3 ways to power the Arduino Uno [3]:

Barrel Jack—The Barrel jack, or DC Power Jack can be used to power your Arduino board. The barrel jack is usually connected to a wall adapter. The board can be powered by 5–20 volts but the manufacturer recommends to keep it between 7–12 volts. Above 12 volts, the regulators might overheat, and below 7 volts, might not suffice.

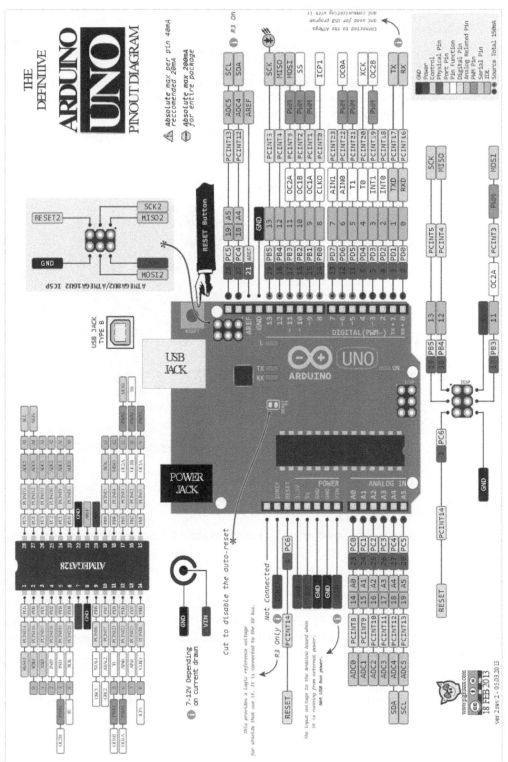

Figure 3: Arduino pin lay-out

VIN Pin—This pin is used to power the Arduino Uno board using an external power source. The voltage should be within the range mentioned above.

USB cable—when connected to the computer, provides 5 volts at 500 mA.

There is a polarity protection diode (see Figure 4) connecting between the positive of the barrel jack to the VIN pin, rated at 1 Ampere.

The power source you use determines the power you have available for your circuit. For instance, powering the circuit using the USB limits you to 500 mA. Take into consideration that this is also used for powering the MCU, its peripherals, the on-board regulators, and the components connected to it. When powering your circuit through the barrel jack or VIN, the maximum capacity available is determined by the 5 and 3.3 volts regulators on-board the Arduino.

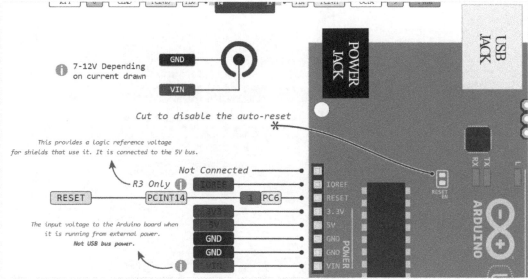

Figure 4: Arduino Power Diagram

5v and 3v3

They provide regulated 5 and 3.3v to power external components according to manufacturer specifications.

GND

In the Arduino Uno pinout, you can find 5 GND pins, which are all interconnected.

The GND pins are used to close the electrical circuit and provide a common logic reference level throughout your circuit. Always make sure that all GNDs (of the Arduino, peripherals and components) are connected to one another and have a common ground.

RESET—resets the Arduino

IOREF—This pin is the input/output reference. It provides the voltage reference with which the microcontroller operates.

Serial Communication

Serial communication is used to exchange data between the Arduino board and another serial devices such as computers, displays, sensors and more. Each Arduino board has at least one serial port. Serial communication occurs on digital pins 0 (RX) and 1 (TX) as well as via USB. Arduino also supports serial communication through digital pins with the Software Serial Library. This allows the user to connect multiple serial-enabled devices and leave the main serial port available for the USB.

I^2C

- SCL/SDA pins are the dedicated pins for I2C communication. On the Arduino Uno they are found on Analog pins A4 and A5.
- I2C is a communication protocol commonly referred to as the "I2C bus". The I2C protocol was designed to enable communication between components on a single circuit board. With I2C, there are 2 wires referred to as SCL and SDA.
- SCL is the clock line which is designed to synchronize data transfers.
- SDA is the line used to transmit data.
- Each device on the I2C bus has a unique address, up to 255 devices can be connected on the same bus.

Tinyduino Humidity Sensor (SI7021)

The TinyDuino Humidity Sensor specification can be available at [3,4]: https://tinycircuits.com/blogs/learn/139739207-temperature-humidity-tinyshield-tutorial.

Arduino library for SI7020 and SI7021 environmental sensors Install as usual in your Arduino/libraries folder, restart IDE.[3]

SI702—Header file

The Header file defines environmental sensors data structure:

```
typedef struct si7021_env {
int celsiusHundredths;
int fahrenheitHundredths;
unsigned int humidityBasisPoints;
} si7021_env;
```

The structure contains Celsius & Fahrenheit Hundredth Second value of temperature had Humidity basis values. Similarly, si7021_thc contains temperature in Celsius and relative humidity values (see Table 1).

---

[3]   https://github.com/LowPowerLab/SI7021

Table 1: Code Listing Si7021

```
/*
Copyright 2014 Marcus Sorensen <marcus@electron14.com>
This program is licensed, please check with the copyright
holder for terms
Updated: Jul 16, 2015: TomWS1: eliminated Byte constants, fixed
'sizeof' error in _command(), added getTempAndRH() function to
simplify calls for C & RH only
*/
#ifndef si7021_h
#definesi7021_h

#if (ARDUINO >= 100)
#include"Arduino.h"
#else
#include"WProgram.h"
#endif

#ifdef __AVR_ATtiny85__
#include"TinyWireM.h"
#define Wire TinyWireM
#elifdefined(ARDUINO_ARCH_ESP8266) || defined(ARDUINO_ARCH_ESP32)
#include<Wire.h>
#else
#if (defined(__AVR__)) || defined(ARDUINO_ARCH_NRF5)
#include<avr/pgmspace.h>
#else
#include<pgmspace.h>
#endif
#include<Wire.h>
#endif
typedefstructsi7021_env {
int celsiusHundredths;
int fahrenheitHundredths;
unsignedint humidityBasisPoints;
} si7021_env;

// same as above but without fahrenheit parameter and RH %
typedefstructsi7021_thc {
int celsiusHundredths;
unsignedint humidityPercent;
} si7021_thc;

classSI7021
{
public:
    SI7021();
#ifdefined(ARDUINO_ARCH_ESP8266) || defined(ARDUINO_ARCH_ESP32)
bool begin(int SDA, int SCL);
#else
bool begin();
#endif
```

Table 1 contd. ...

...Table 1 contd.

```
bool sensorExists();
int getFahrenheitHundredths();
int getCelsiusHundredths();
unsignedint getHumidityPercent();
unsignedint getHumidityBasisPoints();
structsi7021_env getHumidityAndTemperature();
structsi7021_thc getTempAndRH();
int getSerialBytes(byte * buf);
int getDeviceId();
void setPrecision(byte setting);
void setHeater(boolon);
private:
void _command(byte cmd, byte * buf );
void _writeReg(byte * reg, intreglen);
int _readReg(byte * reg, intreglen);
int _getCelsiusPostHumidity();
};

#endif
```

SI720—Implementation File

To retrieve Sensor values (temperature and Humidity) and pass it to connectivity, the following command locations are used (see Table 2):

```
#define I2C_ADDR 0x40

// I2C commands
#define RH_READ              0xE5
#define TEMP_READ            0xE3
#define POST_RH_TEMP_READ    0xE0
#define RESET                0xFE
#define USER1_READ           0xE7
#define USER1_WRITE          0xE6
```

Embedded System Architecture

To get into embedded software architecture and its relationship to design of Internet of Things products and enablement of mobile operation systems (Android & iOS) as edge devices for IoT, the following concepts are very important in design of hardware products that power the IoT.

Embedded systems are programmed much like any other software project. You work within an IDE; write code in, usually, a high-level language; link in support through libraries and frameworks; and, somehow, download the code to the specific piece of hardware.

What does an embedded systems program contain—that is, what would it look like? At its most fundamental, an embedded program contains (1) setup code, (2) support libraries, (3) interrupt handlers, and (4) a processing loop. The heart of embedded software is a processing loop.

Table 2: Temperature and Humidity Listings

```
/*
Copyright 2014 Marcus Sorensen <marcus@electron14.com>
This program is licensed, please check with the copyright holder for terms
Updated: Jul 16, 2015: TomWS1:
        eliminated Byte constants,
        fixed 'sizeof' error in _command(),
        added getTempAndRH() function to simplify calls for C & RH only
*/
#include "Arduino.h"
#include "SI7021.h"
#include <Wire.h>

#define I2C_ADDR 0x40

// I2C commands
#define RH_READ             0xE5
#define TEMP_READ           0xE3
#define POST_RH_TEMP_READ   0xE0
#define RESET               0xFE
#define USER1_READ          0xE7
#define USER1_WRITE         0xE6

// compound commands
byte SERIAL1_READ[]     ={ 0xFA, 0x0F };
byte SERIAL2_READ[]     ={ 0xFC, 0xC9 };

bool _si_exists = false;

SI7021::SI7021() {
}

#if defined(ARDUINO_ARCH_ESP8266) || defined(ARDUINO_ARCH_ESP32)
bool SI7021::begin(int SDA, int SCL) {
    Wire.begin(SDA,SCL);
#else
bool SI7021::begin() {
    Wire.begin();
#endif
    Wire.beginTransmission(I2C_ADDR);
if (Wire.endTransmission() == 0) {
        _si_exists = true;
    }
return _si_exists;
}

bool SI7021::sensorExists() {
return _si_exists;
}

int SI7021::getFahrenheitHundredths() {
int c = getCelsiusHundredths();
return (1.8 * c) + 3200;
}
```

Table 2 contd. ...

...Table 2 contd.

```
intSI7021::getCelsiusHundredths() {
    byte tempbytes[2];
    _command(TEMP_READ, tempbytes);
long tempraw = (long)tempbytes[0] << 8 | tempbytes[1];
return ((17572 * tempraw) >> 16) - 4685;
}

intSI7021::_getCelsiusPostHumidity() {
    byte tempbytes[2];
    _command(POST_RH_TEMP_READ, tempbytes);
long tempraw = (long)tempbytes[0] << 8 | tempbytes[1];
return ((17572 * tempraw) >> 16) - 4685;
}

unsignedintSI7021::getHumidityPercent() {
    byte humbytes[2];
    _command(RH_READ, humbytes);
long humraw = (long)humbytes[0] << 8 | humbytes[1];
return ((125 * humraw) >> 16) - 6;
}
unsignedintSI7021::getHumidityBasisPoints() {
    byte humbytes[2];
    _command(RH_READ, humbytes);
long humraw = (long)humbytes[0] << 8 | humbytes[1];
return ((12500 * humraw) >> 16) - 600;
}

voidSI7021::_command(byte cmd, byte * buf ) {
    _writeReg(&cmd, sizeofcmd);
#ifdefined(ARDUINO_ARCH_ESP8266) || defined(ARDUINO_ARCH_ESP32)
    delay(25);
#endif
    _readReg(buf, 2);
}

voidSI7021::_writeReg(byte * reg, intreglen) {
    Wire.beginTransmission(I2C_ADDR);
for(int i = 0; i <reglen; i++) {
reg += i;
        Wire.write(*reg);
    }
    Wire.endTransmission();
}
intSI7021::_readReg(byte * reg, intreglen) {
    Wire.requestFrom(I2C_ADDR, reglen);
//while(Wire.available() < reglen); //remove redundant loop-wait per https://
github.com/LowPowerLab/SI7021/issues/12
for(int i = 0; i <reglen; i++) {
reg[i] = Wire.read();
    }
```

Table 2 contd. ...

```
return 1;
}
intSI7021::getSerialBytes(byte * buf) {
  byte serial[8];
  _writeReg(SERIAL1_READ, sizeof SERIAL1_READ);
  _readReg(serial, 8);
//Page23 - https://www.silabs.com/Support%20Documents%2FTechnicalDocs%2F
Si7021-A20.pdf
buf[0] = serial[0]; //SNA_3
buf[1] = serial[2]; //SNA_2
buf[2] = serial[4]; //SNA_1
buf[3] = serial[6]; //SNA_0
  _writeReg(SERIAL2_READ, sizeof SERIAL2_READ);
  _readReg(serial, 6);
buf[4] = serial[0]; //SNB_3 - device ID byte
buf[5] = serial[1]; //SNB_2
buf[6] = serial[3]; //SNB_1
buf[7] = serial[4]; //SNB_0
return 1;
}
intSI7021::getDeviceId() {
//0x0D=13=Si7013
//0x14=20=Si7020
//0x15=21=Si7021
  byte serial[8];
  getSerialBytes(serial);
int id = serial[4];
return id;
}

// 0x00 = 14 bit temp, 12 bit RH (default)
// 0x01 = 12 bit temp, 8 bit RH
// 0x80 = 13 bit temp, 10 bit RH
// 0x81 = 11 bit temp, 11 bit RH
voidSI7021::setPrecision(byte setting) {
    byte reg = USER1_READ;
    _writeReg(&reg, 1);
    _readReg(&reg, 1);

    reg = (reg & 0x7E) | (setting& 0x81);
    byte userwrite[] = {USER1_WRITE, reg};
    _writeReg(userwrite, sizeof userwrite);
}
// NOTE on setHeater() function:
// Depending if 'on' parameter it sets it to:
// - 0x3A (0011_1010 - POR default) for OFF
// - 0x3E (0011_1110) for ON
// This resets the resolution bits to 00 in both cases - which is maximum
resolution (12bit RH and 14bit Temp)
// For User1 register usage see p26 (section 6) in DSheet https://www.silabs.
com/documents/public/data-sheets/Si7021-A20.pdf
voidSI7021::setHeater(boolon) {
```

Table 2 contd. ...

...Table 2 contd.

```
    byte userbyte;
if (on) {
        userbyte = 0x3E;
    } else {
        userbyte = 0x3A;
    }
    byte userwrite[] = {USER1_WRITE, userbyte};
    _writeReg(userwrite, sizeof userwrite);
}

// get humidity, then get temperature reading from humidity measurement
structsi7021_envSI7021::getHumidityAndTemperature() {
si7021_env ret = {0, 0, 0};
    ret.humidityBasisPoints      = getHumidityBasisPoints();
    ret.celsiusHundredths        = _getCelsiusPostHumidity();
    ret.fahrenheitHundredths     = (1.8 * ret.celsiusHundredths) + 3200;
return ret;
}

// get temperature (C only) and RH Percent
structsi7021_thcSI7021::getTempAndRH()
{
si7021_thc ret;

    ret.humidityPercent   = getHumidityPercent();
    ret.celsiusHundredths = _getCelsiusPostHumidity();
return ret;

}
```

```
while(true) {

statusData = checkStatus();            // Check if any interrupts happened
processInterrupts(statusData);         // Run any code to handle whatever happened
performPeriodicMaintenance();          // timers, LEDs, polling, etc.

}
```

The processing loop continuously checks for the interrupt status, runs the code to handle the interrupt and performs periodic calibration of timer or other maintenances.

Dairy IoT Sensor

Lifesaving diagnostic IoT Sensor
The following is case study of Dairy IoT Sensor for diagnostic aid purposes: Generally, CLASS 10 classification is reserved to Medical Apparatus such as: Surgical, medical, dental and veterinary apparatus and instruments, artificial limbs, eyes and teeth; orthopedic articles; suture materials. Specifically, in Dairy IoT Sensor case, it has been assigned **"Class 10: Wearable veterinary sensor for use in capturing a cow's vital signs, providing data to the farmer to monitor the cow's milk productivity, and improving its overall health."**

For deploying adaptive edge analytics, we have developed IoT Embedded device (Edge) with following Sensor stack, comprising of hardware, connectivity and sensors.

1. BLE Chip (top)
2. Accelerometer
3. Timer Sensor
4. Proto Terminal Block
5. EEPROM
6. Microcontroller (ATmega328P)
7. Power Supply (bottom)

We have based the sensor design on reference architecture provided by Hanumayamma Innovations and Technologies, Inc., IoT Dairy Sensor.[4]

The embedded system, Figure, is built on top of a high-performance 8-bit AVR RISC-based microcontroller that combines 32KB ISP flash memory with read-while-write capabilities, 1024B EEPROM, 2KB SRAM, 23 general purpose I/O lines, 32 general purpose working registers, three flexible timer/counters with compare modes, internal and external interrupts and serial programmable USART. The device operates between 1.8–5.5 volts.

The embedded system consists of following components: Battery Power Supply, Microcontroller, EPROM, Terminal Blocks, Timer Sensor, Accelerometer, Bluetooth Low Energy (BLE) and Temperature & Humidity Sensors.

Battery Power Supply

Three AA batteries power the system. The total input voltage of 5V.

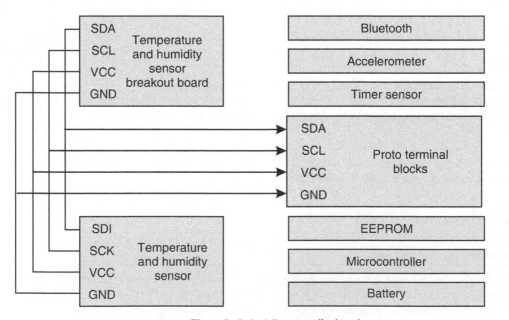

Figure 5: Dairy Microcontroller board

---

4   Hanumayamma Innovations and Technologies, Inc., http://hanuinnotech.com/dairyanalytics.html

Microcontroller

The embedded system is built on top of high-performance 8-bit AVR RISC[5]-based microcontroller [4] that combines 32KB ISP flash memory (see Figure 5: Dairy Microcontroller board) with read-while-write capabilities, 1024B EEPROM, 2KB SRAM, 23 general purpose I/O lines, 32 general purpose working registers, three flexible timer/counters with compare modes, internal and external interrupts and serial programmable USART. The device operates between 1.8–5.5 volts.

EEPROM

EEPROM (see Figure 5: Dairy Microcontroller board) provides nonvolatile data storage support. 0–255 record counters will be stored in EEPROM location zero. 256–336 record counter with maximum of 2 weeks at 24 records/day will be stored in EEPROM location one. Records will start at EEPROM location two. Additionally, the data arrays store on EEPROM and retrieve command over Bluetooth Low Energy (BLE).

Bluetooth

The Bluetooth module (see Table 1) provides an interface to receive commands from mobile application. In addition, the EEPROM stored values are transferred back to the mobile application via Bluetooth transfer protocol.

```
void setup(void)
{
        Wire.begin();
        SerialMonitorInterface.begin(9600);
        // initialize record counters here for testing - must be removed or
power cycling loses all data
        EEPROMwrite(0, 0);
        EEPROMwrite(1, 0);
        delay(500);
        if (!htu.begin()) {
                Serial.println("Couldn't find HTU21 sensor!");
                while (1);
        }
        if (!bme.begin()) {
                Serial.println("Could not find a valid BME280 sensor, check wir-
ing!");
                while (1);
        }
        //Case LED Support.
        pinMode(A3, OUTPUT);
        int tickEvent = t.every(3600000, HourlySensorReadAndStore); // hourly
data grab 3600000 is an hour
        accel.begin(BMA250_range_2g, BMA250_update_time_64ms);//This sets up
the BMA250 accelerometer
        RTC.start(); // ensure RTC oscillator is running, if not already
        BLEsetup();
}
```

Figure 6: Bluetooth setup

[5] AVR RISC – Advanced Virtual reduced instruction set computer (http://www.atmel.com/products/microcontrollers/avr/)

Setup

The Sensor Module kicks-in by running setup command (see Figure 7). As part of the setup, Serial Monitor Interface with 9600 milliseconds, initiates humidity and temperature sensor, initiates BME 280 Sensor module, starts real-time clock, and initiates Bluetooth.

Continuous Loop

Sensor is a state machine with continuous event loop (see Table 3) capture Cow movement, temperature, Humidity and Data & Time Stamp. The State machine (see Figure 7) initiates the call as in the following diagram:

- Initiate Bluetooth Radio (call to aci_loop())
- Update Motion Detection (UpdateMotionDetection())
- Check Mobile Request for the Connection (Check module Command)
- Transfer Data (Data Transfer)

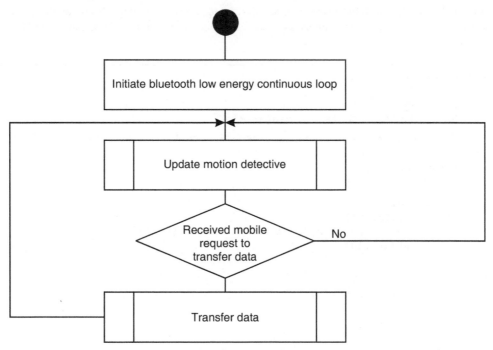

Figure 7: State Diagram

Table 3: Loop Code

```
void loop() {
      aci_loop();//Process any ACI commands or events from the NRF8001- main
BLE handler, must run often. Keep main loop short.
      UpdateMotionDetection();
      t.update(); // time to grab sensor snapshot and store? if so call Hour-
lySensorReadAndStore.
      Send Command Data();
```

Libraries

Dairy IoT Sensor is a real-time RTOS based Sensor with Temperature, Humidity, and Activity Level detection for Cattle. It's an agriculture sensor for Veterinarian purposes.

The following libraries used (see Table 4):

- Bluetooth Low Energy Sensor

Table 4: Dairy Sensor Libraries code

```
#include <Event.h>
#include <Timer.h>

//Bluetooth Section

//when using this project in the Arduino IDE, set the following to false
and rename file UART.h to UART.ino
#define CODEBENDER false

#if defined(ARDUINO_ARCH_SAMD)
#define SerialMonitorInterface SerialUSB
#else
#define SerialMonitorInterface Serial
#endif

#define BLE_DEBUG false

#include <SPI.h>
#include "lib_aci.h"
#include "aci_setup.h"
#include "uart_over_ble.h"
#include "services.h"
#include "kalman.h"

#if CODEBENDER
#include "UART.ino"
#endif
// Sensor Support Section
#include <Wire.h>
#include "Adafruit_HTU21DF.h"
#include <Adafruit_Sensor.h>
#include <Adafruit_BME280.h>
#define SEALEVELPRESSURE_HPA (1013.25) //necessary to know at local loca-
tion for precision elevation measurement.
Adafruit_BME280 bme; // I2C
Adafruit_HTU21DF htu = Adafruit_HTU21DF();

#include "BMA250.h"
BMA250 accel;

        //RTC board support
#include "DSRTCLib.h"
DS1339 RTC = DS1339();

// Periodic hourly timer support for data storage, non blocking delay.
//#include "Timer.h"  //http://github.com/JChristensen/Timer
Timer t;                 //instantiate the timer object
```

- Adafruit BME 280 I2C or SPI Temperature and Humidity Sensors
- BMA250 [5] Digital, triaxial acceleration Sensor[6]
- DS 1339 I2C [6] Serial Real-Time Clock[7]

Bluetooth Low Energy Sensor

Dairy IoT Sensor needs to communicate the stored sensors data to the Farmer mobile (Android & iOS). The Sensor has on-board Bluetooth Module [7] and is used Tiny Circuits Bluetooth.[8] Bluetooth Low Energy (BLE) TinyShield combines both master and slave modes into one module!

The Dairy Sensor employs Low Energy Communication Sensors for saving Farmer's battery.

The Bluetooth communication is initiated by calling aci_loop().

Application Control Interface (ACI):

The flow control is credit-based and the credit is initially given using the "device started" event. A credit of more than 1 is given to the application MCU. These credits are used only after the "ACI Connected Event" is sent to the application MCU.

Every send_data that is used decrements the credit available by 1. This is to be tracked by the application MCU. When the credit available reaches 0, the application MCU shall not send any more send_data. Credit is returned using the "credit event", this returned credit can then be used to send more send_data. This flow control is not necessary and not available for Broadcast. The entire credit available with the external MCU expires when a "disconnected" event arrives.

When a command is sent over the ACI, the next command shall not be sent until after a response for the command sent (see Table 5) [8,9] has arrived.[9]

Update Motion Detection

Updates Cattle motion based on most recent accelerometer values (see Table 6).

The swap of old temperature and humidity values are required to preserve the historical values.

Transfer Data

The Transfer (see Figure 8 & Table 7) to Mobile is performed based on commands sent to the Sensor by the phone.

The following commands will result in the corresponding actions:

- Send Ambient Temperature
- Send Ambient Humidity
- Send Cattle Temperature
- Send Cattle Humidity
- Send Cattle Activity
- Send Record Count
- Send Read Pointer

---

[6]   BMA Acceleration Sensor - http://www1.futureelectronics.com/doc/BOSCH/BMA250-0273141121.pdf

[7]   Real-Time Clock - https://www.maximintegrated.com/en/products/digital/real-time-clocks/DS1339.html

[8]   Tiny Circuits BLE Module - https://tinycircuits.com/products/bluetooth-low-energy-tinyshield

[9]   ACI - https://devzone.nordicsemi.com/f/nordic-q-a/6487/

Table 5: ACI_LOOP Code

```
void aci_loop()
{
        static bool setup_required = false;

        // We enter the if statement only when there is a ACI event available
to be processed
        if (lib_aci_event_get(&aci_state, &aci_data))
        {
                aci_evt_t * aci_evt;
                aci_evt = &aci_data.evt;

                switch (aci_evt->evt_opcode)
                {
                case ACI_EVT_DEVICE_STARTED: // As soon as you reset the
nRF8001 you will get an ACI Device Started Event
                    {
                        aci_state.data_credit_available = aci_evt->params.de-
vice_started.credit_available;
                        switch (aci_evt->params.device_started.device_mode)
                        {
                        case ACI_DEVICE_SETUP: //When the device is in the setup
mode
                                aci_state.device_state = ACI_DEVICE_SETUP;
                                Serial.println(F("Evt Device Started: Setup"));
                                setup_required = true;
                                break;
                        case ACI_DEVICE_STANDBY:
                                aci_state.device_state = ACI_DEVICE_STANDBY;

                        Serial.println(F("Evt Device Started: Standby"));
                                if (aci_evt->params.device_started.hw_error)
                                {
                                        delay(20); //Magic number used to make
sure the HW error event is handled correctly.
                                }
                                else
                                {
                                        lib_aci_connect(180/* in seconds */,
0x0100 /* advertising interval 100ms*/);
                                        Serial.println (F("Advertising started"));
                                }
                                break;
                        }
                }
        break; //ACI Device Started Event
                    case ACI_EVT_CMD_RSP:
                                //If an ACI command response event comes with an error
-> stop
                                if (ACI_STATUS_SUCCESS != aci_evt->params.cmd_rsp.cmd_
status)
                                {
                                        //ACI ReadDynamicData and ACI WriteDynamicData
will have status codes of
                                        //TRANSACTION_CONTINUE and TRANSACTION_COMPLETE
                                        //all other ACI commands will have
status code of ACI_STATUS_SCUCCESS for a successful command
                                        Serial.print(F("ACI Status of ACI Evt Cmd Rsp"));
```

Table 5 contd. ...

...Table 5 contd.

```
                              Serial.println(aci_evt->params.cmd_rsp.cmd_status,
HEX);
                              Serial.print(F("ACI Command 0x"));
                              Serial.println(aci_evt->params.cmd_rsp.cmd_opcode,
HEX);
                              Serial.println(F("Evt Cmd respone: Error. Arduino
is in a while(1); loop"));
                              while (1);
                    }
                    break;
            case ACI_EVT_PIPE_STATUS:
                    Serial.println(F("Evt Pipe Status"));
                    // Check if the peer has subscribed to any particleBox
characterisitcs for notifications
                    if (lib_aci_is_pipe_available(&aci_state, PIPE_AIR_QUAL-
ITY_SENSOR_TEMPERATURE_MEASUREMENT_TX)
                            && (false == timing_change_done))
                    {
                            // Request a change to the link timing as set in
the GAP -> Preferred Peripheral Connection Parameters
                            lib_aci_change_timing_GAP_PPCP();
                            timing_change_done = true;
                    }
                    if (lib_aci_is_pipe_available(&aci_state, PIPE_AIR_QUAL-
ITY_SENSOR_RELATIVE_HUMIDITY_TX)
                            && (false == timing_change_done))
                    {
                            // Request a change to the link timing as set in
the GAP -> Preferred Peripheral Connection Parameters
                            lib_aci_change_timing_GAP_PPCP();
                            timing_change_done = true;
                    }
                    if (lib_aci_is_pipe_available(&aci_state, PIPE_AIR_QUAL-
ITY_SENSOR_CARBON_MONOXIDE_LEVEL_TX)
                            && (false == timing_change_done))
                    {
                            // Request a change to the link timing as set in
the GAP -> Preferred Peripheral Connection Parameters
                            lib_aci_change_timing_GAP_PPCP();
                            timing_change_done = true;
                    }
                    if (lib_aci_is_pipe_available(&aci_state, PIPE_AIR_QUAL-
ITY_SENSOR_PM10_CONCENTRATION_TX)
                            && (false == timing_change_done))
                    {
                            // Request a change to the link timing as set in
the GAP -> Preferred Peripheral Connection Parameters
                            lib_aci_change_timing_GAP_PPCP();
                            timing_change_done = true;
                    }
                    if (lib_aci_is_pipe_available(&aci_state, PIPE_AIR_QUAL-
ITY_SENSOR_PM25_CONCENTRATION_TX)
                            && (false == timing_change_done))
                    {
                            // Request a change to the link timing as set in
the GAP -> Preferred Peripheral Connection Parameters
```

Table 5 contd. ...

...Table 5 contd.

```
                              lib_aci_change_timing_GAP_PPCP();
                              timing_change_done = true;
                        }
                        break;
              case ACI_EVT_TIMING: // Link timing has changed
                        Serial.print(F("Timing changed: "));
                        Serial.println(aci_evt->params.timing.conn_rf_interval,
HEX);
                        break;
              case ACI_EVT_CONNECTED:
                        radio_ack_pending = false;
                        timing_change_done = false;
                        aci_state.data_credit_available = aci_state.data_credit_
total;
                        Serial.println(F("Evt Connected"));
                        break;
              case ACI_EVT_DATA_CREDIT:
                        aci_state.data_credit_available = aci_state.data_credit_
available + aci_evt->params.data_credit.credit;
                        /**
                        Bluetooth Radio ack received from the peer radio for the
data packet sent.
                        This also signals that the buffer used by the nRF8001
for the data packet is available again.
                        */
                        radio_ack_pending = false;
                        break;
              case ACI_EVT_PIPE_ERROR:
                        /**
                        Send data failed. ACI_EVT_DATA_CREDIT will not come.
                        This can happen if the pipe becomes unavailable by the
peer unsubscribing to the Heart Rate
                        Measurement characteristic.
                        This can also happen when the link is disconnected after
the data packet has been sent.
                        */
                        radio_ack_pending = false;
                        //See the appendix in the nRF8001 Product Specication
for details on the error codes
                        Serial.print(F("ACI Evt Pipe Error: Pipe #:"));
                        Serial.print(aci_evt->params.pipe_error.pipe_number,
DEC);
                        Serial.print(F(" Pipe Error Code: "));
                        Serial.println(aci_evt->params.pipe_error.error_code,
HEX);
                        //Increment the credit available as the data packet was
not sent.
                        //The pipe error also represents the Attribute protocol
Error Response sent from the peer and that should not be counted
                        //for the credit.
                        if (ACI_STATUS_ERROR_PEER_ATT_ERROR != aci_evt->params.
pipe_error.error_code)
                        {
                                aci_state.data_credit_available++;
                        }
                        break;
```

Table 5 contd. ...

```
                case ACI_EVT_DISCONNECTED:
                        // Advertise again if the advertising timed out
                        if (ACI_STATUS_ERROR_ADVT_TIMEOUT == aci_evt->params.
disconnected.aci_status)
                        {
                                Serial.println(F("Evt Disconnected -> Advertising
timed out"));
                                {
                                        Serial.println(F("nRF8001 going to
sleep"));
                                        lib_aci_sleep();
                                        aci_state.device_state = ACI_DEVICE_SLEEP;
                                }
                        }
                        if (ACI_STATUS_EXTENDED == aci_evt->params.disconnected.
aci_status)
                        {
                                Serial.print(F("Evt Disconnected -> Link lost.
Bluetooth Error code = "));
                                Serial.println(aci_evt->params.disconnected.btle_
status, HEX);
                                lib_aci_connect(180/* in seconds */, 0x0050 /*
advertising interval 50ms*/);
                                Serial.println(F("Advertising started"));
                        }
                        break;
                case ACI_EVT_HW_ERROR:
                        Serial.print(F("HW error: "));
                        Serial.println(aci_evt->params.hw_error.line_num, DEC);
                        for (uint8_t counter = 0; counter <= (aci_evt->len - 3);
counter++)
                        {
                                Serial.write(aci_evt->params.hw_error.file_
name[counter]); //uint8_t file_name[20];
                        }
                        Serial.println();
                        lib_aci_connect(30/* in seconds */, 0x0100 /* advertis-
ing interval 50ms*/);
                        Serial.println(F("Advertising started"));
                        break;
                }
        }
        else
        {
                // Serial.println(F("No ACI Events available"));
                // No event in the ACI Event queue and no event in ACI comman
queue
                // Arduino can go to sleep now
        }
        /* setup_required is set to true when the device starts up and enters
setup mode.
        * It indicates that do_aci_setup() should be called. The flag should be
cleared if
        * do_aci_setup() returns ACI_STATUS_TRANSACTION_COMPLETE.
        */
        if (setup_required)
        {
```

Table 5 contd. ...

...Table 5 contd.

```
              if (SETUP_SUCCESS == do_aci_setup(&aci_state))
              {
                      setup_required = false;
              }
       }
}
```

Table 6: Update Motion Code

```
// Cow Motion detection and running count. Y-axis values increment a counter
only
// when above or below a threshold (sensitivity var) of the last reading.
void UpdateMotionDetection() {
      accel.read();
      newTempY = abs(accel.Y);
      if ((abs(newTempY - oldTempY)) > (sensitivity)) {
              activityCount = activityCount + 1;
              oldTempY = newTempY;
      }
}
```

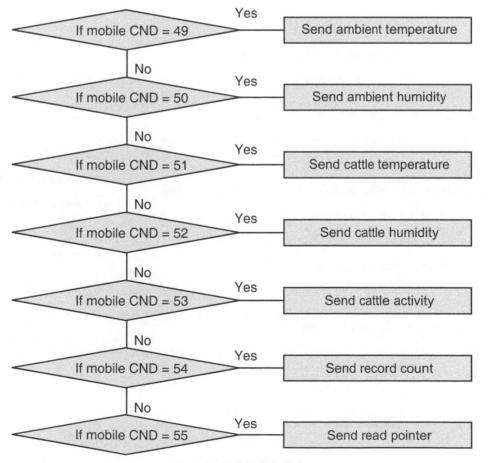

Figure 8: Transfer Data Code

Table 7: Transfer Code

```
//Receive from Celphone App after connect and command decode section.
        if (ble_rx_buffer_len) {//Check if data is available
                moduleCommand = ((char*)ble_rx_buffer[0]); //look at first char
in buffer
                if (moduleCommand == 49) { // cmd "1"
                        SendKelman();
                        SendAmbientTemp();
                }
                if (moduleCommand == 50) { // cmd "2"
                        SendAmbientHumdity();
                }
                if (moduleCommand == 51) { // cmd "3"
                        SendCowTemp();
                }
                if (moduleCommand == 52) { // cmd "4"
                        SendCowHumidity();
                }
                if (moduleCommand == 53) { // cmd "5"
                        SendCowActivity();
                }
                if (moduleCommand == 54) { // cmd "6"
                        SendDateTime();
                }
                if (moduleCommand == 55) { // cmd "7"
                        SendRecordCount(); // count of current number of records
stored
                }
                if (moduleCommand == 56) { // cmd "8"
                        SendRecordPartOne(); //send timedate and 4 sensor values
part of record, MUST be followed by cmd "9"
                }
                if (moduleCommand == 57) { // cmd "9", MUST be preceeded by cmd
"8"
                        SendRecordPartTwo(); //send timedate and cow activity
value part of record
                }
                if (moduleCommand == 48) { // cmd "0" // zero
                        ResetCurrentRecordCount(); // explicit clear of number
of records available to return
                                                        // and reset both
eeprom read and write address counters to beginning
                                                        // read and write
addresses of 02 which is first record in eeprom.
                                                        // This is the cor-
rect way to end a reading session and log will
                                                        // now begin to
overwrite starting at first record.
                }
                if (moduleCommand == 82) { // cmd "R" // capital letter R -
used only for testing it
                        ResetRecordPointer(); // resets base address of record
storage so can read again from first
                                                        // record without
clearing current record count. Does not affect write address counter.
                }
                if (moduleCommand == 75) { // cmd "K"
                        SendKelman();
                }
                ble_rx_buffer_len = 0;//clear after reciept
                DataTransferBlink();
        }
```

Table 7 contd. ...

Chapter Summary:
- Covers in-depth Arduino board and should enable the design of the Arduino based Sensors
- Write code to initiate Bluetooth & Temperature/Humidity value reads
- Should enable design of Microcontroller & sensor board

References

1. Facchinm. Download the Arduino IDE. May 21, 2018, https://www.arduino.cc/en/Main/Software, Access date: September 18, 2018
2. Anat Zait. An Introduction to Arduino Uno Point. April 22, 2018, https://www.circuito.io/blog/arduino-uno-pinout/ Access Date: September 10, 2018
3. Ken Burns, Ben Rose and Tiny Circuits Staff. Temperature/Humidity Tiny Shield. March 22 2016, https://tinycircuits.com/blogs/learn/139739207-temperature-humidity-tinyshield-tutorial, Access Date: September 10, 2018
4. Mina Khan and Atmel Software Developers. AVR RISC – Advanced Virtual reduced instruction set computer. March 2016, http://www.atmel.com/products/microcontrollers/avr/Access Date: September 10, 2018
5. Bosch Product Group, BMA 250 – Digital, Triaxial Accelerometer, 03 March 2011, http://www1.futureelectronics.com/doc/BOSCH/BMA250-0273141121.pdf, Access Date: September 10, 2018
6. Maxim Software Group, I²C Serial Real-Time Clock, Feb 2018, https://www.maximintegrated.com/en/products/digital/real-time-clocks/DS1339.html, Access date: August 06 2018
7. Mouser Development Group, Bluetooth Development Tool Kit, December 2017, https://www.mouser.com/ProductDetail/TinyCircuits/ASD2116-R?qs=5aG0NVq1C4whKPU%252bTAaEBw%3D%3D, Access date: August 06 2018
8. Carltonlee. 2015. Nordic, Multiple notifications on multiple characteristics, https://devzone.nordicsemi.com/f/nordic-q-a/6487/multiple-notifications-on-multiple-characteristics, Access Date: September 08, 2018
9. Tiny Circuits, Bluetooth Low Energy Tinyshield (ST), https://tinycircuits.com/products/bluetooth-low-energy-tinyshield

---

[i] ATmega328P - http://www.microchip.com/wwwproducts/en/ATmega328P

## CHAPTER 4

IoT Data Sources

This Chapter Covers:
- Basic Sensors for IoT Applications
- Sensors handling in the Android and iOS Platform
- Other Sensor data types
  - Audio Data
  - Video Data
  - Geospatial Data

The purpose of the chapter is to provide an overview of several different types of IoT devices that generate data at various velocities and that generate with several different formats and that would generate with different frequencies. The Devices are diverse, and there are no rules about size, form factor or origin. For instance, some devices will be the size of small chip, some will be as large as a connected vehicle or Oil drilling equipment. The devices include a wide range of hardware with various capabilities. Additionally, the rate of new devices developed by entrepreneurs and IoT by manufacturer over the world is steadily increasing. In general, as per Cisco IoT reference architecture,[1] [1] the IoT device must contain (see Figure 1):

- Analog to Digital (A/D) Conversion, as required
- Generating data
- Being queried/Controlled Over the net

The data sources include basic sensors, mobile devices, cameras, audio systems, video systems, geospatial sensors, industrial equipment, satellite imagery, edge devices that are closes to earth and telecommunication equipment. The velocity includes data generation frequencies in microsecond, milliseconds, and seconds. The format includes intrinsic sensor values (Humidity, temperature, etc.) and compound data values (Audio & video signal).

---

[1] Cisco IoT Reference Model - http://cdn.iotwf.com/resources/71/IoT_Reference_Model_White_Paper_June_4_2014.pdf

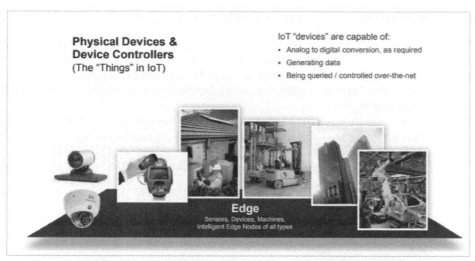

Figure 1: IoT Edge

Enterprise IoT Data Sources

The World generates more than 2 Exabytes of data every day [1]. Connected Smart Sensor Objects (IoT) generate Big Data.[2] For instance, 46 million Smart meters in US alone generate 1.1 billion data points per day [1]. Similarly, a single CPG (Consumer Packaged Good) manufacturing machine generates 13 MB data samples per day [2] (see Figure 2).

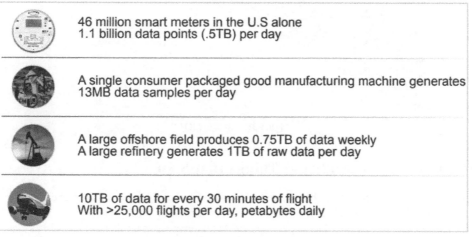

Figure 2: IoT Data Pay loads [2]

---

[2]   Connected Objects generate Big Data - https://www.cisco.com/web/offer/emear/38586/images/Presentations/P11.pdf

Connected Sensors & Objects Generate Huge Data

Given, today's data centric world that we live in, enterprises apply several tools and technologies in order to handle the data that is needed for business growth or sustainability purposes. The data, importantly, has become a competitive differentiator. No one tool nor one technology fulfills the needs of storing, processing and reporting of data. Small or large organizations need a data platform to transit from traditional businesses to data driven organizations.[3]

It would be very tough to pin-point enterprise IoT Data Sources. As listed in Table—Enterprise IoT Data Sources (see Figure 4), it has seven major sources, all are required in order to assess the value generated by IoT to the enterprises (see Figure 3) [4].

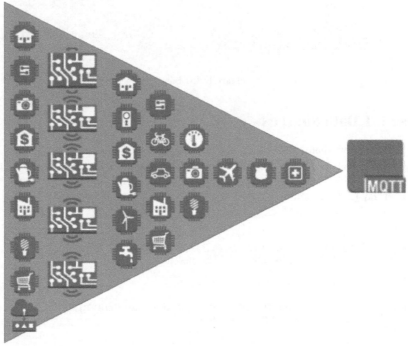

Figure 3: Assets to MQTT

Physical Asset Perspective—"Things that Spin"

As per Physical Asset Perspective championed by General Electrical Industrial Internet of Things (IIOT),[4] the Industrial IoT sources (see Figure 4) include sensors & rotating machinery.

All the sensors that capture data as per Physical Asset Perspective could be employed as data sources (see Table 1):

---

[3] Building a Great Data Platform - https://www.mckinsey.com/industries/electric-power-and-natural-gas/our-insights/building-a-great-data-platform

[4] GE IIOT - https://www.ge.com/docs/chapters/Industrial_Internet.pdf

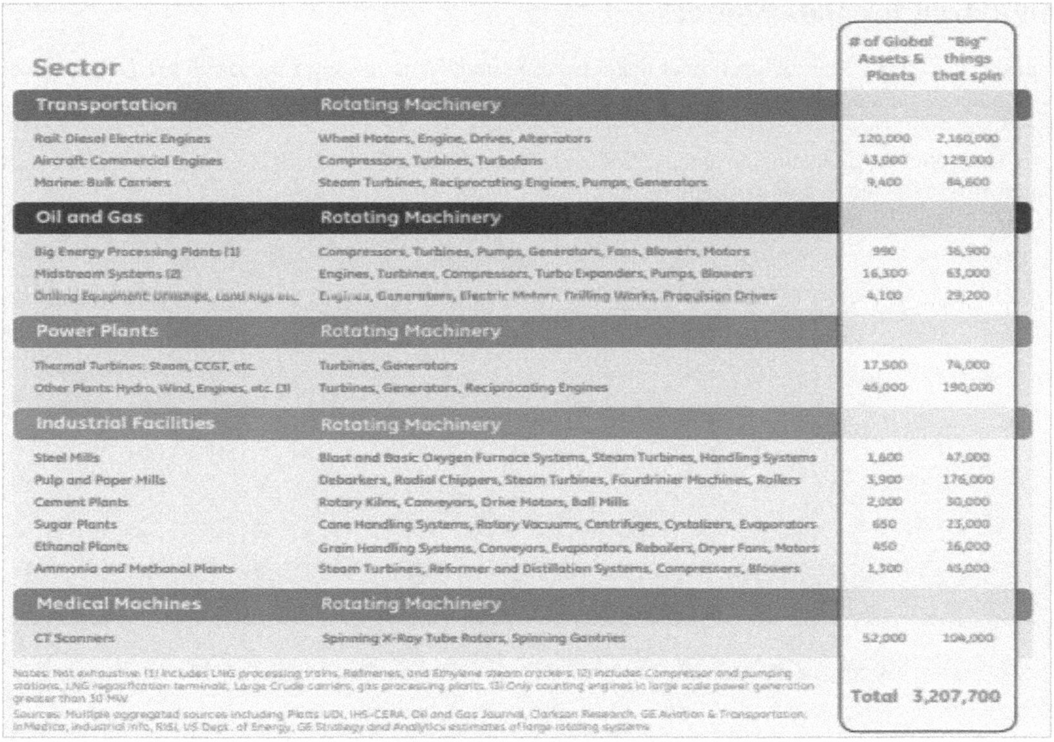

Sector	Rotating Machinery	# of Global Assets & Plants	"Big" things that spin
Transportation	**Rotating Machinery**		
Rail: Diesel Electric Engines	Wheel Motors, Engine, Drives, Alternators	120,000	2,160,000
Aircraft: Commercial Engines	Compressors, Turbines, Turbofans	43,000	129,000
Marine: Bulk Carriers	Steam Turbines, Reciprocating Engines, Pumps, Generators	9,400	84,600
Oil and Gas	**Rotating Machinery**		
Big Energy Processing Plants (1)	Compressors, Turbines, Pumps, Generators, Fans, Blowers, Motors	990	36,900
Midstream Systems (2)	Engines, Turbines, Compressors, Turbo Expanders, Pumps, Blowers	16,300	63,000
Drilling Equipment: Drillships, Land Rigs etc.	Engines, Generators, Electric Motors, Drilling Works, Propulsion Drives	4,100	29,200
Power Plants	**Rotating Machinery**		
Thermal Turbines: Steam, CCGT, etc.	Turbines, Generators	17,500	74,000
Other Plants: Hydro, Wind, Engines, etc. (3)	Turbines, Generators, Reciprocating Engines	46,000	190,000
Industrial Facilities	**Rotating Machinery**		
Steel Mills	Blast and Basic Oxygen Furnace Systems, Steam Turbines, Handling Systems	1,600	47,000
Pulp and Paper Mills	Debarkers, Radial Chippers, Steam Turbines, Fourdrinier Machines, Rollers	3,900	176,000
Cement Plants	Rotary Kilns, Conveyors, Drive Motors, Ball Mills	2,000	30,000
Sugar Plants	Cane Handling Systems, Rotary Vacuums, Centrifuges, Cystalizers, Evaporators	650	23,000
Ethanol Plants	Grain Handling Systems, Conveyors, Evaporators, Reboilers, Dryer Fans, Motors	450	16,000
Ammonia and Methanol Plants	Steam Turbines, Reformer and Distillation Systems, Compressors, Blowers	1,300	45,000
Medical Machines	**Rotating Machinery**		
CT Scanners	Spinning X-Ray Tube Rotors, Spinning Gantries	52,000	104,000
		Total	**3,207,700**

Notes: Not exhaustive (1) Includes LNG processing trains, Refineries, and Ethylene steam crackers (2) Includes Compressor and pumping stations, LNG regasification terminals, Large Crude carriers, gas processing plants (3) Only counting engines in large scale power generation greater than 30 MW

Sources: Multiple aggregated sources including Platts UDI, IHS-CERA, Oil and Gas Journal, Clarkson Research, GE Aviation & Transportation, InMedica, Industrial Info, RISI, US Dept. of Energy, GE Strategy and Analytics estimates of large rotating systems

Figure 4: GE IIOT

Table 1: Enterprise IoT Data Sources

Data Sources	Example	Data Format	Case Study
Industrial Control Systems[5]	SCADA - Supervisory control and data acquisition - Systems	Time-Series Historical Data Machine Data	Preventive maintenance
Business Applications	CRM Data ERP Data	Structure Data	Customer Centric
Wearables	Employee Badge Data RFID Data	Location Data	Employee Performance
Sensors & Devices	Real-Time Sensor Data: Temperature, Humidity, Location, Access Control, Surveillance Data	Basic Sensors data, GPS Latitude/ Longitude, Time & Resource Access	Employee Engagement
Open & Web Data	Open Source Data like NYC Open Data,[6] Traffic & Geospatial Data	Structured/ Unstructured	Click Stream and Safety Data
Media	Camera Data	Unstructured	Safety & Security
Location	Gel Location, GPS	Location	User gets location data

5 7 Types of Industrial IoT Data - http://xmpro.com/7-types-industrial-iot-data-sources/

6 NYC Open Data - https://data.cityofnewyork.us/Transportation/Real-Time-Traffic-Speed-Data/qkm5-nuaq

Industrial IoT Data Sources

Industrial IoT Data Sources itself does not solve case study. It is fusion of Industrial IoT Data Sources with other disparate sources would enable solutioning of several business cases. As an example, please find NYC Real-Time [6] (see Figure 5) Traffic Speed Data.[7]

The common denominator among "Industrial Data Sources", GE's IIOT, and Seven Sources of Enterprise IoT is the Sensor that captures data.

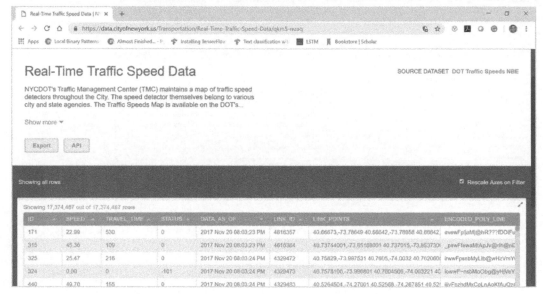

Figure 5: New York Data Sources

Sensors

A Sensor is a device that converts a physical phenomenon into an electrical signal. Generally, a Sensor represents the part of interface between the physical world and the world of electrical or electronic devices, such as computers and/or mobile phones.

In the diagram (see Figure 6), the physical world (left side) activities are translated into electrical signals and fed to electronic devices (Phone, Computer, etc.) through the Sensors.

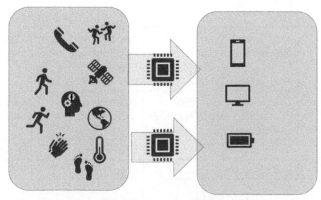

Figure 6: Sensor activities

---

[7]   NYC Data - https://data.cityofnewyork.us/Transportation/Real-Time-Traffic-Speed-Data/qkm5-nuaq

In an IoT environment, high sensitivity sensors result in more number of data events and demand a low latency closed loop behavior.

Case Study: Investment in IoT is investment in our future generations' Safety & Security

IoT Data sources are the best means to protect our planet from global warming and Climate change affects. Through IoT Devices data collection, one can deduce the impacts of global warming and can take proactive steps to building engineering products, policies, and government regulations that can help humanity move forward.

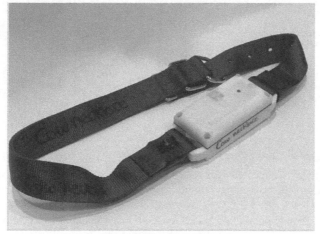

Figure 7: Dairy IoT Sensor

Consider Dairy IoT Sensor (see Figure 7):

Climate change is impacting milk production worldwide. For instance, increased heat stress in cows is causing average-sized dairy farms to lose thousands of gallons of milk each year; drastic climate change, especially in developing countries, is pushing small farmers below the poverty line and triggering suicides due to economic stress and social stigma. It's profoundly clear that current dairy agriculture practices are falling short in countering the impacts of climate change. What we need are innovative and intelligent dairy farming techniques that employ the best of traditional practices with data infused insights that can counter negative effects of climate change. However, there weren't any heat stress-to-climate models, nor the data. To overcome the data challenge, we have developed low-cost ($20 per cattle in India) dairy sensors and deployed in Punjab and Telangana states and collected, at every 5-minute interval from 09/18/2015 to 11/22/2017, cattle activity, body temperature, and humidity data. The application of data science and machine learning technique is known as Decision Tree (ID3 Iterative Dichotomiser) on sensor collected data (around 168 million records) with medication, geospatial (ambient temperature/ humidity) data has helped us to develop cattle heat stress-to-climate decision models with an accuracy of around 73%. Similarly, the use of linear regression and Analysis of variance (ANOVA) techniques on the collected data helped us to develop water consumption and milk productivity forecasting models with an accuracy of 65%. The deployment of these data science algorithms into sensors, making it a precision sensor, enabled us to deliver early warning heat-stress related notifications to small farmers. This has led small farmers to take proactive actions, such as cooling dairy cattle through watering or improving dairy farm ventilation air-flow systems. The water consumption and milk productivity forecasting models helped us to improve overall financials and productivity outcomes for small farmers. Finally, we staunchly believe that perpetual learning, by man and machine, and dissemination of information to small farmers is the best defense against the negative effects of climate change and, in this regard, we would like to provide our precision dairy sensors to small farmers all over the world.

Industrial Case Study: Connected Bus and Mass Transportation[8]

In the connected Bus, the sources of sensors are: GPS Radio, Telemetry Transmitter, Sensors, Digital screens and Digital Cameras [2]

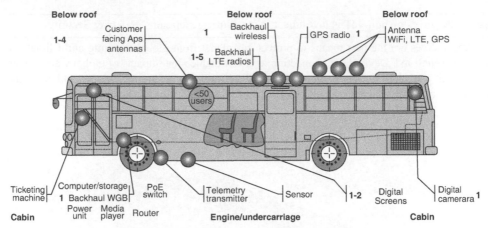

Figure 8: The Connected Bus

The Sensors integrated Connected Bus:

The Sensors integrated Connected Bus (see Figure 8) encompasses various sensors that include environmental sensors (temperature, Humidity), Geo-positional sensors (GPS), activity sensors (Accelerometer) and Video cameras. The data collected by these sensors are processed on premises (Bus) and analyzed for any immediate closed loop actions. The major sensor modules include:

1. Backhaul Work Group Bridge (WGB)[9]
2. Digital Screens
3. Customer Facing APIs/Antennas
4. Backhaul/LTE Radios

The Connected Bus enables case study that define Smart City and Smart Transportation. The principal advantage of Smart Connected Bus is that it enables Safety & Security and Fleet management for the cities.

Sensors Performance and Characteristic Definitions

The following are some of the important characteristics of sensors:

Transfer Function

A Sensor translates an input signal into electrical signal (see Figure 9). Transfer function represents the relationship between input signal and output electrical signal.

In the figure below, the transfer function relates the output signal Volts to input signal Temperature (a classical temperature sensor—please see Figure 10).

---

[8]   Cisco Connected IoT Case Studys - https://www.cisco.com/web/offer/emear/38586/images/Presentations/P11.pdf

[9]   Cisco Mobility - https://www.cisco.com/c/en/us/support/docs/wireless/4400-series-wireless-lancontrollers/111902-outdoor-mobi-guide-00.pdf

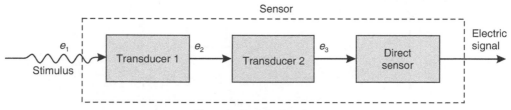

Figure 9: Transfer Function

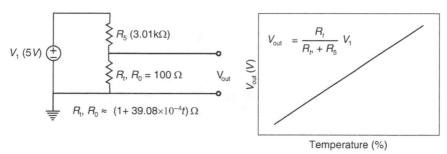

Figure 10: Temperature Circuit

Sensitivity

The relationship between input physical signal and output electrical signal is defined in terms of sensitivity. In other words, the small change in electrical signal to the change in the input signal. Mathematically, a derivative function defines the sensitivity. For instance, a high sensitivity temperature sensor results into larger reading (Voltage) for a small temperature change.

	In an IoT environment, high sensitivity sensors result in a greater number of data events and demand a low latency closed loop behavior.

Span or Dynamic Range

A dynamic rage or span is the range of input signals that are converted into electrical signals by the sensor. The wider the span, the greater the number of events being generated.

	In an IoT environment, a high sensitivity & high span sensor result in a greater number of data events with higher magnitude of output signal.

Hysteresis

Some sensors do not return to the same output value when the input signal is scaled up or down. Hysteresis is the width of the output signal error in terms of measured quantity.

For instance, in the figure below, presence of Hysteresis error is clearly depicted in terms of the width of graph of pressure difference (see Figure 11).

	In an IoT deployment, the Hysteresis error results in error in data and huge number of service calls!

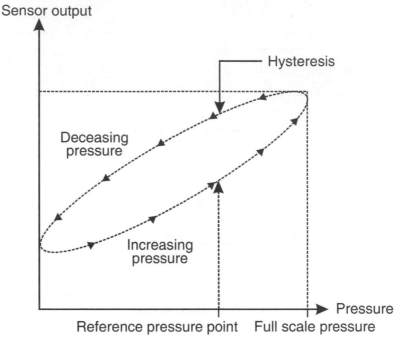

Figure 11: Hysteresis Curve

Case study

Hanumayamma Dairy IoT Sensor

Hanumayamma Dairy IoT Sensors are a CLASS 10 Veterinary Diagnostics Cow Sensor. We have observed the reporting of humidity values greater than 100% during the five month period of October 30, 2016 to January 18, 2017. The x-axis 0:00 represents midnight (12:00 am) and value 9.36 data at morning 9:36 am (see Figure 12).

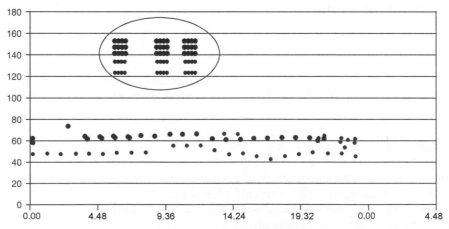

Figure 12: Sensor residual values

Later we discovered that the residual values of humidity on the Sensor casing had resulted in accumulation of hysteresis error range and triggered a huge number of service calls from Farmers.

Other important Sensor attributes include Noise, Bandwidth and Resolution.

Given various types of input signals, the sensors come in various types and form factors to convert physical activities to electrical signals.

Type of Sensors

Given the wide range input physical activities and the corresponding sensors to translate into electrical signals, classifying sensors is a very complex process. On a broad case study level, in this book, we have divided sensors based on their availability to the public. A good example would be the Smartphone mobile world (see Figure 13) [7]. The Mobile Smartphone is the universal edge device that is the center of the IoT, connecting both individual aspirations and enterprise business opportunities.[10]

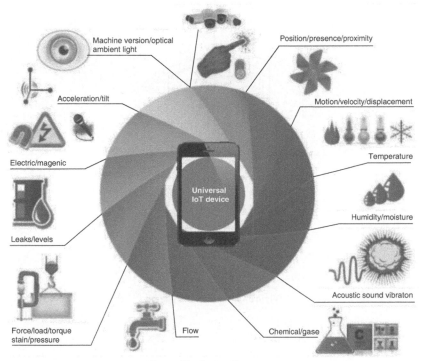

Figure 13: Sensor Ecosystem

Basic Sensors

The Smartphones are powered by built-in sensors that measure motion, activity, wellness, various environmental factors, and near-field & Geolocation details of the Users. Broadly, the sensors in Smartphones could be classified in the following broad categories:

- Motion Sensors
- Position Sensors
- Environmental Sensors
- Audio & Visual Sensors (Microphone and Camera)

Motion Sensors

Acceleration forces and rotational forces along three axes are measured by motion sensors. Example of sensors include: accelerometer, gravity sensors, gyroscope, and rotational sensors. Sensor Framework in popular Smartphone systems, Android and iOS, use standard three coordinate system to express the relative position of the device. The relative position is relative to the default position of the phone to the screen.

- When device is held in its default orientation (as shown in the Figure 14), the x axis is horizontal and points to the right
- The Y axis is vertical and points up
- And the Z axis is points to outward of the screen face.

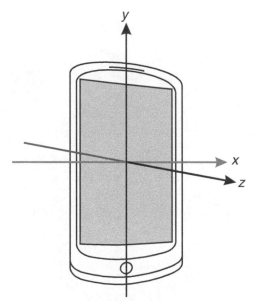

Figure 14: Device Orientation

Sensor Types Supported in Android:[11]

Sensor	Type	Description	Common Case Study
TYPE_ACCELEROMETER	Hardware	Measures acceleration in m/s^2 that is applied to the device in all three axes	Motion Detection
TYPE AMBIENT_ TEMPERATURE	Hardware	Measures room or ambient temperature in Celsius	Monitoring air temperatures
TYPE_GRAVITY	Software or Hardware	Measure force of gravity m/s^2 that is applied to the device in all three axes	Motion detection
TYPE_GYROSCOPE	Hardware	Device's rate of rotation in rad/sec	Rotation detection (spin, turn, etc.)

---

[11]   Sensor Types in Android - https://developer.android.com/guide/topics/sensors/sensors_overview

TYPE_LIGHT	Hardware	Measures the ambient light	Controlling the screen brightness
TYPE_LINEAR_ ACCELEROMETER	Software or Hardware	Excluding force of gravity, acceleration in m/s² that is applied to the device in all three axes	Monitoring acceleration along single axis
TYPE_MAGNETIC_FIELD	Hardware	Ambient geomagnetic field for all three physical axes in µT	Determining device position
TYPE_PRESURE	Hardware	Measures the ambient pressure	Monitoring air pressure
TYPE_PROXIMITY	Hardware	Measures proximity of an object in cm	Phone position during a call
TYPE_RELATIVE_ HUMIDITY	Hardware	Measures relative humidity in %	Monitoring dew point, absolute and relative humidity

Please find all the list of Sensors for Android at: https://developer.android.com/guide/topics/sensors/sensors_overview

The workings of the Sensor in Android are well understood if we know the architecture of Android Operating System [8].

Android Architecture[12]

The following are the major components in Android architecture [9] (please see Figure 15):

System Apps

The System applications are shipped during the device manufacture or delivered as part of Original Equipment Manufacturer rollout (OEM). The System Apps include Email, Calendar or Camera.

Java API Framework

Originally called, Application Framework, the role of Java API Framework is to expose the capabilities of Android Operating System capabilities to Application Developers.

Content Provider

Content Providers enable Android Applications to access data from other applications (Calendar, Contacts) or to share their own data.

Activity Manager

The role of Activity Manager is to handle life cycle of Activities (please see Figure 16).

---

[12]　Android Architecture - https://developer.android.com/guide/platform/

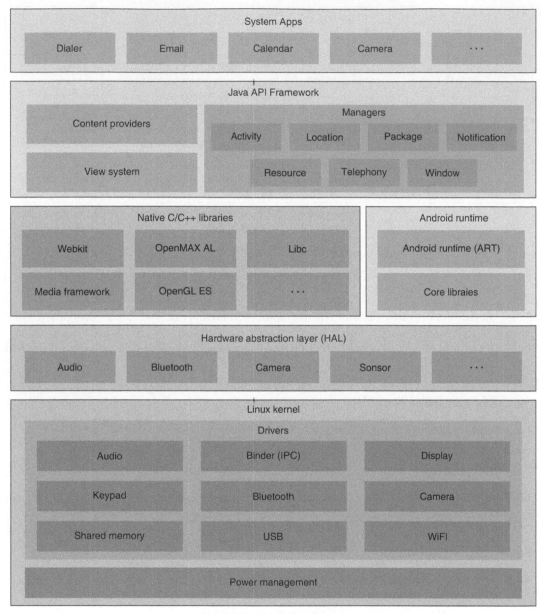

Figure 15: Android Architecture

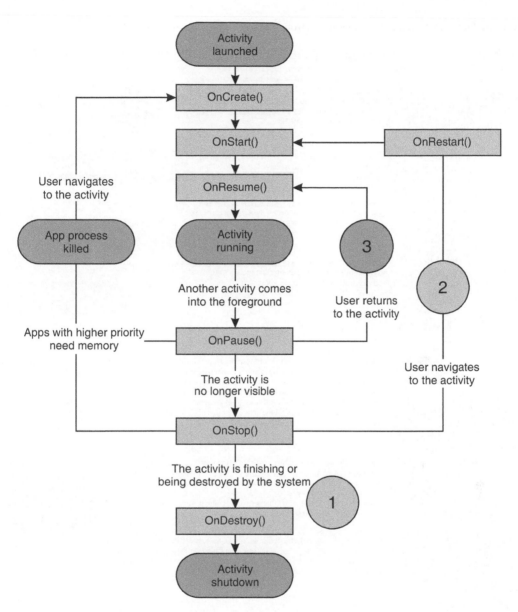

Figure 16: Activity Lifecycle

Activity Heap: In general, the movement through an activity's lifecycle looks like this:

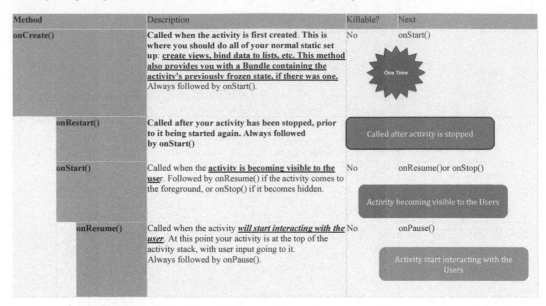

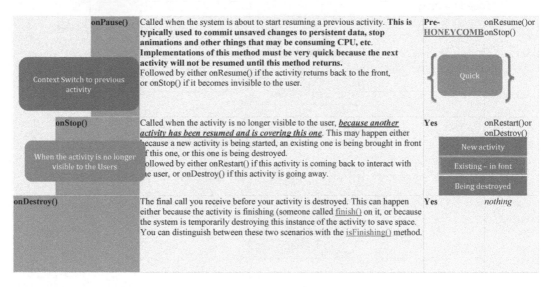

Hardware Abstraction Layer (HAL)

As the name implies, the HAL is an abstraction layer over the hardware. It plays an important role of bridging the calls between application framework and Hardware. For instance, when an App needs to access Camera capabilities, it calls Camera API that will be bridged to Camera device driver to load camera functionality in memory, marshal/unmarshall, the calls between the Camera and the App.

The Sensor management plays an important role in IoT Space—collecting data from device or Sensors of phone and either process locally at phone or move the data to the Cloud.

Sensor Framework

The Sensor Framework in Android enables to setup listener for supported sensors and process the sensor events by handling OnSensorChanged() call back handler [10].

The following steps are required to get data from sensors:

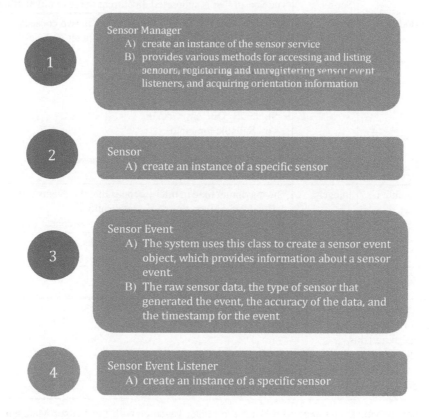

Sensor Service and Sensor Values

registeredListener

```
public boolean registerListener (SensorEventListener listener,
            Sensor sensor,
            int samplingPeriodUs,
            int maxReportLatencyUs,
            Handler handler)
```

Registered Listener API [13] configures [10] and setup sensor event listener.

Parameters		
Listener		SensorEventListener is the object that receives installed sensor values.
Sensor	Sensor	The Sensor that is supported on current the Android platform.
samplingPeriodUs	Integer	The configurable sampling period between two consecutive sensor events. The possible values for this parameter includes: • SENSOR_DELAY_NORMAL 　o Rate suitable for orientation change • SENSOR_DELAY_UI 　o Suitable for User Interface • SENSOR_DELAY_GAME 　o Suitable for games • SENSOR_DELAY_FATEST 　o Gets the sensor values as fast as possible • Custom delay in Microseconds
MaxReportLatencyUs	Integer	The maximum time in microseconds that the events can be held or delayed before reporting to the application. The values include Custom value in microseconds.
Handler	Call back handler	Sensor Event Handler to which the Events are delivered.

Code example:

```
SensorManager sm = null;
sm = (SensorManager) getSystemService(SENSOR_SERVICE);
sm.registerListener(this,
            SensorManager.SENSOR_ORIENTATION |
                SensorManager.SENSOR_ACCELEROMETER,
            SensorManager.SENSOR_DELAY_NORMAL);
```

To retrieve Sensor value, you need to register the sensor Listener with the Sensor Manager and when complete, on listening, unregister the sensor by calling UnregisterListener.

Un Register Sensor

`public void unregisterListener (SensorEventListener listener)`

	Parameters
Listener	SensorEventListener: a SensorListener object

UnRegisterListener API[14] unregisters all the Sensors [11]. If you would like to un register a specific sensor, please call following API:

public void unregisterListener (SensorEventListener listener, `Sensor sensor`)

---

[13]   Register Listener - https://developer.android.com/reference/android/hardware/SensorManager#registerListener(android. hardware.SensorEventListener,%20android.hardware.Sensor,%20int,%20int,%20android.os.Handler)

[14]   Un   register   all   sensors   -   https://developer.android.com/reference/android/hardware/SensorManager. html#unregisterListener(android.hardware.SensorEventListener)

List All Sensors App—Sensor Code in Android

List All Sensors App code is very straightforward code. It provides a detailed view of interfacing with the Android Sensor manager and retrieves all the sensors in the user's Smartphone (please see Figure 17).

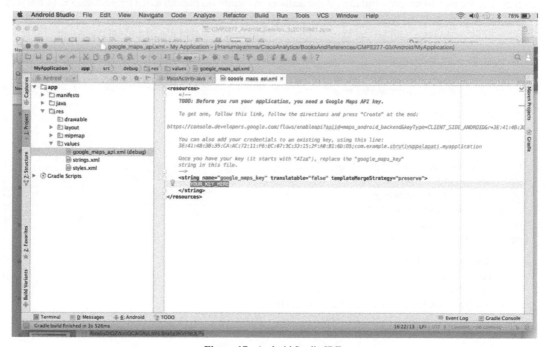

Figure 17: Android Studio IDE

Android Manifest

Android Manifest file (see Table 2) is one central place holder that contains all the versions, features, Sensors and securities used by the application.

Table 2: Android Manifest File

```xml
<?xml version="1.0" encoding="utf-8"?>
<manifest xmlns:android="http://schemas.android.com/apk/res/android"
     package="com.androidbook.sensors.list"
     android:versionCode="1"
     android:versionName="1.0">
     <uses-sdk android:minSdkVersion="8" android:targetSdkVersion="19" />
     <application android:icon="@drawable/icon" android:label="@string/
       app_name">
          <activity android:name=".MainActivity"
             android:label="@string/app_name">
            <intent-filter>
               <action android:name="android.intent.action.MAIN" />
               <category android:name="android.intent.category.LAUNCHER" />
            </intent-filter>
          </activity>
     </application>
</manifest>
```

The above manifest file has default SDK and Main Activity.

Sensor Code

The code performs reference to SYSTEM Services and retrieves a reference to SENSOR_SERVICE object. The Sensor Manager emulates all the sensors available on the device. The next step is to retrieve all ACTIVE sensors in the phone by calling getSensorList of ALL Sensor Type. After this call, the Sensors LIST will be populated with all the Sensors on the Android phone that are in current usage (see Table 3).

Table 3: List All Sensors Code

```java
package com.androidbook.sensors.list;

//This file is MainActivity.java
import java.util.HashMap;
import java.util.List;

import android.annotation.SuppressLint;
import android.app.Activity;
import android.hardware.Sensor;
import android.hardware.SensorManager;
import android.os.Bundle;
import android.widget.TextView;

public class MainActivity extends Activity {
    @SuppressLint("NewApi")
        @Override
    public void onCreate(Bundle savedInstanceState) {
        super.onCreate(savedInstanceState);
        setContentView(R.layout.main);

        TextView text = (TextView)findViewById(R.id.text);
        SensorManager mgr = (SensorManager) this.getSystemService(SENSOR_SERVICE);
        List<Sensor> sensors = mgr.getSensorList(Sensor.TYPE_ALL);

        StringBuilder message = new StringBuilder(2048);
        message.append("The sensors on this device are:\n");
    for(Sensor sensor : sensors) {
        message.append(sensor.getName() + "\n");
        message.append(" Type: " + sensorTypes.get(sensor.getType()) + "\n");
        message.append(" Vendor: " + sensor.getVendor() + "\n");
        message.append(" Version: " + sensor.getVersion() + "\n");
        try {
        message.append(" Min Delay: " + sensor.getMinDelay() + "\n");
        } catch(NoSuchMethodError e) {} // ignore if FIFO method not found
        try {
        message.append(" FIFO Max Event Count: " + sensor.getFifoMaxEventCount() + "\n");
        } catch(NoSuchMethodError e) {} // ignore if FIFO method not found
        message.append(" Resolution: " + sensor.getResolution() + "\n");
        message.append(" Max Range: " + sensor.getMaximumRange() + "\n");
        message.append(" Power: " + sensor.getPower() + " mA\n");
    }
    text.setText(message);
}
private HashMap<Integer, String> sensorTypes = new HashMap<Integer, String>();
```

Table 3 contd. ...

...Table 3 contd.

```
        { // Initialize the map of sensor type values and names
        sensorTypes.put(Sensor.TYPE_ACCELEROMETER, "TYPE_ACCELEROMETER"); // 1
        sensorTypes.put(Sensor.TYPE_AMBIENT_TEMPERATURE, "TYPE_AMBIENT_TEMPERA-
TURE"); // 13
        sensorTypes.put(Sensor.TYPE_GAME_ROTATION_VECTOR, "TYPE_GAME_ROTATION_
VECTOR"); // 15
        sensorTypes.put(Sensor.TYPE_GEOMAGNETIC_ROTATION_VECTOR, "TYPE_GEOMAG-
NETIC_ROTATION_VECTOR"); // 20
        sensorTypes.put(Sensor.TYPE_GRAVITY, "TYPE_GRAVITY"); // 9
        sensorTypes.put(Sensor.TYPE_GYROSCOPE, "TYPE_GYROSCOPE"); // 4
        sensorTypes.put(Sensor.TYPE_GYROSCOPE_UNCALIBRATED, "TYPE_GYROSCOPE_UN-
CALIBRATED"); // 16
        sensorTypes.put(Sensor.TYPE_LIGHT, "TYPE_LIGHT"); // 5
        sensorTypes.put(Sensor.TYPE_LINEAR_ACCELERATION, "TYPE_LINEAR_ACCELERA-
TION"); // 10
        sensorTypes.put(Sensor.TYPE_MAGNETIC_FIELD, "TYPE_MAGNETIC_FIELD"); // 2
        sensorTypes.put(Sensor.TYPE_MAGNETIC_FIELD_UNCALIBRATED, "TYPE_MAGNET-
IC_FIELD_UNCALIBRATED"); // 14
        sensorTypes.put(Sensor.TYPE_ORIENTATION, "TYPE_ORIENTATION (deprecated)"); // 3
        sensorTypes.put(Sensor.TYPE_PRESSURE, "TYPE_PRESSURE"); // 6
        sensorTypes.put(Sensor.TYPE_PROXIMITY, "TYPE_PROXIMITY"); // 8
        sensorTypes.put(Sensor.TYPE_RELATIVE_HUMIDITY, "TYPE_RELATIVE_HUMIDITY"); //
12
        sensorTypes.put(Sensor.TYPE_ROTATION_VECTOR, "TYPE_ROTATION_VECTOR"); // 11
        sensorTypes.put(Sensor.TYPE_SIGNIFICANT_MOTION, "TYPE_SIGNIFICANT_MO-
TION"); // 17
        sensorTypes.put(Sensor.TYPE_STEP_COUNTER, "TYPE_STEP_COUNTER"); // 19
        sensorTypes.put(Sensor.TYPE_STEP_DETECTOR, "TYPE_STEP_DETECTOR"); // 18
        sensorTypes.put(Sensor.TYPE_TEMPERATURE, "TYPE_TEMPERATURE (deprecated)"); // 7
    }
}
```

 getSensorList() will retrieve all the Sensors on the current phone hardware. It's not required that the current hardware has all the sensors as supported by the Android.

Now iterate through each Sensors list and display sensor attributes that include:

- Sensor Name
- Sensor Type
- Vendor
- Version
- Delay
- Resolution
- Maximum Range
- And Power

Screen Capture

The output of the application is below (see Figure 18):

```
Name: BMI160 accelerometer(android.sensor.accelerometer) minDelay=2500 maxDelay=160000
getReportingMode=0 getMaximumRange=156.9064 getResolution=0.0047884034
Name: BMI160 gyroscope(android.sensor.gyroscope) minDelay=2500 maxDelay=160000 getReportingMode=0
getMaximumRange=17.453293 getResolution=5.326322E-4
Name: BMM150 magnetometer(android.sensor.magnetic_field) minDelay=20000 maxDelay=320000
getReportingMode=0 getMaximumRange=1300.0 getResolution=0.0
Name: BMP280 pressure(android.sensor.pressure) minDelay=100000 maxDelay=10000000 getReportingMode=0
getMaximumRange=1100.0 getResolution=0.005
Name: BMP280 temperature(com.google.sensor.internal_temperature) minDelay=40000 maxDelay=10000000
getReportingMode=0 getMaximumRange=85.0 getResolution=0.01
Name: RPR0521 Proximity Sensor(android.sensor.proximity) minDelay=200000 maxDelay=10000000
getReportingMode=1 getMaximumRange=5.0 getResolution=1.0
Name: RPR0521 Light Sensor(android.sensor.light) minDelay=200000 maxDelay=10000000
getReportingMode=1 getMaximumRange=43000.0 getResolution=10.0
Name: Orientation(android.sensor.orientation) minDelay=5000 maxDelay=80000 getReportingMode=0
getMaximumRange=360.0 getResolution=1.0
Name: BMI160 Step detector(android.sensor.step_detector) minDelay=0 maxDelay=0 getReportingMode=3
getMaximumRange=1.0 getResolution=1.0
Name: Significant motion(android.sensor.significant_motion) minDelay=-1 maxDelay=0
getReportingMode=2 getMaximumRange=1.0 getResolution=1.0
Name: Gravity(android.sensor.gravity) minDelay=5000 maxDelay=80000 getReportingMode=0
getMaximumRange=1000.0 getResolution=1.0
Name: Linear Acceleration(android.sensor.linear_acceleration) minDelay=5000 maxDelay=80000
getReportingMode=0 getMaximumRange=1000.0 getResolution=1.0
Name: Rotation Vector(android.sensor.rotation_vector) minDelay=5000 maxDelay=80000
getReportingMode=0 getMaximumRange=1000.0 getResolution=1.0
```

Figure 18: All Sensors List Output

Nexus 5X Sensors List

The Sensor list application on the Nexus 5X[15] hardware generates the following Sensor Sampling and performance details (please see Table 4).

Table 4: Nexus 5X Sensor Data

Data Source	Sampling PeriodUs (ms)	MaxReport LatencyUs (ms)	Sensitivity	Range Of Events	Reporting Mode
Accelerometer	2,500	160,000	0.0047884034	156.9064	Continuous
Accelerometer (uncalibrated)	2,500	160,000	0.0047884034	156.9064	Continuous
Ambient Temperature	n/a[1]	n/a[1]	n/a[1]	n/a[1]	n/a[1]
Device Orientation	0	0	1.0	3.0	On Change
Double Tap	0	0	1.0	1.0	Special
Double Twist	0	0	1.0	1.0	Special
Game Rotation Vector	5,000	80,000	1.0	1000.0	Continuous
Geomagnetic Rotation Vector	5,000	80,000	1.0	1000.0	Continuous
Gravity	5,000	80,000	1.0	1000.0	Continuous
Gyroscope	2,500	160,000	5.326322E-4	17.453293	Continuous
Gyroscope (uncalibrated)	2,500	160,000	5.326322E-4	17.453293	Continuous

Table 4 contd. ...

---

15   Google Nexus 5X - https://www.google.com/search?q=nexus+5x+sensors&rlz=1C1GGRV_enUS808US808&oq=Nexus+5X+Sensors&aqs=chrome.0.0l6.2149j0j8&sourceid=chrome&ie=UTF-8

Table 4 contd. ...					
Data Source	Sampling PeriodUs (ms)	MaxReport LatencyUs (ms)	Sensitivity	Range Of Events	Reporting Mode
Heart beat	n/a[1]	n/a[1]	n/a[1]	n/a[1]	n/a[1]
Heart rate	n/a[1]	n/a[1]	n/a[1]	n/a[1]	n/a[1]
Light	200,000	10,000,000	10.0	43000.0	On Change
Linear Acceleration	5,000	80,000	1.0	1000.0	Continuous
Low latency off-body detect sensor	n/a[1]	n/a[1]	n/a[1]	n/a[1]	n/a[1]
Magnetic field	20,000	320,000	0.0	1300.0	Continuous
Magnetic field (uncalibrated)	20,000	320,000	0.0	1300.0	Continuous
Motion detect	n/a[1]	n/a[1]	n/a[1]	n/a[1]	n/a[1]
Orientation	5,000	80,000	1.0	360.0	Continuous
Pose	n/a[1]	n/a[1]	n/a[1]	n/a[1]	n/a[1]
Pickup gesture	−1	0	1.0	1.0	One Shot
Pressure	100,000	10,000,000	0.005	1100.0	Continuous
Proximity	200,000	10,000,000	1.0	5.0	On Change
Relative Humidity	n/a[1]	n/a[1]	n/a[1]	n/a[1]	n/a[1]
Rotation vector	5,000	80,000	1.0	1000.0	Continuous
Significant motion	−1	0	1.0	1.0	One Shot
Stationary detect	n/a[1]	n/a[1]	n/a[1]	n/a[1]	n/a[1]
Sensors Sync	0	0	1.0	1.0	Special
Step counter	0	0	1.0	1.0	One Change
Step detector	0	0	1.0	1.0	Special
Temperature	40,000	10,000,000	0.01	85.0	Continuous
Tilt detector	0	0	1.0	1.0	Special

Please note: 1 Sensor is not available on the Nexus 5X
Android doesn't support the generation of events with delay minor less than 1 ms (1000 Hz).

Accelerometer App

The purpose of the App is to measures the ambient geomagnetic field for all three physical axes (x, y, z) in micro Tesla (μT) units. The magnetometer sensor (used in Android device) is crucial for detecting the orientation of your device relative to the Earth's magnetic north. In Android platform, the Magnetometer sensor is a Hardware Sensor.

Let's look at the Magnetometer App. It has three parts.

1. Android Manifest Code
2. Sensor Configuration Code
3. Sensor processing & application logic code

Android Manifest Code

The Manifest code (see Table 5) sets up the Application Main activity. Please note the App does not require any special privileges:

Table 5: Accelerometer Manifest File

```xml
<?xml version="1.0" encoding="utf-8"?>
<manifest xmlns:android="http://schemas.android.com/apk/res/android"
  package="com.hanuinnotech.android.accelerometer">

  <uses-feature android:name="android.hardware.sensor.
   accelerometer" />

  <application
      android:allowBackup="true"
      android:icon="@mipmap/ic_launcher"
      android:label="@string/app_name"
      android:roundIcon="@mipmap/ic_launcher_round"
      android:supportsRtl="true"
      android:theme="@style/AppTheme">
      <activity android:name=".MainActivity">
        <intent-filter>
            <action android:name="android.intent.action.MAIN" />
            <category android:name="android.intent.category.
              LAUNCHER" />
        </intent-filter>
      </activity>
  </application>
</manifest>
```

Android UI

The UI contains following (see Figure 19):

Accelerometer App Code

The Code (see Table 6) calls getSystemServices context Sensor Service to get the system sensor services. Next, it will get the default sensor for the TYPE_MAGNETIC_FIELD.

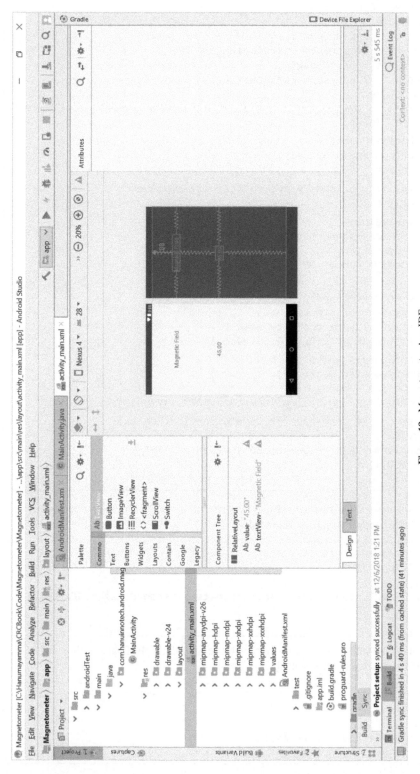

Figure 19: Magnetometer App IDE

Table 6: Code listing Magnetometer

```java
package com.hanuinnotech.android.magnetometer;
import android.content.Context;
import android.hardware.Sensor;
import android.hardware.SensorEvent;
import android.hardware.SensorEventListener;
import android.hardware.SensorManager;
import android.support.v7.app.AppCompatActivity;
import android.os.Bundle;
import android.widget.TextView;
import android.widget.Toast;

import java.text.DecimalFormat;
import java.text.DecimalFormatSymbols;
import java.util.Locale;

public class MainActivity extends AppCompatActivity implements
SensorEventListener {

    private TextView value;
    private SensorManager sensorManager;
    private Sensor sensor;
    public static DecimalFormat DECIMAL_FORMATTER;

    @Override
    protected void onCreate(Bundle savedInstanceState) {
       super.onCreate(savedInstanceState);
       setContentView(R.layout.activity_main);
       value = (TextView)findViewById(R.id.value);

       DecimalFormatSymbols symbols = new
       DecimalFormatSymbols(Locale.US);
       symbols.setDecimalSeparator('.');
       DECIMAL_FORMATTER = new DecimalFormat("#.000", symbols);

       sensorManager = (SensorManager)getSystemService(Context.
       SENSOR_SERVICE);
       sensor = sensorManager.getDefaultSensor(Sensor.TYPE_
       MAGNETIC_FIELD);
    }
```

To activate the Sensor data collection, the sensor register listener is tied to OnResume as this system callback function is called whenever App life cycle triggers:

```java
//register the accelerometer to listen to the events
protected void onResume(){
 super.onResume();
 sensorManager.registerListener(this, accelerometer, SensorManager.SENSOR_DE-
LAY_NORMAL);
}
```

We are retrieving Magnetic values on a Normal retrieval level. Similarly, the Pause method unregisters the Sensor to conserve the battery and other valuable resources.

```java
//unregister the accelerometer to stop listening to the events
protected void onPause(){
 super.onPause();
 sensorManager.unregisterListener(this);
}
```

The Accelerometer Sensor values are captured on Sensor Changed event value (Eq. 1). The net magnetic field value is computed by square root and all the sums of square of X, Y and Z values.

$$Magnetic\ Value = \sqrt[2]{x^2} + y^2 + z^2 \tag{1}$$

Accelerometer App Full Code

Please find the complete integrated code (please see Table 7):

Table 7: Magnetometer Full Code

```java
package com.hanuinnotech.android.accelerometer;

import android.content.Context;
import android.hardware.Sensor;
import android.hardware.SensorEvent;
import android.hardware.SensorEventListener;
import android.hardware.SensorManager;
import android.support.v7.app.AppCompatActivity;
import android.os.Bundle;
import android.view.View;
import android.widget.Button;
import android.widget.TextView;

public class MainActivity extends AppCompatActivity implements
SensorEventListener, View.OnClickListener{

  private SensorManager sensorManager;
  private Sensor accelerometer;

  private float deltaX = 0;
  private float deltaY = 0;
  private float deltaZ = 0;

  private TextView currentX, currentY, currentZ;
  private Button btnstart, btnstop;

  @Override
  protected void onCreate(Bundle savedInstanceState) {
    super.onCreate(savedInstanceState);
    setContentView(R.layout.activity_main);
    initializeView();
    btnstart = (Button)findViewById(R.id.btn_start);
    btnstart.setOnClickListener(this);
    btnstop = (Button)findViewById(R.id.btn_stop);
    btnstop.setOnClickListener(this);
    sensorManager = (SensorManager)getSystemService(Context.
SENSOR_SERVICE);
    if (sensorManager.getDefaultSensor(Sensor.TYPE_
    ACCELEROMETER) != null) //we have accelerometer
    {
      accelerometer = sensorManager.getDefaultSensor(Sensor.
        TYPE_ACCELEROMETER);
```

Table 7 contd. ...

...Table 7 contd.

```java
        sensorManager.registerListener(this, accelerometer,
        SensorManager.SENSOR_DELAY_NORMAL);
    }
    else {
        //no accelerometer
    }
}

public void initializeView(){
    currentX = (TextView)findViewById(R.id.x_value);
    currentY = (TextView)findViewById(R.id.y_value);
    currentZ = (TextView)findViewById(R.id.z_value);
}
//register the accelerometer to listen to the events
protected void onResume(){
    super.onResume();
    sensorManager.registerListener(this, accelerometer,
     SensorManager.SENSOR_DELAY_NORMAL);
}
//unregister the accelerometer to stop listening to the events
protected void onPause(){
    super.onPause();
    sensorManager.unregisterListener(this);
}
@Override
public void onSensorChanged(SensorEvent sensorEvent) {
    deltaX = Math.abs(sensorEvent.values[0]);
    deltaY = Math.abs(sensorEvent.values[1]);
    deltaZ = Math.abs(sensorEvent.values[2]);
    displayCurrentValues();
}
@Override
public void onAccuracyChanged(Sensor sensor, int accuracy) {
}
//display current X, Y, Z accelerometer values
public void displayCurrentValues(){
    currentX.setText(Float.toString(deltaX));
    currentY.setText(Float.toString(deltaY));
    currentZ.setText(Float.toString(deltaZ));
}
@Override
public void onClick(View v) {
    switch (v.getId()){
      case R.id.btn_start:
        sensorManager.registerListener(this, accelerometer,
         SensorManager.SENSOR_DELAY_NORMAL);
        break;
      case R.id.btn_stop:
        sensorManager.unregisterListener(this);
        break;
    }
}
}
```

Accelerometer Play App—a Precursor to Digital Twin Apps

This App[16] (see Figure 20) demonstrates how to use an accelerometer sensor as input for a physics-based view. The input from the accelerometer is used to simulate a virtual surface, and a number of free-moving objects placed on top of it [12].

Figure 20: Accelerometer Play

Any effects from the device's acceleration vector (including both gravity and temporary movement) will be translated to the on-screen particles.

Source Code

Please see Table 8 for critical laws of motion compute code.

Table 8: Accelerometer Play Code

```
public void computePhysics(float sx, float sy, float dT) {

  final float ax = -sx/5;
  final float ay = -sy/5;

  mPosX += mVelX * dT + ax * dT * dT / 2;
  mPosY += mVelY * dT + ay * dT * dT / 2;
mVelX += ax * dT;
  mVelY += ay * dT;

} /*

* Resolving constraints and collisions with the Verlet integrator
```

Table 8 contd. ...

---

[16]   Accelerometer Play - https://github.com/googlesamples/android-AccelerometerPlay

...Table 8 contd.

```
 * can be very simple, we simply need to move a colliding or
 * constrained particle in such way that the constraint is
 * satisfied.
 */
public void resolveCollisionWithBounds() {
  final float xmax = mHorizontalBound;
  final float ymax = mVerticalBound;
  final float x = mPosX;
  final float y = mPosY;
  if (x > xmax) {
      mPosX = xmax;
      mVelX = 0;
  } else if (x < -xmax) {
      mPosX = -xmax;
      mVelX = 0;
  }
  if (y > ymax) {
      mPosY = ymax;
      mVelY = 0;
  } else if (y < -ymax) {
      mPosY = -ymax;
      mVelY = 0;
  }
 }
}
```

Magnetometer App

The purpose of the App is to measures the ambient geomagnetic field for all three physical axes (x, y, z) in micro Tesla (µT) units. The magnetometer sensor (used in Android device - please see Figure 21) is crucial for detecting the orientation of your device relative to the Earth's magnetic north. In Android

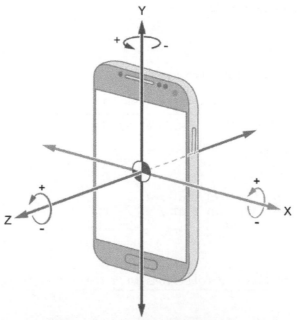

Figure 21: Magnetometer Device Vectors

platform, the Magnetometer sensor is a Hardware Sensor [13]. This illustration shows the orientation of the X, Y, and Z axes for a typical Android mobile phone.

This illustration shows the orientation of the X, Y, and Z axes for a typical Android tablet (please see Figure 22).

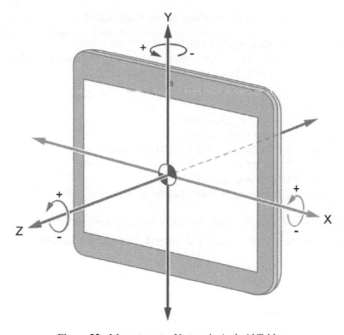

Figure 22: Magnetometer Vectors in Android Tablet

The App reads the strength of the magnetic field from the built-in magnetometer sensor on your Android device. The block outputs the magnetic field as a 1-by-3 vector. The Data Display block accepts the vector and displays the values on your Android device (please see Figure 23).

Sensor Code

Let's look at the Magnetometer App. It has three parts.
4. Android Manifest Code
5. Sensor Configuration Code
6. Sensor processing & application logic code

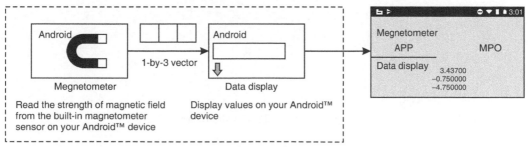

Figure 23: Simulink Model for Android Devices

Android Manifest Code

The Manifest code (see Table 9) sets up the Application Main activity. Please note that the App does not require any special privileges:

Table 9: Magnetometer Manifest Code

```xml
<?xml version="1.0" encoding="utf-8"?>
<manifest xmlns:android="http://schemas.android.com/apk/res/android"
  package="com.hanuinnotech.android.magnetometer">

  <application
    android:allowBackup="true"
    android:icon="@mipmap/ic_launcher"
    android:label="@string/app_name"
    android:roundIcon="@mipmap/ic_launcher_round"
    android:supportsRtl="true"
    android:theme="@style/AppTheme">
    <activity android:name=".MainActivity">
      <intent-filter>
        <action android:name="android.intent.action.MAIN" />

        <category android:name="android.intent.category.LAUNCHER" />
      </intent-filter>
    </activity>
  </application>
</manifest>
```

Android UI

The development of view of the Magnetometer (please see Figure 24):

Figure 24: Android Studio Magnetometer

Magnetometer App Code

The Code (see Table 10) calls getSystemServices context Sensor Service to get the system sensor services. Next, it will get the default sensor for the TYPE_MAGNETIC_FIELD.

Table 10: Magnetometer App Code

```
package com.hanuinnotech.android.magnetometer;

import android.content.Context;
import android.hardware.Sensor;
import android.hardware.SensorEvent;
import android.hardware.SensorEventListener;
import android.hardware.SensorManager;
import android.support.v7.app.AppCompatActivity;
import android.os.Bundle;
import android.widget.TextView;
import android.widget.Toast;

import java.text.DecimalFormat;
import java.text.DecimalFormatSymbols;
import java.util.Locale;

public class MainActivity extends AppCompatActivity implements SensorEventListener {

  private TextView value;
  private SensorManager sensorManager;
  private Sensor sensor;
  public static DecimalFormat DECIMAL_FORMATTER;

  @Override
  protected void onCreate(Bundle savedInstanceState) {
    super.onCreate(savedInstanceState);
    setContentView(R.layout.activity_main);
    value = (TextView)findViewById(R.id.value);

    DecimalFormatSymbols symbols = new
    DecimalFormatSymbols(Locale.US);
    symbols.setDecimalSeparator('.');
    DECIMAL_FORMATTER = new DecimalFormat("#.000", symbols);

    sensorManager = (SensorManager)getSystemService(Context.
    SENSOR_SERVICE);
    sensor = sensorManager.getDefaultSensor(Sensor.TYPE_MAGNETIC_FIELD);
  }
```

To activate the Sensor data collection, the sensor register listener is tied to OnResume as this system callback function is called whenever App life cycle triggers (see Table 11):

Table 11: Magnetometer OnResume()

```
public void onResume()
{
    super.onResume();
    if (sensor != null)
    {
        sensorManager.registerListener(this, sensor, SensorManager.SENSOR_
        DELAY_NORMAL);
    }
    else {
        Toast.makeText(this, "Not Supported", Toast.LENGTH_SHORT).show();
        finish();
    }
}
```

We are retrieving Magnetic values on a Normal retrieval level. Similarly, the Pause method (see Table 12) will unregisters the Sensor in order to conserve the battery and other valuable resources.

Table 12: Magnetometer OnPause ()

```
public void onPause()
{
  super.onPause();
  sensorManager.unregisterListener(this);
}
```

The Magnetic Sensor values are captured on Sensor Changed event value (Eq. 1). The net magnetic field value computed by square root and all the sums of square of X, Y and Z values. The values are displaced on the UI (see Figure 25).

$$Magnetic\ Value = \sqrt[2]{x^2} + y^2 + z^2 \tag{1}$$

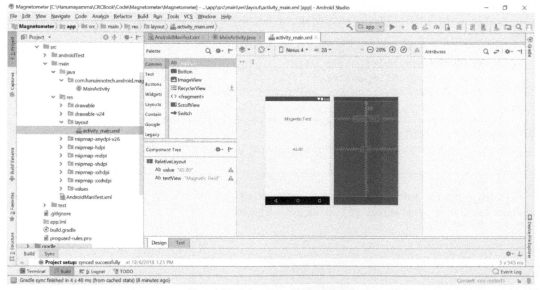

Figure 25: Magnetometer Activity UI

Magnetometer App Full Code

Please find the complete integrated code (see Table 13):

Table 13: Magnetometer Complete Code

```java
package com.hanuinnotech.android.magnetometer;

import android.content.Context;
import android.hardware.Sensor;
import android.hardware.SensorEvent;
import android.hardware.SensorEventListener;
import android.hardware.SensorManager;
import android.support.v7.app.AppCompatActivity;
import android.os.Bundle;
import android.widget.TextView;
import android.widget.Toast;

import java.text.DecimalFormat;
import java.text.DecimalFormatSymbols;
import java.util.Locale;

public class MainActivity extends AppCompatActivity implements SensorEventListener {
    private TextView value;
    private SensorManager sensorManager;
    private Sensor sensor;
    public static DecimalFormat DECIMAL_FORMATTER;
    @Override
    protected void onCreate(Bundle savedInstanceState) {
      super.onCreate(savedInstanceState);
      setContentView(R.layout.activity_main);
      value = (TextView)findViewById(R.id.value);

      DecimalFormatSymbols symbols = new DecimalFormatSymbols(Locale.US);
      symbols.setDecimalSeparator('.');
      DECIMAL_FORMATTER = new DecimalFormat("#.000", symbols);

      sensorManager = (SensorManager)getSystemService(Context.SENSOR_SERVICE);
      sensor = sensorManager.getDefaultSensor(Sensor.TYPE_MAGNETIC_FIELD);

    }
    @Override
    public void onResume()
    {

        super.onResume();
        if (sensor != null)
    {
     sensorManager.registerListener(this, sensor, SensorManager.SENSOR_DELAY_NORMAL);
     }
     else {
     Toast.makeText(this, "Not Supported", Toast.LENGTH_SHORT).show();
     finish();
        }
     }
    @Override
    public void onPause()
    {
```

Table 13 contd. ...

...Table 13 contd.

```
super.onPause();
sensorManager.unregisterListener(this);
}
@Override
    public void onSensorChanged(SensorEvent event) {
    float azimuth = Math.round(event.values[0]);
    float pitch = Math.round(event.values[1]);
    float roll = Math.round(event.values[2]);

    double tesla = Math.sqrt((azimuth*azimuth) + (pitch*pitch) + (roll*roll));
    value.setText(DECIMAL_FORMATTER.format(tesla) + "\u00B5Tesla");
}
@Override
public void onAccuracyChanged(Sensor sensor, int accuracy) {
}
}
```

The Role of Sensors in Healthcare & Wellness

The Sensors are playing a very vital and useful role in fueling health and fitness-based Mobile Apps. Sensors capture useful User activity related data points that are directly related to the health and fitness of the User.

Case Study: A System to Detect Mental Stress using Machine Learning and Mobile Development

Stress can be defined as a physical response to the excessive amount of pressure faced by an individual. The stress could be induced by any psychological or social scenario [1]. When excessive pressure is induced onto an individual, this could lead to multiple psychological disorders. These might include being depressed or, if worsened, suffering cardiac arrests.

According to American Psychological Association (APA) 2012 study, "Stress is costing organizations a Fortune", in some cases as much as $300 billion a year. The Yerkes-Dodson curve (see Figure 26) clearly demonstrates high stress results in low-performance for individual and organizations.

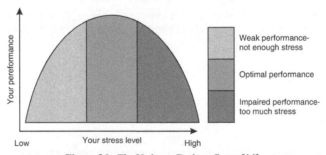

Figure 26: The Yerkes—Dodson Curve [14]

Timely stress detection and the individual propensity/level to healthily operate under stress "the Stress Hump [14] see Figure 27" are very important in managing and curing physiological disorders. However, the traditional methods of assessing stress levels, such as:
(I) interviewing the individual and asking stress related questions in order to gain a better understanding of their condition and (II) Observing the facial gestures—people under stress react by giving different facial expressions, i.e., the eyebrows shape differently, their pupils dilate, or the blinking rate might differ, are limited [14] as they may miss stress episodes. Thus, we argue that our novel approach employ detection using sensors, machine learning, de-identified collaborative insights and mobile devices.

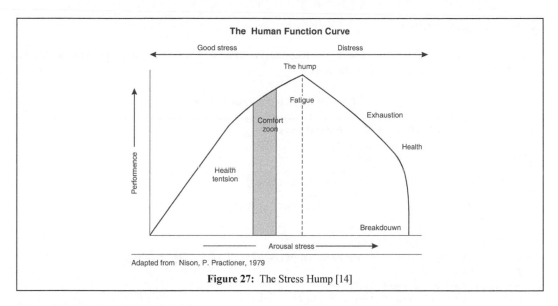

Figure 27: The Stress Hump [14]

Health Fitness IoT App

Step counter sensor is very useful for tracking and forming a prognosis of the physical activity levels of Users. Interweaving step counter activity-based sensor with Electronic Health Records would provide new clinical pathways. This not only helps individual users but also provides a great benefit to humanity.

Android Manifest File

Since the health fitness App requires Hardware Step Counter details, we need to add permissions of using Step counter sensor as part of the manifest file (see Table 14).

Table 14: Step Counter Manifest File

```xml
<?xml version="1.0" encoding="utf-8"?>
    <manifest xmlns:android="http://schemas.android.com/apk/res/android"
    package="com.example.shruti.stepcounter">

    <uses-feature android:name="android.hardware.sensor.stepcounter"
    android:required="true"/>
    <uses-feature android:name="android.hardware.sensor.stepdetector"
    android:required="true"/>
    <application
      android:allowBackup="true"
      android:icon="@mipmap/ic_launcher"
      android:label="@string/app_name"
      android:roundIcon="@mipmap/ic_launcher_round"
      android:supportsRtl="true"
      android:theme="@style/AppTheme">
    <activity android:name=".MainActivity">
        <intent-filter>
            <action android:name="android.intent.action.MAIN" />

            <category android:name="android.intent.category.LAUNCHER" />
        </intent-filter>
    </activity>
    </application>

</manifest>
```

Step Counter Value

The Step Counter sensor is set during the On Resume() method and values of the Steps are calculated as part of OnSensorChanged event (see Table 15).

Table 15: Step Counter Activity Code

```java
package com.example.shruti.stepcounter;
import android.content.Context;
import android.hardware.Sensor;
import android.hardware.SensorEvent;
import android.hardware.SensorEventListener;
import android.hardware.SensorManager;
import android.support.v7.app.AppCompatActivity;
import android.os.Bundle;
import android.widget.TextView;
import android.widget.Toast;
public class MainActivity extends AppCompatActivity implements
SensorEventListener{
  private TextView textView;
  private SensorManager sensorManager = null;

  boolean running = false;

  @Override
  protected void onCreate(Bundle savedInstanceState) {
      super.onCreate(savedInstanceState);
      setContentView(R.layout.activity_main);

      textView = (TextView)findViewById(R.id.steps);
      sensorManager = (SensorManager)getSystemService(Context.SENSOR_SERVICE);
  }
  @Override
  protected void onResume() {
      super.onResume();
      Sensor countSensor = sensorManager.getDefaultSensor(Sensor.TYPE_STEP_
COUNTER);
      if (countSensor != null){
          sensorManager.registerListener(this, countSensor, SensorManager.
SENSOR_DELAY_UI);
      }
      else {
          Toast.makeText(this, "Sensor not found", Toast.LENGTH_SHORT).show();
      }
  }
  @Override
  protected void onPause() {
      super.onPause();
      running= false;
  }
  @Override
  public void onSensorChanged(SensorEvent event) {
      if (running)
      {
          textView.setText(String.valueOf(event.values[0]));
      }
  }
  @Override
  public void onAccuracyChanged(Sensor sensor, int accuracy) {

  }
}
```

App UI

Sensor Batching

Android 4.4 introduces Sensor batching in an effort to help preserve power for Sensor operations. With the sensor batching, Hardware collects the sensor event data based on MaxReportLatencyUs.

 | Sensor batching is a useful optimization technique for long running IoT Case Studys, such as fitness, tracking, continuous updating, and high activity sensor Apps.

In the following application, let's say your App is listening for Sensor events from an Accelerometer sensor in order to detect the movement of the User.

Without Sensor Batching (see Figure 28), the Application OnSensorChanged event method will be called whenever User moves. The events are delivered to OS queue through HAL and from OS Queue the events are delivered to Application.

With Sensor Batching (see Figure 28), we can optimize the rate at which events are delivered to the application. For instance, we can set MaxReportLatencyUs to batch the events in the sensor and have them delivered to the Application. Please note that the hardware collects the events but will not wake up the CPU to deliver events to your application.

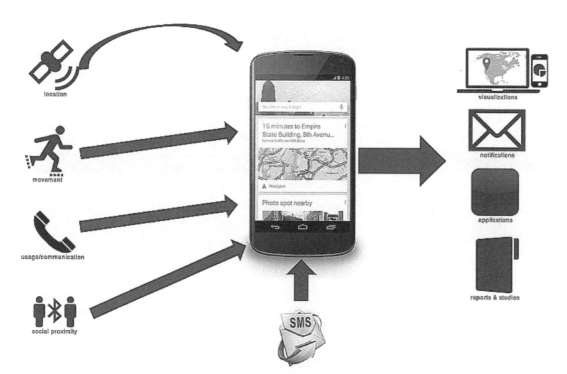

Figure 28: Sensor Hub

---

[17] Sensor Batch - https://github.com/googlesamples/android-BatchStepSensor/

Android BatchStepSensor Sample

This sample[17] demonstrates the use of the two step sensors (step detector and counter) and sensor batching.

It shows how to register a SensorEventListener with and without batching and shows how these events are received.

The Step Detector sensor (see Figure 29) fires an event when a step is detected, while the step counter returns the total number of steps since a listener was first registered for this sensor.

Both sensors only count steps while a listener is registered. This sample only covers the basic case, where a listener is only registered while the app is running. Likewise, batched sensors can be used in the background (when the CPU is suspended), which requires manually flushing the sensor event queue before it overflows, which is not covered in this sample.

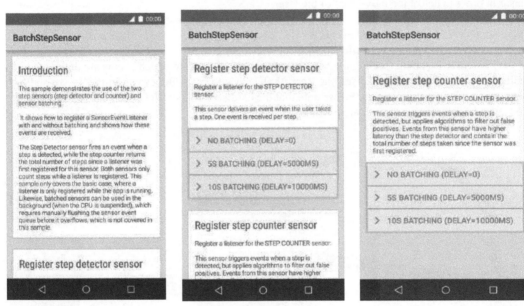

Figure 29: Batch Step Sensor

The Source Code

Please see Table 16 for viewing the complete code.

Table 16: Step Counter Code

```
package com.example.shruti.stepcounter;

import android.content.Context;
import android.hardware.Sensor;
import android.hardware.SensorEvent;
import android.hardware.SensorEventListener;
import android.hardware.SensorManager;
import android.support.v7.app.AppCompatActivity;
import android.os.Bundle;
import android.widget.TextView;
import android.widget.Toast;

public class MainActivity extends AppCompatActivity implements SensorEventListener{

    private TextView textView;
    private SensorManager sensorManager = null;

    boolean running = false;

    @Override
    protected void onCreate(Bundle savedInstanceState) {
      super.onCreate(savedInstanceState);
      setContentView(R.layout.activity_main);

      textView = (TextView)findViewById(R.id.steps);
      sensorManager = (SensorManager)getSystemService(Context.SENSOR_SERVICE);
    }

    @Override
    protected void onResume() {
      super.onResume();
      Sensor countSensor = sensorManager.getDefaultSensor(Sensor.TYPE_STEP_COUNTER);
      if (countSensor != null){
        sensorManager.registerListener(this, countSensor, SensorManager.SENSOR_DELAY_UI);
      }
      else {
        Toast.makeText(this, "Sensor not found", Toast.LENGTH_SHORT).show();
      }
    }

    @Override
    protected void onPause() {
      super.onPause();
      running= false;
    }

    @Override
    public void onSensorChanged(SensorEvent event) {
      if (running)
      {
        textView.setText(String.valueOf(event.values[0]));
      }
    }

    @Override
    public void onAccuracyChanged(Sensor sensor, int accuracy) {

    }
}
```

Full Sensor & Data List

Following (see Table 17) provides Sensor values in the Android platform.[18]

Table 17: Sensor Values in Android

Sensor	Sensor event data	Description	Units of measure
TYPE_ ACCELEROMETER	SensorEvent.values[0]	Acceleration force along the x axis (including gravity).	m/s²
	SensorEvent.values[1]	Acceleration force along the y axis (including gravity).	
	SensorEvent.values[2]	Acceleration force along the z axis (including gravity).	
TYPE_GRAVITY	SensorEvent.values[0]	Force of gravity along the x axis.	m/s²
	SensorEvent.values[1]	Force of gravity along the y axis.	
	SensorEvent.values[2]	Force of gravity along the z axis.	
TYPE_GYROSCOPE	SensorEvent.values[0]	Rate of rotation around the x axis.	rad/s
	SensorEvent.values[1]	Rate of rotation around the y axis	
	SensorEvent.values[2]	Rate of rotation around the z axis	
TYPE_STEP_ COUNTER	SensorEvent.values[0]	Number of steps taken by the user since the last reboot while the sensor was activated.	Steps

Core Motion Framework in iOS

The Core Motion framework is responsible for handling all User and device movements in iOS.

iOS App Anatomy

The Apple iOS User Interface Kit has three major objects: UI Application, System Events, and Application Delegate Code. The UI Application represents the User Interface of the App. The System events could be application user bound, for instance App Click, and system bound, for instance battery low event.

App Architecture

The iOS architecture following Model-View-Controller design pattern (see Figure 30) with clear separation of roles of Controller (handles events), Model (lifecycle of data objects) and View (User Interface elements). The Application Delegate Object creates the Window Object. Under the Window Object, View and View Controller Objects are created. The data is stored in the Data Object that connects to Application and View objects in order to send or receive the data [15].

---

[18] Android Platform - https://developer.android.com/guide/topics/sensors/sensors_motion

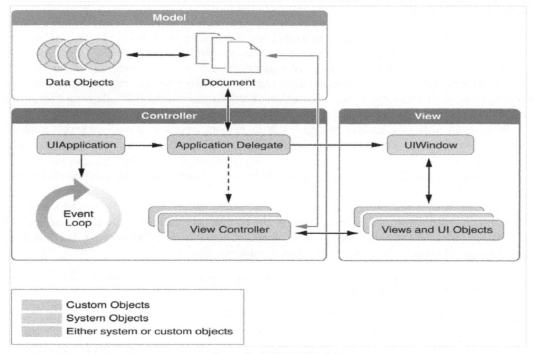

Figure 30: iOS MVC Model

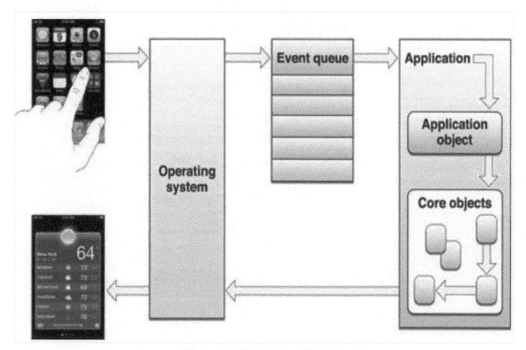

Figure 31: iOS App Event Handling

Main Event Loop

In the main event loop, an application continuously routes incoming events to objects for handling and, because of that handling, updates its appearance and state. An event loop is simply a run loop: An event-processing loop for scheduling work and coordinating the receipt of events from various input sources attached to the run loop. Every thread has access to a run loop. In all but the main thread, the run loop must be configured and run manually by your code. In Cocoa applications, the run loop for the main thread—the main event loop—is run automatically by the application object.

What distinguishes the main event loop is that its primary input source receives events from the operating system that are generated by user actions—for example, tapping a view or entering text using a keyboard.

For instance, in the following figure (Figure 31) the User launches the Application by clicking the App icon on the Phone Home page.

The Click event is enqueued into Operating System Queue. When Operating System allocated time to process the event, in a round-robin scheduling, the event is dequeued and processed. The result of the process creates the application and Show the application, in this case, Weather App. The Scheduler maintains several queues: Idle Queue, Dead Queue, and Ready Queue.[19]

Microsoft Windows Message Pump

Windows application programming follows similar message processing systems link iOS:

1. System or Thread send a message to the applications (see Figure 32)
2. Operating System Enqueues the Message in the System Queue and each event is processed by Message Pump.
3. Message Pump, based on the Window Handle, sends or dispatches to targeted Window Procedure.
4. The Window processes and updates the User Interface (UI).

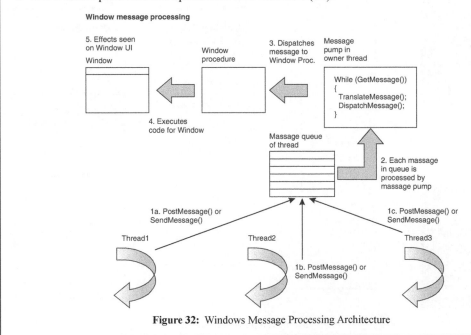

Figure 32: Windows Message Processing Architecture

---

19   Inside iOS - https://cdn.ttgtmedia.com/searchNetworking/downloads/InsideIOS.pdf

```
int WINAPI WinMain(HINSTANCE hInstance, HINSTANCE hPrevInstance, LPSTR lpCmdLine, int
nCmdShow)
{
   MSG msg;
   BOOL bRet;
   while(1)
   {
      bRet = GetMessage(&msg, NULL, 0, 0);
      if (bRet > 0) // (bRet > 0 indicates a message that must be processed.)
      {
         TranslateMessage(&msg);
         DispatchMessage(&msg);
      }
      else if (bRet < 0) // (bRet == -1 indicates an error.)
      {
         // Handle or log the error; possibly exit.
         // ...
      }
      else // (bRet == 0 indicates "exit program".)
      {
         break;
      }
   }
   return msg.wParam;
}
```

Users generate motion events when they move, shake, or tilt the device. These motion events are detected by the device hardware, specifically, the accelerometer and the gyroscope. The Core Motion framework lets your application receive motion data from the device hardware and process that data.

Core Motion provides following device events:

- Accelerometer
- Gyroscope
- Pedometer
- Magnetometer
- Barometer

Like Android Sensor Framework, iOS Core Motion Framework[20] in iOS provides and reports motion and environmental related data from onboard hardware of iOS devices (see Figure 33), including from the accelerometers and gyroscopes, and from the pedometer, magnetometer, and barometer.

---

[20] Core Motion Framework - https://developer.apple.com/documentation/coremotion

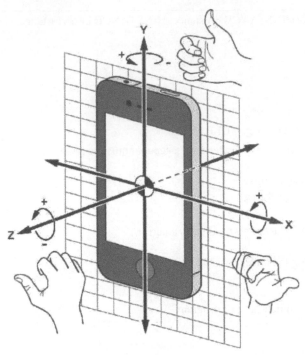

Figure 33: iOS Core Motion

Topics	Classes	
First Steps	Class CMMotionManager	Core Motion Manager that manages the object for creating Motion Manager
Device Motion	CMDeviceMotion	Delivers device motion, rotation rage, and acceleration of device
	CMAttitude	The device's orientation relative to know reference at a point in time
Accelerometer data	CMAccelerometerData	Device's Accelerometer Data
	CMRecordedAccelerometerData	Recorded accelerometer data
	CMSensorRecorder	Gathering and retrieval of Accelerometer data from the device
	CMSensorDataList	A list of the accelerometer data from the device

Core Motion Classes

The Core Motion Framework (CM) has the following classes (see Figure 34):

Core Motion iOS Code

The core motion iOS App uses Sensor Manager to capture Device Motion, Gyroscope and, Accelerometer data. The data can be fetched at user configured rate (as shown in Figure 35).

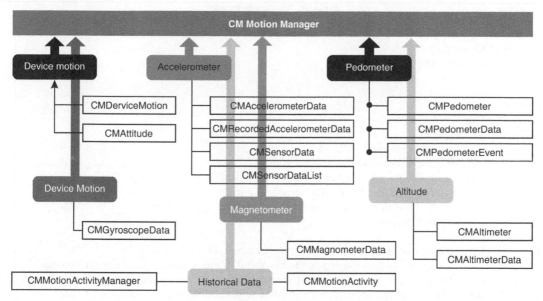

Figure 34: iOS Core Motion Framework

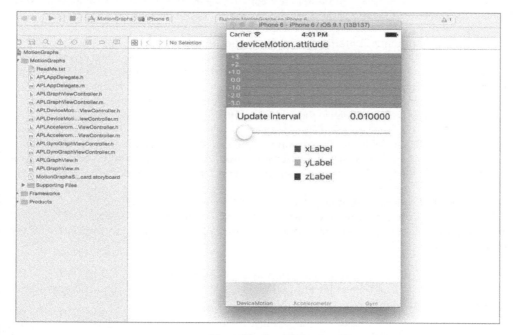

Figure 35: Core Motion App UI

Code

App Delegate Code

The App Delegate (see Table 18) handles all the code transition: Transition from foreground to background, active to pause mode, and active to background mode.

Table 18: App Delegate Code

```
//
// AppDelegate.swift
// CoreMotionExample
//

//
import UIKit
@UIApplicationMain
class AppDelegate: UIResponder, UIApplicationDelegate {
        var window: UIWindow?
        func application(_ application: UIApplication, didFinishLaunchingWithOptions
launchOptions: [UIApplication.LaunchOptionsKey : Any]? = nil) -> Bool {
                // Override point for customization after application launch.
                return true
        }
        func applicationWillResignActive(_ application: UIApplication) {
                // Sent when the application is about to move from active to inactive state.
This can occur for certain types of temporary interruptions (such as an incoming phone call
or SMS message) or when the user quits the application and it begins the transition to the
background state.
                // Use this method to pause ongoing tasks, disable timers, and throttle down
OpenGL ES frame rates. Games should use this method to pause the game.
        }
        func applicationDidEnterBackground(_ application: UIApplication) {
                // Use this method to release shared resources, save user data, invalidate
timers, and store enough application state information to restore your application to its current
state in case it is terminated later.
                // If your application supports background execution, this method is called
instead of applicationWillTerminate: when the user quits.
        }
        func applicationWillEnterForeground(_ application: UIApplication) {
                // Called as part of the transition from the background to the inactive state;
here you can undo many of the changes made on entering the background.
        }
        func applicationDidBecomeActive(_ application: UIApplication) {
                // Restart any tasks that were paused (or not yet started) while the
application was inactive. If the application was previously in the background, optionally refresh
the user interface.
        }
        func applicationWillTerminate(_ application: UIApplication) {
                // Called when the application is about to terminate. Save data if appropriate.
See also applicationDidEnterBackground:.
        }
}
```

View Controller Code

The View Controllers arranges Story Boards for Accelerometer, Gyro, and Device Motions Views (see Table 19).

Table 19: Core Motion Accelerometer Updates

```
//
// ViewController.swift
// CoreMotionExample
//
//
import UIKit
import CoreMotion
class ViewController: UIViewController {
        let motionManager = CMMotionManager()
        var timer: Timer!

        override func viewDidLoad() {
                super.viewDidLoad()

                motionManager.startAccelerometerUpdates()
                motionManager.startGyroUpdates()
                motionManager.startMagnetometerUpdates()
                motionManager.startDeviceMotionUpdates()

                timer  =  Timer.scheduledTimer(timeInterval:  3.0,  target:  self,  selector:
#selector(ViewController.update), userInfo: nil, repeats: true)
        }
        @objc func update() {
                if let accelerometerData = motionManager.accelerometerData {
                        print(accelerometerData)
                }
                if let gyroData = motionManager.gyroData {
                        print(gyroData)
                }
                if let magnetometerData = motionManager.magnetometerData {
                        print(magnetometerData)
                }
                if let deviceMotion = motionManager.deviceMotion {
                        print(deviceMotion)
                }
        }
}
```

Output

```
x 0.082092 y -0.638794 z -0.770035 @ 118190.336395
x -0.613495 y 0.366119 z -0.493500 @ 118190.431574
x -7.839874 y -18.345612 z -65.320282 @ 118190.435309
QuaternionX 0.337296 QuaternionY 0.038609 QuaternionZ -0.008262 QuaternionW
0.940570 UserAccelX -0.007219 UserAccelY -0.000872 UserAccelZ 0.000194
RotationRateX -0.022522 RotationRateY 0.010460 RotationRateZ -0.000189
MagneticFieldX 0.000000 MagneticFieldY 0.000000 MagneticFieldZ 0.000000
MagneticFieldAccuracy -1 Heading -1.000000 @ 118190.436584
x 0.251816 y -0.750671 z -0.894073 @ 118193.332021
x 35.390411 y -10.392548 z 10.837311 @ 118193.337030
x 18.206879 y -16.890945 z -61.640625 @ 118193.327468
QuaternionX 0.385979 QuaternionY 0.054317 QuaternionZ -0.052126 QuaternionW
0.919431 UserAccelX 0.111695 UserAccelY -0.046572 UserAccelZ -0.197934
RotationRateX 0.605888 RotationRateY -0.177316 RotationRateZ 0.197591
MagneticFieldX 0.000000 MagneticFieldY 0.000000 MagneticFieldZ 0.000000
MagneticFieldAccuracy -1 Heading -1.000000 @ 118193.342040
x 0.096817 y -0.679611 z -0.996552 @ 118196.337638
x 112.781448 y 102.204865 z 49.687347 @ 118196.332628
x 15.080399 y -13.295288 z -60.180542 @ 118196.334916
QuaternionX 0.359207 QuaternionY 0.006589 QuaternionZ -0.009305 QuaternionW
0.933187 UserAccelX 0.073929 UserAccelY 0.008747 UserAccelZ -0.260938
RotationRateX 1.956633 RotationRateY 1.787879 RotationRateZ 0.875666
MagneticFieldX 0.000000 MagneticFieldY 0.000000 MagneticFieldZ 0.000000
MagneticFieldAccuracy -1 Heading -1.000000 @ 118196.337638
x 0.234695 y -0.696487 z -0.841339 @ 118199.333216
x 43.316010 y 52.855972 z 21.723465 @ 118199.338226
x 27.676117 y -16.922943 z -61.158340 @ 118199.341703
QuaternionX 0.392182 QuaternionY 0.143840 QuaternionZ 0.039591 QuaternionW
0.907705 UserAccelX 0.004620 UserAccelY 0.026874 UserAccelZ -0.190333
RotationRateX 0.896214 RotationRateY 1.405376 RotationRateZ 0.458711
MagneticFieldX 0.000000 MagneticFieldY 0.000000 MagneticFieldZ 0.000000
MagneticFieldAccuracy -1 Heading -1.000000 @ 118199.338226
x -0.113968 y -0.540314 z -0.805084 @ 118202.338797
x 28.752853 y -81.964859 z -6.410049 @ 118202.333788
x 11.785004 y -8.191986 z -60.685181 @ 118202.311253
RotationRateZ 0.084201 MagneticFieldX 0.000000 MagneticFieldY 0.000000
MagneticFieldZ 0.000000 MagneticFieldAccuracy -1 Heading -1.000000 @
118217.341415
x -0.010315 y 0.009766 z -1.008194 @ 118220.331888
x 1.210968 y -0.627075 z -1.095947 @ 118220.336897
x 11.762009 y -7.462921 z -39.715225 @ 118220.317951
QuaternionX -0.007884 QuaternionY -0.007102 QuaternionZ -0.422925 QuaternionW
0.906097 UserAccelX 0.009224 UserAccelY 0.001486 UserAccelZ -0.008419
RotationRateX 0.009441 RotationRateY -0.006884 RotationRateZ -0.010602
MagneticFieldX 0.000000 MagneticFieldY 0.000000 MagneticFieldZ 0.000000
MagneticFieldAccuracy -1 Heading -1.000000 @ 118220.341906
x -0.009247 y 0.005402 z -0.998627 @ 118223.337378
x 1.703171 y -0.557861 z -0.544861 @ 118223.332369
x 12.780258 y -8.298157 z -41.708649 @ 118223.325466
QuaternionX -0.007103 QuaternionY -0.006643 QuaternionZ -0.493429 QuaternionW
0.869726 UserAccelX 0.008860 UserAccelY 0.001083 UserAccelZ -0.003317
RotationRateX 0.018035
```

```
RotationRateY -0.005676 RotationRateZ -0.000981 MagneticFieldX 0.000000
MagneticFieldY 0.000000 MagneticFieldZ 0.000000 MagneticFieldAccuracy -1
Heading -1.000000 @ 118223.337378
x -0.010590 y 0.007446 z -1.002930 @ 118226.332846
x 0.729385 y -0.248611 z -0.303421 @ 118226.337855
x 13.334152 y -8.945099 z -43.797546 @ 118226.331997
QuaternionX -0.007251 QuaternionY -0.005849 QuaternionZ -0.553885 QuaternionW
0.832535 UserAccelX 0.007183 UserAccelY 0.001852 UserAccelZ -0.003103
RotationRateX -0.000046 RotationRateY 0.001940 RotationRateZ 0.000011
MagneticFieldX 0.000000 MagneticFieldY 0.000000 MagneticFieldZ 0.000000
MagneticFieldAccuracy -1 Heading -1.000000 @ 118226.337855
x -0.010132 y 0.006378 z -1.004257 @ 118229.338330
x 0.666885 y -0.437088 z -0.485672 @ 118229.333321
x 13.297012 y -8.620132 z -43.263718 @ 118229.338106
```

Pedometer App

The purpose of Pedometer App is to detect Step Counts and type of activity by the User (see Figure 36). The App keep tracks of number of user steps and types of activity.

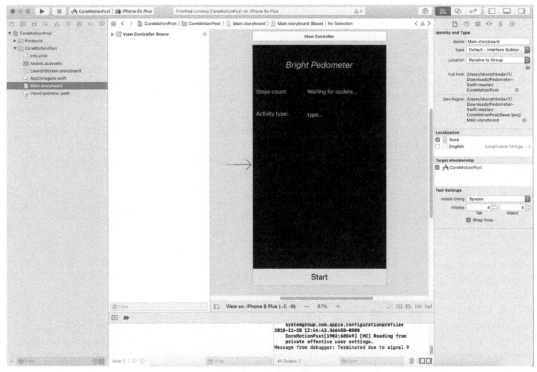

Figure 36: Pedometer App

CMPedometerData

Another class that should catch our attention is CMPedometerData. This class represents data that will be sent with every update in the above functions. It contains a lot of useful information like:

numberOfSteps: NSNumber?

distance: NSNumber?

currentPace: NSNumber?

floorsAscended: NSNumber?

floorsDescended: NSNumber?

CMMotionActivityManager

If we want to start counting steps, it would be good to know about what kind of activity our user is doing now. Here, with some help, comes the CMMotionActivityManager class (please see Table 20). Using the instance of this class we can get updates about the user activity type. To do this, we should call: The permissions UI will grant necessary privileges to access the device core features.

Table 20: Pedometer App Full Code

CMPedometerData
Another class that should catch our attention is CMPedometerData. This class represents data that will be sent with every update in the above functions. It contains a lot of useful information like:
numberOfSteps: NSNumber? distance: NSNumber? currentPace: NSNumber? floorsAscended: NSNumber? floorsDescended: NSNumber? CMMotionActivityManager
If we want to start counting steps, it will be good to know about what kind of activity our user is doing at the moment. Here with some help comes the CMMotionActivityManager class. Using the instance of this class we are able to get updates about the user activity type. In order to do this we should call:
open func startActivityUpdates(to queue: OperationQueue, withHandler handler: @escaping CoreMotion.CMMotionActivityHandler) and the result of that is getting updates with CMMotionActivity which represents data for a single motion event update. This data is a pack of bool values:
stationary: Bool walking: Bool running: Bool automotive: Bool cycling: Bool unknown: Bool Code step by step… 1. Add NSMotionUsageDescription to your info.plist As we can read in Apple docs
Important An iOS app linked on or after iOS 10.0 must include usage description keys in its Info. plist file for the types of data it needs. Failure to include these keys will cause the app to crash. To >access motion and fitness data specifically, it must include NSMotionUsageDescription.
So add to your info.plist NSMotionUsageDescription key modifying plain file:
<key>NSMotionUsageDescription</key> <string>In order to count steps I need an access to your pedometer</string> or adding new key via Xcode
info plist motion usage description

Table 20 contd. ...

...Table 20 contd.

```
2. Create an CMMotionActivityManager and CMPedometer instances
private let activityManager = CMMotionActivityManager()
private let pedometer = CMPedometer()

3. Create a method for tracking activity events
private func startTrackingActivityType() {
    activityManager.startActivityUpdates(to: OperationQueue.main) {
        [weak self] (activity: CMMotionActivity?) in

        guard let activity = activity else { return }
        DispatchQueue.main.async {
          if activity.walking {
            self?.activityTypeLabel.text = "Walking"
          } else if activity.stationary {
            self?.activityTypeLabel.text = "Stationary"
          } else if activity.running {
            self?.activityTypeLabel.text = "Running"
          } else if activity.automotive {
            self?.activityTypeLabel.text = "Automotive"
          }
        }
      }
}

4. Create a method for steps counting updates
private func startCountingSteps() {
  pedometer.startUpdates(from: Date()) {
    [weak self] pedometerData, error in
    guard let pedometerData = pedometerData, error == nil else { return }

    DispatchQueue.main.async {
      self?.stepsCountLabel.text = pedometerData.numberOfSteps.stringValue
    }
  }
}

5. Start getting updates
private func startUpdating() {
 if CMMotionActivityManager.isActivityAvailable() {
    startTrackingActivityType()
 }
  if CMPedometer.isStepCountingAvailable() {
    startCountingSteps()
 }
}
```

App in Action

The deployed app can detect User activities and updates the UI.

Case Study: California Reservoir Continuous Water Monitoring

Worldwide, on a daily basis, sensors in water reservoirs collect water level values and upload them to a central repository. The water level values enable the local governments to better manage citizenry water needs. The following are reservoirs in California.

Continuous Monitoring Stations

The California Departments of Water Resources (DWR) Division of Operations and Maintenance (O&M) currently maintains 16 continuous water quality monitoring stations located throughout the State Water Project. Data from these automated stations are uploaded to the California Data Exchange Center (CDEC) website. The parameters monitored at these stations are:

- Electrical conductivity
- Water temperature
- Turbidity
- pH
- Fluorometry
- UVA-254 absorption

Current and historic hourly and daily data for individual stations can be viewed by selecting the corresponding station link in the Stations and Sensors table below.

Water quality monitoring station at Banks Pumping Plant[21]—This is a caption display style to be used for describing various media items including photos and videos.

Stations and Sensors

This table (see Table 21) contains[22] information on the location and water quality parameters monitored at continuous, automated stations operated by O&M.

Station Pictures and Sensor Information

Instruments at each station[23] collect data 24 hours a day, 7 days a week. Measurements of water quality parameters are taken every 5 seconds and then combined into hourly averages by onsite data loggers (see Figure 37). These data loggers are then automatically polled every hour by our server in Sacramento. Once the data is downloaded to the server, each measurement is automatically compared against broad screening criteria in order to eliminate erroneous data, and the preliminary data are then transferred to the California Data Exchange Center (CDEC) website. These data are also regularly reviewed by DWR scientists to ensure their validity and accuracy.[24]

1. Spectrophotometer: Measures UVA in u/cm
2. Stepped filtration: Filters incoming water

---

[21] Continuous monitoring station - https://water.ca.gov/Programs/State-Water-Project/Water-Quality/Continuous-Monitoring-Stations#

[22] Station and Sensors - https://water.ca.gov/Programs/State-Water-Project/Water-Quality/Continuous-Monitoring-Stations#

[23] Station Pictures and Sensor Information - https://water.ca.gov/Programs/State-Water-Project/Water-Quality/Continuous-Monitoring-Stations/Station-Pictures-and-Information

[24] Station Pictures and Sensor Information - https://water.ca.gov/Programs/State-Water-Project/Water-Quality/Continuous-Monitoring-Stations/Station-Pictures-and-Information

Table 21: Water Quality parameters

#	CDEC ID	Station Name	Lat	Long	County	Turb	EC	Temp	pH	UVA-254	Fluoro
1	BKS	Barker Sough Pumping Plant	38.275900	-121.796500	Solano	X	X	X	X		
2	C13	Check 13	37.074200	-121.015100	Merced	X	X	X	X	X	
3	C21	Check 21	36.014200	-119.976900	Kings	X	X	X	X		
4	C41	Check 41	34.834000	-118.711900	Kern	X	X	X	X		
5	C66	Check 66	34.336000	-117.303600	San Bernardino	X	X	X			
6	CLC	Clifton Court	37.829800	-121.557400	Contra Costa	X	X	X	X		X
7	CPP	Cordelia Pumping Plant	38.227600	-122.134700	Solano	X	X	X			
8	CSO	Castaic Lake Outlet	34.527000	-118.611000	Los Angeles	X	X	X	X		
9	DC2	Devil Canyon 2nd Afterbay	34.230500	-117.340500	San Bernardino	X	X	X	X		
10	DCO	Del Valle COW	37.617800	-121.746800	Alameda	X	X	X	X		
11	DV7	Del Valle Check 7	37.654700	-121.741900	Alameda	X	X	X	X		X
12	DYR	Dyer Reservoir	37.757824	-121.674913	Contra Costa	X	X	X	X		X
13	HBP	Banks Pumping Plant	37.801944	-121.620278	Contra Costa	X	X	X	X	X	X
14	PPP	Pacheco Pumping Plant	37.061900	-121.180600	Merced	X	X	X	X		X
15	TEE	Teerink Pumping Plant	35.032749	-119.008102	Kern	X	X	X	X		
16	VSB	Vallecitos Turnout	37.598500	-121.819300	Alameda	X	X	X	X		

3. Turbidimeter: Turbidity in Nephelometric turbidity units (N.T.U.)

4. Turbidimeter controller: Shows turbidity readings

5. EC and water temperature controller: Shows EC and water temperature readings

6. Datalogger: Installed July 2008

7. EC and water temperature probe: EC in uS/cm and water temperature in degrees C

8. pH probe: Measures pH

9. pH display transmitter: Shows pH readings

10. Fluorometer: Calibrated with Rhodamine dye; reporting units (fluorometry units) are a chlorophyll equivalent.

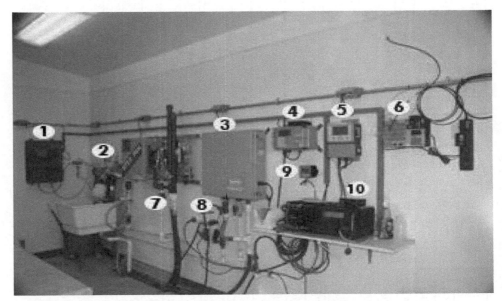

Figure 37: Water quality equipment and functions

Current Reservoir Conditions

The Sensor collected data from different reservoirs are posted on the Web site to report current reservoir conditions[25] (please see Figure 38):

Audio Sensors

Microphone plays an important role in capturing the critical audio data (see Figure 39). The audio data plays an important role in creating mobile applications that are senior citizen assistant apps, in capturing audio signals data that could predict health conditions for agriculture and detecting presence of crowd to detect venue based ambient analytics.

Microphone Sensors

A measurement microphone is like an ordinary microphone in the superficial features: It is typically tubular, with a sensor at one end and a connector at the other, and the sensor itself is a lightweight diaphragm that is

---

[25]   Current Reservoir Conditions - http://cdec.water.ca.gov/reportapp/javareports?name=rescond.pdf

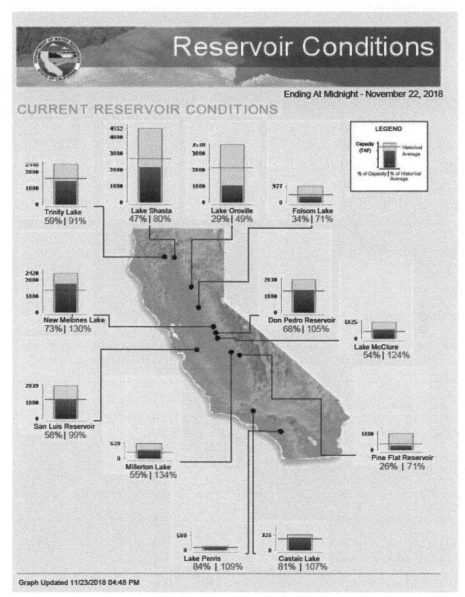

Figure 38: Current Reservoir Conditions

excited by changes in air pressure, responding in a way that can produce an electrical signal. At this point, however, the two microphone types diverge: You won't see a singer's wireless mic measuring loudspeaker drivers in an anechoic chamber, and you won't see a comedian using a measurement microphone for the mic drop at the end of his routine.

A typical sound sensor, microphone (MIC), captures environmental sound variations (see figure Audio Recordings Spectrum). The Case Study for using sound sensors span from agriculture to oil drills (heavy equipment). Sound variation can also predict health factors. The following sound sensor[26] enables capture of environmental noise variations.

---

[26]   Sound Sensor - https://www.osepp.com/electronic-modules/sensor-modules

After a device is activated, it may sometimes transmit the full range of audible sounds (including voices), for example to enable cloud-based speech-to-text translation. However, other devices may not send audio at all, but instead may use the microphone to detect patterns and transmit other information about the user's environment.

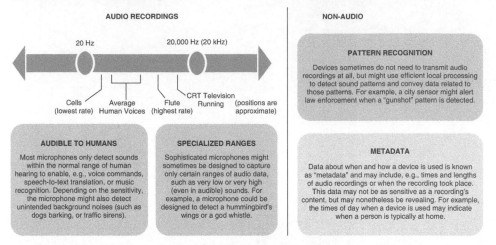

Figure 39: Audio Recording Spectrum

Sensor with Arduino board

The sound sensor is connected to Arduino board and captures the sound signals. The Sound Sensor is connected to the microphone pin and measures the sound input from the microphone sensor.

```
/*
OSEPP Sound Sensor Example
OSEPP example of measure sound input from microphone sensor.
depending on loudness, an LED brightness is changed.
Louder sounds = brighter LED. Test with music to get a strobe
*/
int micPin = A0; // microphone sensor input
int ledPin = 11; // select the pin for the LED
int micValue = 0; // variable to store the value coming from the mic sensor
void setup() {
        Serial.begin(9600);
}
void loop() {
        // read the value from the sensor:
        micValue = analogRead(micPin);
        micValue = constrain(micValue, 0, 100); //set sound detect clamp 0-100

// turn the ledPin on
        int ledbrightness = map(micValue, 0, 100, 0, 255);
        Serial.print("incoming value from microphone =");
        Serial.println(micValue);
        analogWrite(ledPin, ledbrightness);
        delay(100);
}
```

The Circuit board diagram[27]

The MK1/Mic2 (see Figure 40) captures the sound input which is translated into sound electrical analog signal. The waveforms for the sound signals can be rendered in the following:

---

[27]   OSEPP Sound Sensor - https://www.osepp.com/downloads/pdf/SOUND-01-Schematic.pdf

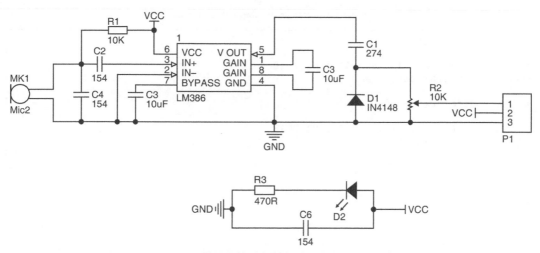

Figure 40: MIC Circuit Board

The sound waveforms can be generated by running:

Case Study—Voice disorders detection using Machine Learning and Artificial Intelligence

The Voice order pathological issues are dominant health issues faced by rural communities in developing countries. Lack of skilled healthcare professionals and delayed detection are major factors in solving the issue. Having a diagnostic app that is readily available in a portal manner to healthcare professional makes a huge positive impact. The S-EHR[28] Voice Detect App team has developed Neural Network that has predicted with 90% accuracy pathological issues in voice detection. The Voice samples were obtained from a voice clinic in a tertiary teaching hospital (Far Eastern Memorial Hospital, FEMH), which included 50 normal voice samples and 150 samples of common voice disorders, including vocal nodules, polyps, and cysts (collectively referred to Phonotrauma); glottis neoplasm; unilateral vocal paralysis. Voice samples of a 3-second sustained vowel sound/a:/were recorded at a comfortable level of loudness, with a microphone-to-mouth distance of approximately 15–20 cm, using a high-quality microphone (Model: SM58, SHURE, IL), with a digital amplifier (Model: X2u, SHURE) under a background noise level between 40 and 45 dBA. The sampling rate was 44,100 Hz with a 16-bit resolution, and data were saved in an uncompressed .wav format.

Wave File: Training\Pathological\Neoplasm\003.wav

Image: The sound wave (see Figure 41) represents the converted voice to digital signal.

Figure 41: Sound wave

Spectrogram Code generator

In order to analyze sound wave for possible feature engineering, the wave file has to be translated into several representations: Spectral Centroid, Mega Centroid, Spectral Mega Centroid, Log Power Spectrogram (please see Table 22 and Figure 42)

Table 22: Spectrogram Librosa Python Code

```
# -*- coding: utf-8 -*-
"""
Created on Mon Oct 15 16 : 41 : 49 2018
@author: cvuppalapati
"""
# -*- coding: utf-8 -*-
"""
Created on Fri Oct 12 12 : 18 : 03 2018
@author: cvuppalapati
"""
# -*- coding: utf-8 -*-
"""
Created on Fri Oct 12 11 : 58 : 12 2018
@author: cvuppalapati
"""
import os
import matplotlib
#matplotlib.use('Agg') # No pictures displayed
import pylab
import librosa
import librosa.display
import numpy as np
np.set_printoptions(precision = 3)
np.set_printoptions(threshold = np.nan)
# https://librosa.github.io/librosa/generated/librosa.feature.melspectrogram.
html#librosa.feature.melspectrogram
sig, sample_rate = librosa.load('C:\\Hanumayamma\\FEMH\\FEMH Data\\Training
Dataset\\Pathological\\Neoplasm\\002.wav')
stft = np.abs(librosa.stft(sig))
print('{0:.15f}'.format(sig[len(sig) - 1]))
# magnitude spectrogram
magnitude = np.abs(stft) # (1 + n_fft / 2, T)
# power spectrogram
power = magnitude * * 2 # (1 + n_fft / 2, T)
mfccs = np.array(librosa.feature.mfcc(y = sig, sr = sample_rate, n_mfcc =
8).T)
mfccs_40 = np.mean(librosa.feature.mfcc(y = sig, sr = sample_rate, n_mfcc =
40).T, axis = 0)
print("\n mfccs_40 \n")
print(mfccs_40)
chroma = np.array(librosa.feature.chroma_stft(S = stft, sr = sample_rate).T)
chroma_mean = np.mean(librosa.feature.chroma_stft(S = stft, sr = sample_
rate).T, axis = 0)
print("\n chroma_mean \n")
print(chroma_mean)
#mel = np.array(librosa.feature.melspectrogram(X = sig, sr = sample_rate).T)
mel_mean = np.mean(librosa.feature.melspectrogram(sig, sr = sample_rate).T,
axis = 0)
print("\n mel_mean \n")
print(mel_mean)
contrast = np.array(librosa.feature.spectral_contrast(S = stft, sr = sample_
rate).T)
contrast_mean = np.mean(librosa.feature.spectral_contrast(S = stft, sr =
sample_rate).T, axis = 0)
print("\n contrast_mean \n")
print(contrast_mean)
tonnetz = np.array(librosa.feature.tonnetz(y = librosa.effects.harmonic(sig),
```

Table 22 contd. ...

...Table 22 contd.

```
sr = sample_rate).T)
tonnetz_mean = np.mean(librosa.feature.tonnetz(y = librosa.ef fects.
harmonic(sig), sr = sample_rate).T, axis = 0)
print("\n tonnetz_mean \n")
print(tonnetz_mean)
#https://librosa.github.io/librosa/generated/librosa.feature.spectral_cen-
troid.html
#Compute the spectral centroid.
cent = librosa.feature.spectral_centroid(y = sig, sr = sample_rate)
print("\n Compute the spectral centroid. \n")
print(cent)
# From spectrogram input :
S, phase = librosa.magphase(librosa.stft(y = sig))
cent_mega = librosa.feature.spectral_centroid(S = S)
print("\n Compute the spectral mega centroid. \n")
print(cent_mega)
# Using variable bin center frequencies :
if_gram, D = librosa.ifgram(sig)
cent_if_gram = librosa.feature.spectral_centroid(S = np.abs(D), freq = if_
gram)
print("\n Compute the spectral mega cent_if_gram centroid. \n")
print(cent_if_gram)
import matplotlib.pyplot as plt
plt.figure()
plt.subplot(4, 1, 1)
plt.semilogy(cent.T, label = 'Spectral centroid')
plt.ylabel('Hz')
plt.xticks([])
plt.xlim([0, cent.shape[-1]])
plt.legend()
plt.subplot(4, 1, 2)
plt.semilogy(cent_mega.T, label = 'spectralcent_if_gram mega centroid.')
plt.ylabel('Hz')
plt.xticks([])
plt.xlim([0, cent_mega.shape[-1]])
plt.legend()
plt.subplot(4, 1, 3)
plt.semilogy(cent_if_gram.T, label = 'spectral mega centroid.')
plt.ylabel('Hz')
plt.xticks([])
plt.xlim([0, cent_if_gram.shape[-1]])
plt.legend()
plt.subplot(4, 1, 4)
librosa.display.specshow(librosa.amplitude_to_db(S, ref = np.max),
        y_axis = 'log', x_axis = 'time')
        plt.title('log Power spectrogram')

        plt.tight_layout()
        plt.show()

        # mel spectrogram
        #S = librosa.feature.melspectrogram(S = power, n_mels = hp.n_mels) #
(n_mels, T)
        #np.transpose(S.astype(np.float32)), np.transpose(magnitude.astype(np.
float32)) # (T, n_mels), (T, 1 + n_fft / 2)
```

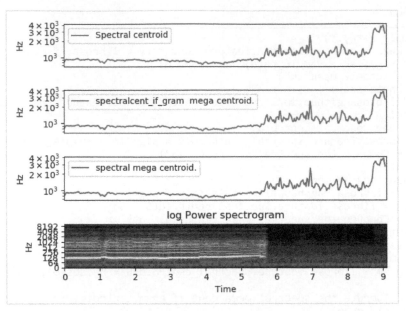

Figure 42: Spectrum representations (Spectral Centroid, Mega Centroid, Spectral Mega Centroid, Log Power Spectrogram)

mfccs_40
[-3.381e+02 1.510e+02 1.213e+01 3.792e+00 -1.770e+01 5.671e+00
 7.133e+00 3.313e+00 4.679e-01 1.043e+00 6.319e+00 -8.136e+00
 1.755e+01 5.969e-01 -8.928e+00 3.334e+00 -1.379e+01 1.120e+00
 -1.356e+00 2.967e-01 -8.713e+00 -3.432e+00 -4.748e+00 -2.915e+00
 -7.313e+00 8.319e-01 -1.378e+00 -3.427e+00 -5.394e+00 -3.428e+00
 -3.671e+00 -3.549e+00 -5.258e+00 -4.364e+00 -1.555e+00 -1.035e+00
 -2.676e+00 -2.589e+00 -1.439e+00 -1.890e+00]

chroma_mean

[0.76 0.737 0.636 0.653 0.513 0.5 0.636 0.532 0.493 0.515 0.554 0.793]

mel_mean

[6.560e-02 9.925e-02 9.475e-02 3.093e+01 1.764e+02 3.447e+01 2.622e+00
 6.215e-01 4.713e+00 1.150e+01 1.963e+00 4.612e-01 3.314e-01 4.742e+00
 5.957e+00 5.211e-01 3.500e-01 6.001e-01 7.821e+00 8.187e+00 1.049e+00
 1.207e+00 5.452e+00 2.276e+01 1.354e+01 1.444e+00 1.408e+00 8.959e+00
 2.608e+01 1.309e+01 1.450e+00 5.430e-01 1.220e+00 3.141e+00 1.014e+00
 2.969e-01 1.195e+00 5.588e+00 1.000e+01 3.956e+00 1.197e+00 3.418e+00
 1.340e+01 5.784e+00 7.605e-01 2.678e-01 4.327e-01 2.547e-01 7.372e-02
 6.208e-02 8.290e-02 3.146e-02 2.345e-02 5.013e-02 2.081e-02 2.065e-02
 5.877e-02 3.494e-02 4.685e-02 8.992e-02 3.469e-02 3.762e-02 2.273e-02
 1.160e-02 1.793e-02 2.614e-03 3.574e-03 5.626e-03 4.474e-03 2.523e-03
 3.231e-03 1.520e-03 1.129e-03 6.526e-04 5.206e-04 3.561e-04 5.273e-04
 6.941e-04 1.071e-03 2.515e-03 4.463e-03 6.817e-03 6.032e-03 4.016e-03
 3.331e-03 2.937e-03 2.446e-03 1.666e-03 1.246e-03 9.489e-04 5.047e-04
 2.592e-04 1.498e-04 9.508e-05 9.836e-05 1.467e-04 9.150e-05 7.593e-05
 1.924e-04 4.853e-04 7.612e-04 7.914e-04 7.920e-04 7.561e-04 6.802e-04
 5.451e-04 4.409e-04 2.875e-04 1.245e-04 3.594e-05 2.505e-05 1.930e-05
 1.408e-05 7.108e-06 4.246e-06 5.054e-06 8.094e-06 9.321e-06 1.008e-05
 4.968e-05 1.616e-04 8.759e-05 1.142e-05 2.858e-05 7.231e-05 4.736e-05
 4.161e-06 4.115e-07]

Audio Case Study: Cow Necklace[29]

Rumination is a window to cow health. Rumination is impacted by external factors as well as internal rumen functioning. Rumination is the regurgitation of fibrous matter from the rumen to the mouth, and the return to the rumen. This biological process is impacted by the forage's nutritional and fiber content. Rumination is not only a requirement for healthy cows; it can be a very early indicator of stress or illness. We can monitor Rumination through the deployment of Internet of Things (IoT) Sensor and can capture the rate of Rumination and enable detection of falling rates of Rumination and notify the farmer proactively any health issues.

Video Sensors

The Camera, a classical video sensor, is replacing other traditional sensors. For example, AT&T replaces IoT Sensors with cameras in order to capture the state of aluminum cans in a refrigerating Units. AT & T moves 186 petabytes across networks and 53% of that traffic is video data.[30] AT & T replaces IoT Sensors with cameras. "Just by measuring the size of the condensation beads on the aluminum cans, you can calculate what the temperature and relative humidity is and taking a high-resolution camera and attaching it to some very sophisticated image recognition software, you can do some amazing things."

For Image and Data analysis, the Video data is a translated series of contiguous images and image processing enables us to find the image patterns.

Image Texture Data

There are several ways to extract the texture of an image and one of the efficient algorithms is Local Binary Pattern (LBP). The purpose of LBP (see Figure 43) is to generate the histogram of an image.

The purpose of Local Binary Pattern (LBP) operator is to transform an image into an array or image of integer labels describing the small scale appearance of the image. In other words, LBP translates an image into a histogram of integer labels and is used for image texture analysis purposes.

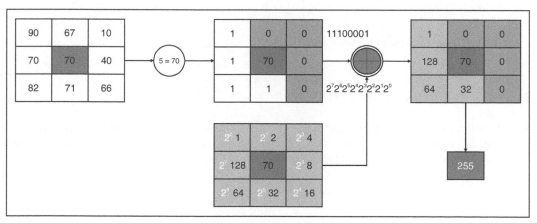

Figure 43: LBP In Action

---

29   Hanumayamma Dairy Sensor - http://hanuinnotech.com/dairyanalytics.html

30   AT&T replaces IoT Sensors with Camera - https://enterpriseiotinsights.com/20171102/news/att-sees-video-replacing-lower-bandwidth-iot-traffic-tag4

The local binary pattern assumes that the image texture has two complimentary features: Pattern and strength. The original version of LBP is based on a 3×3 pixel block of an image.

The LBP performs texture classification based on principle looking points surrounding a central point and tests whether surrounding points are greater than or less than the central point—performs binary transition.

In the figure "LBP In Action", the central pixel value [70] is compared with surrounding pixel values [90,67,10,40,66,71,82,70] and produces the bit pattern. The central pixel is greater than the surrounded pixel the pattern is coded with 1. For example: Surround pixel [90]—Central Pixel [70] = 1. Similarly, Surround pixel [90]—Central Pixel [67] results in 0. The comparison would produce in bit pattern [11100001].

Next step, is to encode bit pattern on binary scale. For instance, the bit pattern [11100001] coded in to binary form: $[(1*2^7) + (1*2^6) + (1*2^5) + (0*2^4) + (0*2^3) + (0*2^2) + (0*2^1) + (1*2^0)]$ and thus producing value of 255.

As shown in below figure "LBP In Action", based on color patterns of surrounding pixels with that of the central one, following types of LBP are possible:

- When surrounding pixels are all white ([11111111]) or all black ([00000000]), then the image is flat or featureless. This is due to the values produced being either 255 or 0 respectively.

- A group of continuous black or white pixels are considered "uniform" patterns and can be interpreted as corners ([00001110]) or edges ([01110000]).

- If surrounding pixels switch back-and-forth between black and white pixels, the pattern is considered non-uniform.[31]

Geospatial Data

Geospatial data is a location relating to or denoting data that is associated with a geographical location. The data contains natural geographical markers and man-made changes. For instance, natural markers include geolocation perimeter and man-made changes include global warming trends & pollution indexes (see figure Hourly Air Quality Index[32] (AQI)).

The Data sources include:

- Pollution Index
- Air Quality
- Geo Location—Latitude and Longitude
- Geofence

Particle Pollution (PM)—PM 2.5

Particle pollution, also called particulate matter or PM, is a mixture of solids and liquid droplets floating in the air. Some particles are released directly from a specific source, while others form in complicated chemical reactions in the atmosphere (see Figure 44).

---

[31]   Texture feature extraction using LBP - Texture Feature Extraction by Using Local Binary Pattern - https://www.researchgate.net/publication/305152373_Texture_Feature_Extraction_by_Using_Local_Binary_Pattern

[32]   AQI - https://airnow.gov/index.cfm?action=airnow.local_city&cityid=312&mapdate=20181205

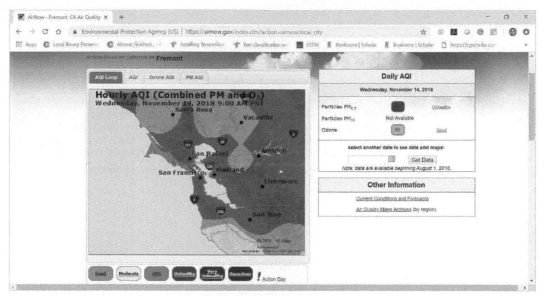

Figure 44: Hourly Air Quality Index(AQI)

Chapter Summary

After reading the chapter:

- Should able to understand IoT Sensor types
- Should able to analyze different data types (Audio, Video, Image, Geospatial)
- Would able to understand how sensors created basic IoT sensor module

References

1. Cisco, The Internet of Things Reference Model. 2014, http://cdn.iotwf.com/resources/71/IoT_Reference_Model_White_Paper_June_4_2014.pdf, Access Date: April 2018
2. Khaled Allam, The Internet of Things – Vertical Solutions, Cisco Connect 2015, https://www.cisco.com/web/offer/emear/38586/images/Presentations/P11.pdf, Access Date: June 2018
3. Adrian Both, Building a Great Data Platform. August 2018, https://www.mckinsey.com/industries/electric-power-and-natural-gas/our-insights/building-a-great-data-platform, Access Date: September 2018
4. Peter C. Evans and Marco Annunziata, Industrial Internet—Pushing the boundaries of Minds and Machines, November 26, 2012, https://www.ge.com/docs/chapters/Industrial_Internet.pdf, Access Date: May 2018
5. Kirsten Schwarzer, 7 Types of Industrial IoT Data Sources (and How to Use them), http://xmpro.com/7-types-industrial-iot-data-sources/, Access Date: September 2018
6. NYC, NYC Open Data. 2018, https://data.cityofnewyork.us/Transportation/Real-Time-Traffic-Speed-Data/qkm5-nuaq, Access Date: August 2018
7. Mehdia Ajana El Khaddar and Mohammed Boulmalf. Smartphone: The Ultimate IoT and IoE Device. November 2nd 2017, Access Date: August 2018
8. Android Developers, Sensors Overview, https://developer.android.com/guide/topics/sensors/sensors_overview, Access Date April 2018
9. Android Developers, Architecture, https://developer.android.com/guide/platform/ Access Date April 2018
10. Android Developers, Sensors Manager, https://developer.android.com/reference/android/hardware/SensorManager#registerListener, Access Date April 2018
11. Android Developers, Sensors Manager, https://developer.android.com/reference/android/hardware/SensorManager#unregisterListener, Access Date April 2018
12. Github. Android AccelerometerPlay Sample, https://github.com/googlesamples/android-AccelerometerPlay
13. Android. Motion Sensors, https://developer.android.com/guide/topics/sensors/sensors_motion. Access Date April 2018

14. Chandrasekar Vuppalapati. A System To Detect Mental Stress Using Machine Learning And Mobile Development, 2018 International Conference on Machine Learning and Cybernetics (ICMLC). IEEE, July 2018

15. Apple. Core Motion Framework, https://developer.apple.com/documentation/coremotion, 2018, Access Date April 2018

16. California. Reservoir Water Condition, http://cdec.water.ca.gov/reportapp/javareports?name=rescond.pdf,

17. Sound Sensor. OSEPP Sound Sensor, https://www.osepp.com/downloads/pdf/SOUND-01-Schematic.pdf, 08/10/2015. Access Date April 2018

18. Martha DeGrasse. AT&T replaces IoT sensors with cameras, November 02 2017, https://enterpriseiotinsights.com/20171102/news/att-sees-video-replacing-lower-bandwidth-iot-traffic-tag4, access date: September 18, 2018

19. EPA. Air Now, https://airnow.gov/index.cfm?action=airnow.local_city&cityid=312&mapdate=20181205

20. Esa Praksa. Texture Feature Extraction by Using Local Binary Pattern, May 2016, https://www.researchgate.net/publication/305152373_Texture_Feature_Extraction_by_Using_Local_Binary_Pattern, Access Date: August 2018

CHAPTER 5

IoT Data Collectors

This Chapter Introduces:
• Data Collectors for the Edge processing
• Introduces Edge & Fog Computing essentials
• Data Collector Algorithms
• Data Circular Buffers
• Audio Signal representation (high level)
• Video Signal representation (high level)

The Data generated by IoT Sensors contain valuable information about the device, the environment, device health, diagnostics, state of firmware, state machine, and more. Generally, the data is time-series with events spaced at an interval. Sensors generate events at a faster pace. Some of the devices might generate the data samples multiple times per second, 24 hours a day, and 365 days a year (see Figure 1). The volume of data per day might be very high. For instance, if a sensor is generating data at every microsecond level by the end of the day, the sensor would have produced 86.4 million (24 x 60 x 60 x 1000) data samples. Not only is the data huge, but the velocity is also very rapid. The basic principle of IoT[1] is that the intelligent system initiates and processes the information as early and as close to the source as possible [1]. This is sometimes to referred as fog computing. In this chapter, we

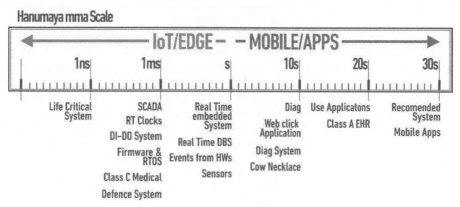

Figure 1: Hanumayamma IoT Scale

---

[1]   IoT Reference Model - http://cdn.iotwf.com/resources/71/IoT_Reference_Model_White_Paper_June_4_2014.pdf

will introduce Fog computing, provide technological imperatives[2] that drives edge computing choices, and the high-performance algorithms that process the data in real-time.

Edge or Fog Computing

IoT may be divided into three main levels: Edge, fog and cloud, as shown in Figure 2. The IoT edge layer typically includes IoT sensors and actuators and perhaps limited sets of getaways. IoT fog layer includes IoT network components including gateways and light weigh applications. The IoT cloud layer includes main severs and applications [2].

Intelligence may be deployed at every level. Edge analytics provides rapid response to the source by collecting and aggregating the data obtained by sensors near to the source instead of transmitting it to the cloud or centralized datacenter. In other words, "IoT processing at the edge pushes the analytics from a central server to devices close to where the data is generated." [3] Edge analytic is a typical requirement for real time systems.

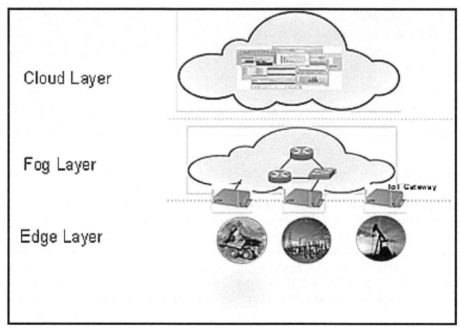

Figure 2: IoT Architecture

The advantage of analyzing the data closer to the source is that Edge Analytics not only provides a rapid response but also aids in the detection of device health markers, data anomalies and abnormalities in order to predict device operational and/or health prognostics and, thus, potentially improve the overall performance and life of the device. "A sensor network can be applied to collect the real-time production line data. The data is then used to analyze the asset health and predict failure or the mean time to failure (MTTF) and suggest possible solutions to minimize disruptive and unscheduled downtime [5]". Edge Analytics with prognostics capabilities, in general, promote core tenets of the Open System Architecture for Condition Based Monitoring (OSA-CBM) architecture—reduces cost & improves the life of the device. As shown in Figure 3, the OSA-CBM architecture consists of seven layers: Data acquisition, data manipulation, state detection, health assessment, prognostics, decision support and presentation [6] and edge analytics, all of which enhance device health capabilities.

---

[2]   The Future Shape of Edge Computing: Five Imperatives - https://www.gartner.com/doc/3880015/future-shape-edge-computing-imperatives

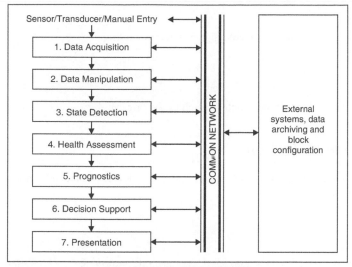

Figure 3: OSA-CBM Model [4]

At Edge or Fog level (see Figure 4), since we are closer to source, generally, the events are generated at a rapid pace. The processing is performed and includes[3] [1]:

- Evaluation
- Formatting
- Expanding
- Distillation or reduction
- Assessment

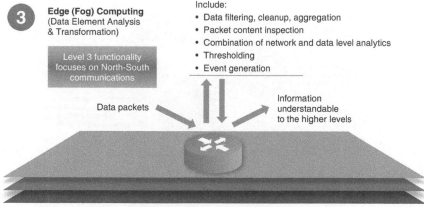

Figure 4: Edge (Fog) Computing

Evaluation

Evaluating data for criteria as to whether the data needs to be processed immediately or it should be processed at a higher level. For instance, to notify or alert a plant floor manager if temperature spiked or increased at rapidly in consecutive time intervals.

---

[3] The IoT Reference Model - http://cdn.iotwf.com/resources/71/IoT_Reference_Model_White_Paper_June_4_2014.pdf

Formatting

The purpose of formatting data is to enable higher level processing.

Expanding

Handling cryptic data with additional context (such as the origin).

Distillation and Reduction

Reducing and/or summarizing data in order to minimize the impact of data and traffic on the network and higher-level processing systems.

Assessment

Determining whether data represents a threshold or alert; this could include redirecting data to additional destinations.

Edge Imperatives

Edge (see Figure 5) computing solution design is based on the performance characteristics of the solution or the Case study that drives it [2]. The following performance or technical parameters play an important role in Edge solution:[4]

- Latency
- Local Interactivity

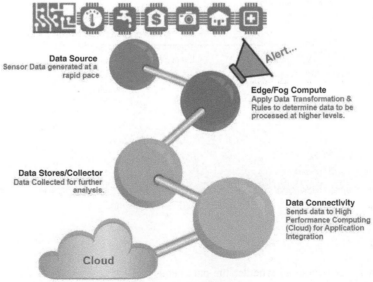

Figure 5: IoT Architecture

---

[4]   The Future Shape of Edge Computing: Five Imperatives - https://www.gartner.com/doc/3880015/future-shape-edge-computing-imperative

- Data/Bandwidth
- Privacy/Security
- Limited Autonomy

Case Study: Note to my great-great Grand Kids: I am Sorry I drank your milk and ate all your milk chocolates!
Climate change is impacting milk production worldwide. If we fail to develop agriculture tools that intelligently counter climate changes through the application of data science, by the end of the century, heat stress in cows could see average-sized dairy farms losing thousands of farms each year; climate change will hit milk products in millions of tonnes, causing unprecedented scarcity for milk-based products and leaving millions of malnourished children worldwide; finally, with the combined impact of climate changes and population growth as well as the scarcity of dairy and agricultural lands, we're leaving the fate of future generations "undefined". Data-driven agriculture tools are the best defense in countering the climate change. Through the application of data science and innovative ways of collecting data, as well as through the development of data sensors, we can claim back the lost productivity in dairy agriculture. Having data enabled dairy farming and application of machine learning infused artificial intelligence and climate models will give hope!

Data Collector Algorithms

The Data at the edge level is very fast paced. To process data at the edge layer, the computation needs to be quick. Holding events longer due to process cost will pile the events at the source and may result in un-wanted system issues or crashes. The following list provides the most frequently used to process the events at the edge:

- Circular Buffers
- Double Buffering

Circular Buffers

Circular Buffers are very useful data structures that are used memory constrained embedded device applications. As the name implies, Circular buffers use predefined fixed size buffers that wrap to the beginning as a close tube (please see Figure 6).

Technically, Circular buffers provide message queue semantics. That is, the data inserted Last in First Out (LIFO) mode.

Circular buffers (also known as ring buffers) are fixed-size buffers that work as if the memory is contiguous & circular in nature. As memory is generated and consumed, data does not need to be reshuffled—rather, the head/tail pointers are adjusted. When data is added, the head pointer advances. When data is consumed, the tail pointer advances. If you reach the end of the buffer, the pointers simply wrap around to the beginning.

Here are design pointers:

- Head pointer—Points to the front of the Queue. Or, in other words, it points to the element to be removed if the dequeue operation is called.
- Tail pointer—Points to the next empty spot in which the new element can be inserted.
- Full—If you tried to fill the queue completely you wouldn't be able to enqueue after the Maximum length or Queue length (Maximum size of the Circular Buffer) index. This is because the Tail has no empty spot to point to after an element is inserted. The queue is considered full and the new elements are inserted restarting from 0. It is for this reason that it is called a circular queue.

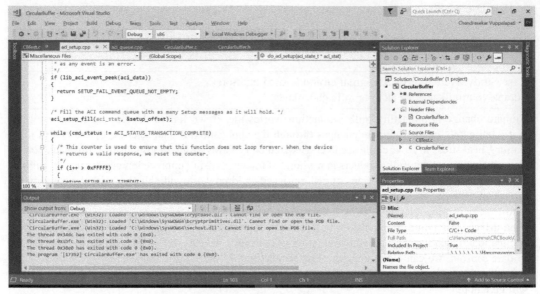

Figure 6: Microsoft Visual Studio

Circular Buffer Animation & Working Semantics [7]

- When Circular Buffer is built, following is the state: Head and Tail points to 0 (see Figure 7).

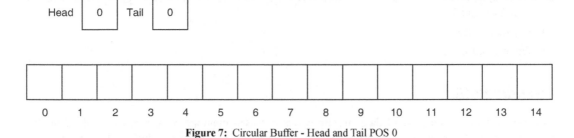

Figure 7: Circular Buffer - Head and Tail POS 0

- Insert first element—enqueue first element, fills index 0 position; head is point to 0 index and tail points to 1 index (see Figure 8).

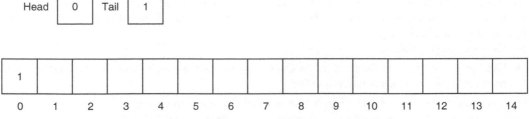

Figure 8: Circular Buffer—Head and Tail POS Insert First Element

- Insert Second Element: Index 1 is populated with new element; Head points to 0 index; tail points 1 index (see Figure 9).

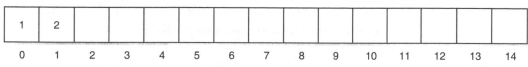

Figure 9: Circular Buffer – Head and Tail POS Insert Second Element

- Insert Third Element: New element is inserted into index 2; head points 0 index; tail points to 2 index (see Figure 10)

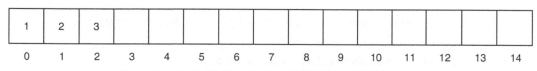

Figure 10: Circular Buffer – Head and Tail POS Insert Third Element

- Dequeue Element: Dequeue first element fetches & removes the element from Index 0. Now, Index 0 is empty; Head points to index 1; Tail points to index 3 (see Figure 11).

Dequeued Value: 1

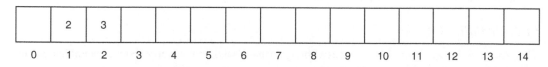

Figure 11: Circular Buffer – Deque Element

- Reached End: Inserting to the size of the circular buffer will make tail points to 0 index and head to 1 index (no change) – see Figure 12.

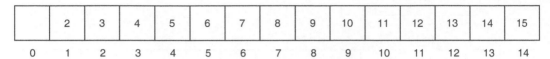

Figure 12: Circular Buffer – Circular Buffer Reached End

- Number of Elements[5]
  - Tail>=Head: Number of elements = Tail - Head. For instance, if Head = 2 and Tail = 5, then the number of elements will be 5 – 2 = 3
  - Head>Tail: Number of elements = (Size of Queue) - (Head-Tail) = (Size of Queue) - Head + Tail. For instance, Head = 9, Tail = 5 and Size of Queue = 10, then the number of elements = 10 – (9 – 5) = 6

Algorithm

1. Initialize the queue, size of the queue (maxSize), head and tail pointers
2. Enqueue:
   a. Check if the number of elements (size) is equal to the size of the queue (maxSize):
   b. If yes, throw error message "Queue Full!"
   c. If no, append the new element and increment the tail pointer
3. Dequeue:
   a. Check if the number of elements (size) is equal to 0:
   b. If yes, throw error message "Queue Empty!"
   c. If no, increment head pointer
4. Size:
   a. If tail>=head, size = tail – head
   b. if head>tail, size = maxSize - (head – tail)

Circular Buffer Python Code

The code (see Table 1) and output (see Figure 13):

Circular Buffer C Code

The generic code (see Figure 14) is used as a library so that User does not worry about code implementation and modification to the library code.

---

[5]   https://www.pythoncentral.io/circular-queue/

Table 1 : Python Code

```
# -*- coding: utf-8 -*-
"""
Created on Mon Nov 5 09:28:19 2018
@author: cvuppalapati
"""
class CircularBuffer:

#Constructor
    def __init__(self):
        self.queue = list()
        self.head = 0
        self.tail = 0
        self.maxSize = 10

#Adding elements to the queue
    def enqueue(self,data):
        if self.size() == self.maxSize-1:
            return ("Circular Buffer Full!")
        self.queue.append(data)
        self.tail = (self.tail + 1) % self.maxSize
        return True

#Removing elements from the queue
    def dequeue(self):
        if self.size()==0:
            return ("Circular Buffer Empty!")
        data = self.queue[self.head]
        self.head = (self.head + 1) % self.maxSize
        return data

#Calculating the size of the queue
    def size(self):
        if self.tail>=self.head:
            return (self.tail-self.head)
        return (self.maxSize - (self.head-self.tail))
# Create a Circular Buffer
q = CircularBuffer()
# Enqueu
print(q.enqueue(1))
print(q.enqueue(2))
print(q.enqueue(3))
print(q.enqueue(4))
print(q.enqueue(5))
print(q.enqueue(6))
print(q.enqueue(7))
print(q.enqueue(8))
print(q.enqueue(9))
#Dequeue
print(q.dequeue())
print(q.dequeue())
print(q.dequeue())
print(q.dequeue())
print(q.dequeue())
print(q.dequeue())
print(q.dequeue())
print(q.dequeue())
print(q.dequeue())
```

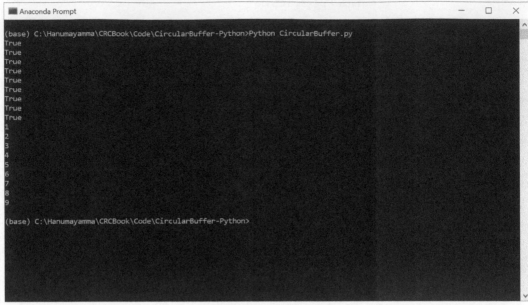

Figure 13: Console Output

Figure 14: Spider IDE

In of our library header, we will forward declare the structure:

```
/// Opaque circular buffer structure
typedef struct circular_buf_t circular_buf_t;
/// Handle type, the way users interact with the API
typedef circular_buf_t* cbuf_handle_t;
```

We have defined Circular buffer and defined access data member cbuf_handle_t;

Circular Buffer Structure

```
// The definition of our circular buffer structure is hidden from the user
struct circular_buf_t {
        uint8_t * buffer;
        size_t head;
        size_t tail;
        size_t max; //of the buffer
        bool full;
};
```

The Circular buffer has following data members: Buffer pointer, head, tail, max and Boolean full flag.

The buffer points to unsigned integer of size 8 bytes. Head and Tail are pointers with size_t data length. In essence, both head and tail points to a circular buffer pointer.

Full flag is Boolean with 0 is empty and 1 with full flag.

Construct and Initialize the Buffer

The buffer is constructed with two parameters: buffer and Size. Please note buffer to be allocated based on the size and number of entries

```
#define EXAMPLE_BUFFER_SIZE 10
uint8_t * buffer = malloc(EXAMPLE_BUFFER_SIZE * sizeof(uint8_t));
printf("\n=== C Circular Buffer Check ===\n");
cbuf_handle_t cbuf = circular_buf_init(buffer, EXAMPLE_BUFFER_SIZE);
```

First order of business, first we need to allocate buffer. The Malloc with number of entries in buffer and size of each entry are needed to allocate buffer.

The following code implements allocation of buffer. First things first, in order to initialize buffer, we need to create circular buffer and assign to cbuf variable. Next, assign each data member of cbuf with the newly allocated buffer.

```
cbuf_handle_t circular_buf_init(uint8_t* buffer, size_t size)
{
        assert(buffer && size);
        cbuf_handle_t cbuf = malloc(sizeof(circular_buf_t));
        assert(cbuf);
        cbuf->buffer = buffer;
        cbuf->max = size;
        circular_buf_reset(cbuf);
        assert(circular_buf_empty(cbuf));
        return cbuf;
}
```

In our case, buffer and max data members. The following calls initialize cbuf data members.

cbuf->buffer = buffer;

cbuf->max = size;

Finally, we need to point head, tail and full members (see Figure 15). Since the buffer is not pointing to any active element, we set other parameter by calling circular buffer reset function.

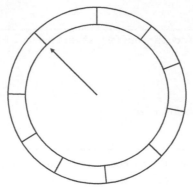

Figure 15: Circular Buffer

```
void circular_buf_reset(cbuf_handle_t cbuf)
{
        assert(cbuf);
        cbuf->head = 0;
        cbuf->tail = 0;
        cbuf->full = false;
}
```

As shown in the Figure 15, both "head" and "tail" points to the same memory address location.

Next, once allocated buffer, calling circular buffer initialize method would allocate the buffer to our structure. After this, we can insert entries into the buffer.

```
printf("Buffer initialized. Full: %d, empty: %d, size: %zu\n",
                circular_buf_full(cbuf),
                circular_buf_empty(cbuf),
                circular_buf_size(cbuf));
```

```
=== C Circular Buffer Check ===
Buffer initialized. Full: 0, empty: 1, size: 0
```

Adding Elements to Buffer

Adding elements to buffer is performed through put operation.

```
printf("\n******\nAdding %d values\n", EXAMPLE_BUFFER_SIZE);
for (uint8_t i = 0; i < EXAMPLE_BUFFER_SIZE; i++)
{
        circular_buf_put(cbuf, i);
        printf("Added %u, Size now: %zu\n", i, circular_buf_size(cbuf));
}
```

After for loop, we have filled all the entries of circular buffer.

```
******
Adding 9 values
Added 0, size now: 1
Added 1, size now: 2
Added 2, size now: 3
Added 3, size now: 4
Added 4, size now: 5
Added 5, size now: 6
Added 6, size now: 7
Added 7, size now: 8
Added 8, size now: 9
Full: 0, empty: 0, size : 9
```

The put operation performs the following: buffer needs to be valid and head points to data.

```
/// Put version 1 continues to add data if the buffer is full
/// Old data is overwritten
/// Requires: cbuf is valid and created by circular_buf_init
void circular_buf_put(cbuf_handle_t cbuf, uint8_t data)
{
        assert(cbuf && cbuf->buffer);
        cbuf->buffer[cbuf->head] = data;
        advance_pointer(cbuf);
}
```

Next, we advance the pointer to point to the next buffer position (see Figure 16). The next buffer position is derived by moving the pointer the size of structure. Please note: In circular buffer, all the entries are allocated in a contiguous manner and by moving the pointer to the size of the preceding structure ensures that we are pointing to a contiguous location.

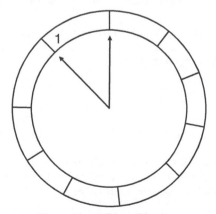

Figure 16: Addition of first item

```
static void advance_pointer(cbuf_handle_t cbuf)
{
        assert(cbuf);
        if (cbuf->full)
        {
                cbuf->tail = (cbuf->tail + 1) % cbuf->max;
        }
        cbuf->head = (cbuf->head + 1) % cbuf->max;
        // We mark full because we will advance tail on the next
time around
        cbuf->full = (cbuf->head == cbuf->tail);
}
```

As data are getting populated, both tail and head move one block at a time (see Figure 17).

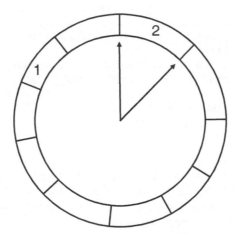

Figure 17: Addition of second item

Please note: head points to new block and tail points first block.

```
static void advance_pointer(cbuf_handle_t cbuf)
{
        assert(cbuf);
        if (cbuf->full)
        {
                cbuf->tail = (cbuf->tail + 1) % cbuf->max;
        }
        cbuf->head = (cbuf->head + 1) % cbuf->max;
        // We mark full because we will advance tail on the next time around
        cbuf->full = (cbuf->head == cbuf->tail);
}
```

Reading Values

Reading Values from Circular buffer is performed by calling Get operation on Circular Buffer.

```
int circular_buf_get(cbuf_handle_t cbuf, uint8_t * data)
{
        assert(cbuf && data && cbuf->buffer);
        int r = -1;
        if (!circular_buf_empty(cbuf))
        {
                *data = cbuf->buffer[cbuf->tail];
                retreat_pointer(cbuf);
                r = 0;
        }
        return r;
}
```

Check Circular buffer is empty first, if not, read the buffer.

```
bool circular_buf_empty(cbuf_handle_t cbuf)
{
        assert(cbuf);
        return (!cbuf->full && (cbuf->head == cbuf->tail));
}
```

Circular buffer empty can be evaluated by checking whether the buffer is full and checking whether head and tail pointers are pointing to the same location.

Driver

```
#include <stdio.h>
#include <stdlib.h>
#include <stdint.h>
#include <stddef.h>
#include <stdbool.h>
#include <assert.h>
#include "CircularBuffer.h"
#define EXAMPLE_BUFFER_SIZE 10
int main(void)
{
        uint8_t * buffer = malloc(EXAMPLE_BUFFER_SIZE * sizeof(uint8_t));
        printf("\n=== C Circular Buffer Check ===\n");
        cbuf_handle_t cbuf = circular_buf_init(buffer, EXAMPLE_BUFFER_SIZE);
        printf("Buffer initialized. Full: %d, empty: %d, size: %zu\n",
                circular_buf_full(cbuf),
                circular_buf_empty(cbuf),
                circular_buf_size(cbuf));
        printf("\n******\nAdding %d values\n", EXAMPLE_BUFFER_SIZE - 1);
        for (uint8_t i = 0; i < (EXAMPLE_BUFFER_SIZE - 1); i++)
        {
                circular_buf_put(cbuf, i);
                printf("Added %u, Size now: %zu\n", i, circular_buf_
size(cbuf));
        }
```

```
        printf("Full: %d, empty: %d, size: %zu\n",
                circular_buf_full(cbuf),
                circular_buf_empty(cbuf),
                circular_buf_size(cbuf));
        printf("\n******\nAdding %d values\n", EXAMPLE_BUFFER_SIZE);
        for (uint8_t i = 0; i < EXAMPLE_BUFFER_SIZE; i++)
        {
                circular_buf_put(cbuf, i);
                printf("Added %u, Size now: %zu\n", i, circular_buf_
size(cbuf));
        }
        printf("Full: %d, empty: %d, size: %zu\n",
                circular_buf_full(cbuf),
                circular_buf_empty(cbuf),
                circular_buf_size(cbuf));
        printf("\n******\nReading back values: ");
        while (!circular_buf_empty(cbuf))
        {
                uint8_t data;
                circular_buf_get(cbuf, &data);
                printf("%u", data);
        }
        printf("\n");
        printf("Full: %d, empty: %d, size: %zu\n",
                circular_buf_full(cbuf),
                circular_buf_empty(cbuf),
                circular_buf_size(cbuf));
        printf("\n******\nAdding %d values\n", EXAMPLE_BUFFER_SIZE + 5);
        for (uint8_t i = 0; i < EXAMPLE_BUFFER_SIZE + 5; i++)
        {
                circular_buf_put(cbuf, i);
                printf("Added %u, Size now: %zu\n", i, circular_buf_
size(cbuf));
        }
        printf("Full: %d, empty: %d, size: %zu\n",
                circular_buf_full(cbuf),
                circular_buf_empty(cbuf),
                circular_buf_size(cbuf));
        printf("\n******\nReading back values: ");
        while (!circular_buf_empty(cbuf))
        {
                uint8_t data;
                circular_buf_get(cbuf, &data);
                printf("%u", data);
        }
        printf("\n");
        printf("\n******\nAdding %d values using non-overwrite version\n", EX-
AMPLE_BUFFER_SIZE + 5);
        for (uint8_t i = 0; i < EXAMPLE_BUFFER_SIZE + 5; i++)
        {
                circular_buf_put2(cbuf, i);
        }
        printf("Full: %d, empty: %d, size: %zu\n",
                circular_buf_full(cbuf),
                circular_buf_empty(cbuf),
                circular_buf_size(cbuf));
        printf("\n******\nReading back values: ");
        while (!circular_buf_empty(cbuf))
```

```
        {
                uint8_t data;
                circular_buf_get(cbuf, &data);
                printf("%u ", data);
        }
        printf("\n");
        free(buffer);
        circular_buf_free(cbuf);
        return 0;
}
```

Case Study: Constrained Devices

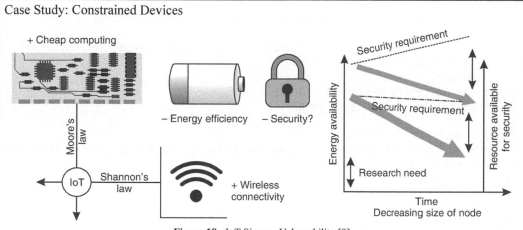

Figure 18: IoT Size vs. Vulnerability [8]

The Internet of things (IoT) devices come in various operating form factors [8]. Some are operated on unconstrained resources by directly connecting to the electrical grid with Cloud Compute driven memory and processing capacities; others are operated on constrained resources by connecting to finite battery sources and limited memory and compute. Unlike securing IoT devices with unconstrained resources (see Figure 18), securing resource constrained IoT devices has many challenges that may include: (a) lack of support of complex and evolving security algorithms due to: (i) limited security compute capabilities, (ii) lack of sufficient higher processing power to support encryption algorithms, and (iii) low CPU cycles vs. effective encryption, (b) lack of built-in connectivity resilience, though designed to operate autonomously in the field, and finally (c) lack of adequate physical security, yet designed to support geographically distributed deployments.

Cow necklace Circular Buffer Code:

```
/************************************************************************/
static bool aci_setup_fill(aci_state_t *aci_stat, uint8_t *num_cmd_offset)
{
 bool ret_val = false;

 while (*num_cmd_offset < aci_stat->aci_setup_info.num_setup_msgs)
 {
        //Board dependent defines
        #if defined (__AVR__)
                //For Arduino copy the setup ACI message from Flash to RAM.
                memcpy_P(&msg_to_send, &(aci_stat->aci_setup_info.setup_
msgs[*num_cmd_offset]),
```

```
                                    pgm_read_byte_near(&(aci_stat->aci_setup_info.
setup_msgs[*num_cmd_offset].buffer[0]))+2);
        #elif defined(ARDUINO_ARCH_SAMD)
                //For Arduino copy the setup ACI message from Flash to RAM.
                memcpy_P(&msg_to_send, &(aci_stat->aci_setup_info.setup_
msgs[*num_cmd_offset]),
                                    pgm_read_byte_near(&(aci_stat->aci_setup_info.
setup_msgs[*num_cmd_offset].buffer[0]))+2);
        #elif defined(__PIC32MX__)
                //In ChipKit we store the setup messages in RAM
                //Add 2 bytes to the length byte for status byte, length for
the total number of bytes
                memcpy(&msg_to_send, &(aci_stat->aci_setup_info.setup_
msgs[*num_cmd_offset]),
                            (aci_stat->aci_setup_info.setup_msgs[*num_cmd_
offset].buffer[0]+2));
        #endif
  //Put the Setup ACI message in the command queue
  if (!hal_aci_tl_send(&msg_to_send))
  {
  //ACI Command Queue is full
  // *num_cmd_offset is now pointing to the index of the Setup command that did
not get sent
  return ret_val;
  }

  ret_val = true;

  (*num_cmd_offset)++;
  }
```

Double Buffers

In general, double buffering algorithm is applied to User Interface, Printing, and Graphic Device (GDI) Interface rendering Case study. A classic example is Screen update for Game programming. The chief reason is continuous user screen or service updates without interruptions. In the same vein, double buffering is also used for Edge processing data payloads that need to be uploaded to central server, involving processing and background uploading on a continuous stream basis. To put it simply, when streams of data are handled between producer (in this Sensor) and consumers (backend server or edge mobile device), double buffering is the best candidate.

As the name implies, double buffering contains two buffers: One buffer gets data from the provider and the other buffer relays data to consumers or a backend data host [9].

The following design factors are critical in designing a double buffer (see Figure 19):

- Data Producer Frequency
- Relay Time Interval
- Capacity or Memory footprint of the device

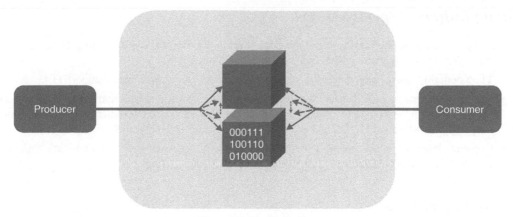

Figure 19: Double Buffers

Thread Synchronization Primitives

When the Double Buffer algorithm is being used, data is being shared while it is being modified. Lack of double buffering can be seen graphically where the screen is not updated properly.
We use Double Buffering in the following case (any one being true):[6]

- We have a state that is being modified incrementally
- The same state may be accessed in the middle of modification
- We want to be able to read the state and we don't want to have to wait while it's being written.

Under the hood, Double Buffer must implement thread synchronization primitives to protect the data. The OS offers following primitive techniques:

- Pthread_join: pthread_join allows one thread to suspend until other thread has terminated.
- Mutex variable: A mutex acts as a mutually exclusive lock, allowing threads to control access to the data.
- Condition variable functions: A condition variable provides a way of naming an event in which the threads have a general interest. Signal Event.
- pthread_once function: Special functions that ensure threads are executed once and only once.
- Reader/Writer exclusion: Multiple threads can read but only one thread is permitted to write.
- Semaphores: A counting semaphore is like a mutex, but it is associated with a counter.

Semaphore vs. Mutex

Mutexes are typically used to serialize access to a section of re-entrant code that cannot be executed concurrently by more than one thread. A mutex object only allows one thread into a controlled section, forcing other threads which attempt to gain access to that section to wait until the first thread has exited from that section.

A semaphore restricts the number of simultaneous users of a shared resource up to a maximum number. Threads can request access to the resource (decrementing the semaphore) and can signal that they have finished using the resource (incrementing the semaphore).

The mutex is like the principles of the binary semaphore with one significant difference: The principle of ownership. Ownership is the simple concept that when a task locks (acquires) only a mutex it can unlock (release) it. If a task tries to unlock a mutex it hasn't locked (i.e., doesn't own) then an error condition is encountered and, most importantly, the mutex is not unlocked. If the mutual exclusion object doesn't have ownership then, irrelevant of what it is called, it is not a mutex.

---

[6] Double Buffer - http://gameprogrammingpatterns.com/double-buffer.html

Audio Buffers—Spectrograms

Audio Signals are processed and stored at the Edge level in the form of Audio spectrograms.
  As indicated in,[7] there are several frames to sounds features [10]:

- Time domain features include: Zero Crossing Rate (ZCR) and Short Time Energy (STE).
- Spectral Features Include: Linear Predictive Coding (LPC) coefficients, Relative Spectral Predictive Linear Coding (RASTA PLP), Pitch, Sone, Spectral Flux (SF) and coefficients from basic time to frequency transforms (FFT, DFT, DWT, CWT and Constant Q-Transform).
- Cepstral domain features are Mel Frequency Cepstral Coefficient (MFCC) and Bark Frequency Cepstral Coefficient (BFCC).

Audio Signal Data

Consider the following simple Audio program: The program is divided into several parts in order to extract the feature extraction. Before going through the example, the WAVE File is part of the data provided FEMH Data as part of IEEE 2018 Big Data Challenge.

Case Study: Aston Martin gets neural network[8]

ASTON MARTIN'S DB9 is having a neural network (see Figure 20) installed in order to make sure the high-performance V-12 knows what it is doing [11].

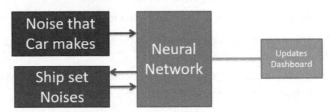

Figure 20: Audio Neural Networks

Step forward neural engineering software, which, similar to the human brain, is based on pattern recognition, learns from experience.

Basically, it listens to all the noises that the car makes and cross-checks them against a list of sounds that the manufacturers think it should make (see figure Audio Neural Networks). If something begins to rattle, that sets off alarm bells for the manufacturer. The network is trained by pushing it into different misfires in different driving conditions. The designers build a critical threshold number of misfires into the system, which, when crossed, lights up the Check Engine light and shuts down the misfiring cylinder in order to avoid damage.

Data Sample

The Voice samples were obtained from a voice clinic in a tertiary teaching hospital (Far Eastern Memorial Hospital, FEMH), these included 50 normal voice samples and 150 samples of common voice disorders, including vocal nodules, polyps, and cysts (collectively referred to Phonotrauma); glottis neoplasm;

---

[7]   Discrimination and retrieval of animal sounds - http://ieeexplore.ieee.org/stamp/stamp.jsp?tp=&arnumber=1651344&is number=34625

[8]   Aston Martin gets Neural Network - https://www.theinquirer.net/inquirer/news/1015284/aston-martin-gets-neural-network

unilateral vocal paralysis. Voice samples of a 3-second sustained vowel sound /a:/ were recorded at a comfortable level of loudness, with a microphone-to-mouth distance of approximately 15–20 cm, using a high-quality microphone (Model: SM58, SHURE, IL), with a digital amplifier (Model: X2u, SHURE) under a background noise level between 40 and 45 dBA. The sampling rate was 44,100 Hz with a 16-bit resolution, and data were saved in an uncompressed.wav format.

```
import os
import matplotlib
#matplotlib.use('Agg') # No pictures displayed
import pylab
import librosa
import librosa.display
import numpy as np
np.set_printoptions(precision=3)
np.set_printoptions(threshold=np.nan)

# https://librosa.github.io/librosa/generated/librosa.feature.melspectrogram.html

print("\n PART I \n")

print("\n mel \n")
sig, sample_rate = librosa.load("002.wav')
mel = librosa.feature.melspectrogram(y=sig, sr=sample_rate)
print(mel)

print("\n mel_mean \n")
mel_mean = np.mean(librosa.feature.melspectrogram(sig, sr=sample_rate).T,axis=0)

print(mel_mean)
```

The output is mels with time.

```
[[1.186e-03 1.041e-02 1.578e-02 5.970e-03 2.206e-03 2.546e-03 1.304e-02
  1.797e-02 8.099e-03 4.751e-03 8.792e-03 3.808e-03 5.589e-03 3.133e-03
  2.640e-03 2.719e-03 1.216e-03 8.780e-04 1.523e-03 3.072e-03 2.842e-03
  4.437e-03 2.941e-03 3.153e-03 8.524e-03 1.147e-02 5.672e-03 5.444e-03
  3.996e-03 1.291e-02 1.375e-02 4.518e-03 2.799e-03 3.401e-03 6.223e-03
  1.000e-02 9.332e-03 2.089e-03 5.287e-03 8.251e-03 4.480e-03 2.338e-03
  2.922e-03 1.799e-03 6.747e-03 1.098e-02 1.683e-02 2.114e-02 1.175e-02
  4.378e-03 2.523e-03 5.747e-03 4.231e-03 1.135e-02 1.224e-02 6.789e-03
  6.733e-03 5.376e-03 5.129e-03 1.186e-02 5.830e-03 8.560e-03 1.402e-02
  1.100e-02 2.488e-02 1.428e-02 2.969e-03 9.693e-03 1.027e-02 2.089e-03
  1.757e-03 6.357e-03 6.484e-03 4.448e-03 6.961e-03 5.251e-03 2.546e-03
  9.974e-03 1.121e-02 3.187e-03 4.820e-03 3.344e-03 1.411e-02 1.843e-02
  3.537e-03 2.055e-03 2.321e-03 7.218e-03 1.335e-02 5.187e-03 2.821e-03
  1.667e-03 7.754e-03 1.795e-02 1.183e-02 5.985e-03 2.768e-03 2.713e-03
  1.502e-03 2.694e-03 1.217e-02 1.775e-02 2.248e-02 1.999e-02 4.830e-03
  1.993e-03 2.412e-03 6.985e-03 1.343e-02 4.970e-03 1.350e-02 6.237e-03
  9.786e-03 7.664e-03 5.587e-03 1.843e-03 6.100e-04 2.255e-03 4.834e-03
  5.024e-03 2.612e-03 1.467e-03 2.075e-03 2.369e-03 5.766e-03 3.161e-03
  6.843e-04 3.259e-03 4.866e-03 9.544e-03 5.245e-03 4.554e-03 4.414e-03
  8.746e-03 2.339e-02 9.972e-03 8.537e-03 1.451e-02 9.108e-03 9.200e-03
  1.717e-02 1.856e-02 5.755e-03 7.440e-03 9.002e-03 7.219e-03 1.311e-03
  5.018e-04 3.065e-03 5.026e-03 2.198e-03 1.566e-02 2.458e-02 6.112e-03
  2.237e-03 6.801e-03 3.511e-03 1.962e-03 1.885e-03 1.369e-03 2.301e-03
  7.659e-03 1.792e-02 9.317e-03 3.680e-03 3.587e-03 2.399e-03 2.679e-03
  1.914e-03 4.853e-03 3.743e-03 4.228e-03 2.255e-03 2.638e-03 5.407e-03
  4.871e-03 7.117e-03 1.417e-02 8.571e-03 4.08]......
```

```
librosa.feature.melspectrogram(y=None, sr=22050, S=None, n_fft=2048, hop_
Length=512, power=2.0, **kwargs)
```

Parameters:	**y:np.ndarray [shape=(n,)] or None** audio time-series **sr:number > 0 [scalar]** sampling rate of *y* **S:np.ndarray [shape=(d, t)]** spectrogram **n_fft:int > 0 [scalar]** length of the FFT window **hop_length:int > 0 [scalar]** number of samples between successive frames. See **librosa.** **core.stft** **power:float > 0 [scalar]** Exponent for the magnitude melspectrogram, e.g., 1 for energy, 2 for power, etc. **kwargs:additional keyword arguments** Mel filter bank parameters. See **librosa.filters.mel** for details.
Returns:	**S:np.ndarray [shape=(n_mels, t)]** Mel spectrogram

The Librosa melspectrogram[9] [12] compute a mel-scaled spectrogram and returns two-dimensional array.

Image Texture Extraction—Histograms

For image data (see Figure 21), there are many ways to extract the features. One method of image data extraction is to identify texture changes. The commonly-used algorithm is Local Binary Pattern (LBP).

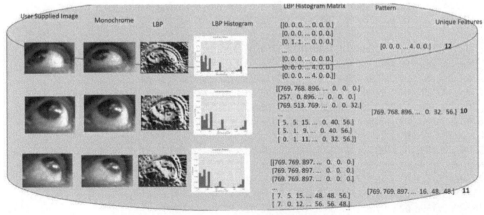

Figure 21: Image Texture Database

---

[9]　Librosa – melspectrogram - https://librosa.github.io/librosa/generated/librosa.feature.melspectrogram.html

```
imagepath = r"sanjeevani_eye_diag.png"
image = cv2.imread(imagepath)
# 2 resize the image
image = resizeImage(image)
(h, w) = image.shape[:2]
cellSize = h/10
# 3 convert the image to grayscale and show it
gray = cv2.cvtColor(image, cv2.COLOR_BGR2GRAY)
cv2.waitKey(10)
# save the image
cv2.imwrite("gray_resized_sanjeevani_eye_diag.png", gray)
# construct the figure
plt.style.use("ggplot")
(fig, ax) = plt.subplots()
fig.suptitle("Local Binary Patterns")
plt.ylabel("% of Pixels")
plt.xlabel("LBP pixel bucket")
# plot a histogram of the LBP features and show it
features = feature.local_binary_pattern(gray, 10, 5, method="default")
cv2.waitKey(10)
# Save figure of lbp_image
cv2.imwrite("lbp_sanjeevani_eye_diag.png", features.astype("uint8"))
ax.hist(features.ravel(), normed=True, bins=20, range=(0, 256))
ax.set_xlim([0, 256])
ax.set_ylim([0, 0.030])
# save figure
```

Chapter Summary:

* Understand basic data collector algorithms for IoT
* Audio signal representations Should enable design of Microcontroller & sensor board
* Video signal representation

References

1. Cisco, The Internet of Things Reference Model, 2014, http://cdn.iotwf.com/resources/71/IoT_Reference_Model_White_Paper_June_4_2014.pdf , Access Date: April 2018
2. Thomas Bittman & Bob Gill, The Future Shape of Edge Computing: Five Imperatives, 28 June 2018, https://www.gartner.com/doc/3880015/future-shape-edge-computing-imperatives
3. Ryan Gillespie and Saurabh Gupta, 2017, Real-time Analytics at the Edge: Identifying Abnormal Equipment Behavior and Filtering Data near the Edge for Internet of Things Applications, SAS Institute Inc., http://support.sas.com/resources/papers/proceedings17/SAS0645-2017.pdf, Access date: June 6, 2017
4. MIMOSA, Open System Architecture for Condition-Based Maintenance, 2017, http://www.mimosa.org/mimosa-osa-cbm/, Access Date: April 2018
5. Dr. David Baglee, Dr Salla Marttonen, and Professor Diego Galar, 2015, International Conference Data Mining, http://worldcompproceedings.com/proc/p2015/DMI8005.pdf , Access Date: June 7, 2017
6. Liangwei Zhang, 2016, Big Data Analytics for Fault Detection and its Application in Maintenance, http://ltu.diva-portal.org/smash/get/diva2:1046794/FULLTEXT01.pdf , Access Date: June 01, 2017
7. Python Development, Circular Queue: An implementation tutorial, August 23 2017, https://www.pythoncentral.io/circular-queue/, Access Date: April 2018

8. Chandrasekar Vuppalapati et al., Cognitive Secure Shield—A Machine Learning enabled threat shield for resource constrained IoT Devices, December 2018, https://www.icmla-conference.org/icmla18/program_2018.pdf, Access Date: December 2018

9. Robert Nystrom, C Programming—Double Buffers, 2014, http://gameprogrammingpatterns.com/double-buffer.html, Access Date: July 2018

10. D. Mitrovic, Discrimination and retrieval of animal sounds, 2006 12th International Multi-Media Modelling Conference, 2006, IEEE, Access Date: March 2018

11. Nick Farrell, Aston Martin gets neural network—Much more useful than an ejector seat, 27 September 2004, https://www.theinquirer.net/inquirer/news/1015284/aston-martin-gets-neural-network, Access Date: April 2018

12. Librosa, Melspectrogram, https://librosa.github.io/librosa/generated/librosa.feature.melspectrogram.html

Data Storage

Data process at Edge device or near to the sensor needs to be very quick. The process requires the analysis of outlier conditions, batching through time-window to detect trends in the data, and filtering the data for prediction and prognosis purposes. The data in motion relied to the backend to the traditional enterprise data stores or to the cloud. Edge devices continuously process the data and generally does not require storage. However, for constrained environments, intermittent network connections or long-periodic connection to the network, the edge devices need to store the data near the source.

The chapter introduces the data stores at the IoT device; It also provides the data store in the universal edge devices; mobile Operating Systems.

Before going into each storage option in the embedded devices, let's look at the data representation for storage. Generally, the data to be persisted is represented in either a single dimensional array, or two-dimensional array or N-dimensional array.

Data in Motion: Data representations

Two-Dimensional array

Let's say we need to represent a two-dimensional data array (zero based index):

	Column 1	Column 2	Column 3	Column 4
Row 1	[0][0]	[0][1]	[0][2]	[0][3]
Row 2	[1][0]	[1][1]	[1][2]	[1][3]
Row 3	[2][0]	[2][1]	[2][2]	[2][3]
Row 4	[3][0]	[3][1]	[3][2]	[3][3]

With zero based index: [0][0] represents the data cell at Row 1 and Column 1; [3][1] represents the row 4 and Column 2 data. The data is of integer:

C—representation:

int x [3][4];

Initializing C two-dimensional array:

/Different ways to initialize two-dimensional array
int x [4][3] = {{1, 3, 0}, {-1, 5, 9}, {4, 0, 6}, {1, 3, 4}};

int x [] [3] = {{1, 3, 0}, {-1, 5, 9}, {4, 0, 6}, {1, 3, 4}};

int x [4][3] = {1, 3, 0, -1, 5, 9, 4, 0, 6, 1, 3, 4};

The program below constructs a simple two-dimensional array and asks the user to populate the data, then prints it out (see Table 1 & Figure 1).

Table 1: Two-Dimensional Array Code

```c
#include <stdio.h>
const int CITY = 2;
const int WEEK = 7;
int main()
{
        int temperature[2][7];
        for (int i = 0; i < CITY; ++i) {
                for (int j = 0; j < WEEK; ++j) {
                        printf("City %d, Day %d: ", i + 1, j + 1);
                        scanf_s("%d", &temperature[i][j]);
                }
        }
        printf("\nDisplaying values: \n\n");
        for (int i = 0; i < CITY; ++i) {
                for (int j = 0; j < WEEK; ++j)
                {
                        printf("City %d, Day %d = %d\n", i + 1, j + 1,
temperature[i][j]);
                }
        }
        return 0;
}
```

Figure 1: Two-Dimensional Array Console Output

Two-Dimensional Matrices Sum

Let's construct a two by two matrix and add them (see Table 2).

Table 2: Two-Dimensional Matrices

```c
#include <stdio.h>
int main()
{
        float a[2][2], b[2][2], c[2][2];
        int i, j;
        // Taking input using nested for loop
        printf("Enter elements of 1st matrix\n");
        for (i = 0; i<2; ++i)
                for (j = 0; j<2; ++j)
                {
                        printf("Enter a%d%d: ", i + 1, j + 1);
                        scanf_s("%f", &a[i][j]);
                }
        // Taking input using nested for loop
        printf("Enter elements of 2nd matrix\n");
        for (i = 0; i<2; ++i)
                for (j = 0; j<2; ++j)
                {
                        printf("Enter b%d%d: ", i + 1, j + 1);
                        scanf_s("%f", &b[i][j]);
                }
        // adding corresponding elements of two arrays
        for (i = 0; i<2; ++i)
                for (j = 0; j<2; ++j)
                {
                        c[i][j] = a[i][j] + b[i][j];
                }
        // Displaying the sum
        printf("\nSum Of Matrix:");
        for (i = 0; i<2; ++i)
                for (j = 0; j<2; ++j)
                {
                        printf("%.1f\t", c[i][j]);
                        if (j == 1)
                                printf("\n");
                }
        return 0;
}
```

Anaconda Prompt - TwoDimensionalMatricesSum.exe   —   □   ×

```
(base) C:\Hanumayamma\CRCBook\Code\Datastorage\TwoDimensionalArray\Debug>TwoDimensionalMatricesSum.exe
Enter elements of 1st matrix
Enter a11: 2
Enter a12: 3
Enter a21: 4
Enter a22: 1
Enter elements of 2nd matrix
Enter b11: 9
Enter b12: 7
Enter b21: 6
Enter b22: _
```

Figure 2: Matrices Sum Output

The above program generates output as follows (see Figure 2)

Three-Dimensional array

Three-Dimensional array contains rows, columns and layers (see Figure 3).

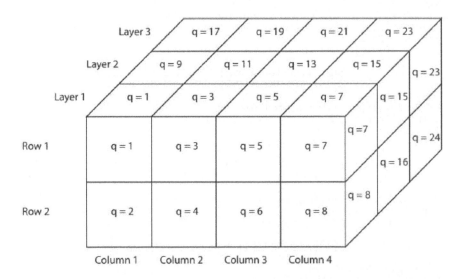

l = # of rows (here l = 2)
m = # of columns (here m = 4)
n = # of layers (here m =3)

Figure 3: 3D Cube

C Style representation:

X [L] [M] [N]

Initializing the three-dimensional array: In the following example, the array X is initialized with 2 rows, 3 columns, and 4 layers.

```
int X [2][3][4] = {
{{3, 4, 2, 3}, {0, -3, 9, 11}, {23, 12, 23, 2}},
{{13, 4, 56, 3}, {5, 9, 3, 5}, {3, 1, 4, 9}}
};
```

Three-Dimensional Array Code: (please see Table 3)

Table 3: Three-Dimensional Array Code

```
#include <stdio.h>
int main()
{
        int i, j, k, test[2][3][2];
        printf("Enter 12 values: \n");
        for (i = 0; i < 2; ++i) {
                for (j = 0; j < 3; ++j) {
                        for (k = 0; k < 2; ++k) {
                                scanf_s("%d", &test[i][j][k]);
                        }
                }
        }
        // Printing values with proper index.
        printf("\nDisplaying values:\n");
        for (i = 0; i < 2; ++i) {
                for (j = 0; j < 3; ++j) {
                        for (k = 0; k < 2; ++k) {
                                printf("test[%d][%d][%d] = %d\n", i, j, k,
test[i][j][k]);
                        }
                }
        }
        return 0;
}
```

The output of the above program—see Figure 4

Arrays in Python

NumPy[1] is a Python library that can be used for scientific and numerical applications and is the tool to use for linear algebra operations. The main data structure in NumPy is the ndarray, which is a shorthand name for N-dimensional array. When working with NumPy, data in an ndarray is simply referred to as an array (see Figure 5) [1].

It is a fixed-sized array in memory that contains data of the same type, such as integers or floating-point values.

---

[1]   https://machinelearningmastery.com/gentle-introduction-n-dimensional-arrays-python-numpy/

```
Anaconda Prompt                                                           —    □    ×
(base) C:\Hanumayamma\CRCBook\Code\Datastorage\TwoDimensionalArray\Debug>ThreeDimensionalArray.exe
Enter 12 values:
10
20
30
44
12
9
4
50
11
34
45
6

Displaying values:
test[0][0][0] = 10
test[0][0][1] = 20
test[0][1][0] = 30
test[0][1][1] = 44
test[0][2][0] = 12
test[0][2][1] = 9
test[1][0][0] = 4
test[1][0][1] = 50
test[1][1][0] = 11
test[1][1][1] = 34
test[1][2][0] = 45
test[1][2][1] = 6

(base) C:\Hanumayamma\CRCBook\Code\Datastorage\TwoDimensionalArray\Debug>
```

Figure 4: 3D Cube Output

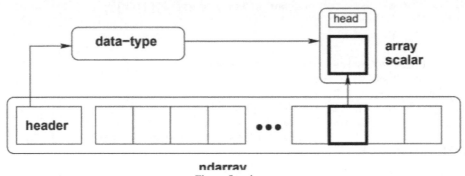

ndarray

Figure 5: ndarray

The data type supported by an array can be accessed via the "dtype" attribute on the array. The dimensions of an array can be accessed via the "shape" attribute that returns a tuple describing the length of each dimension.

Conceptual diagram showing the relationship between the three fundamental objects used to describe the data in an array: (1) the ndarray itself, (2) the data-type object that describes the layout of a single fixed-size element of the array, (3) the array-scalar Python object that is returned when a single element of the array is accessed.

Hierarchy of Objects in Python

Hierarchy of type objects (see Figure 6) representing the array data types.[2] All the number types can be obtained using bit-width names as well [2].

A simple way to create an array from data or simple Python data structures like a list is to use the array () function.

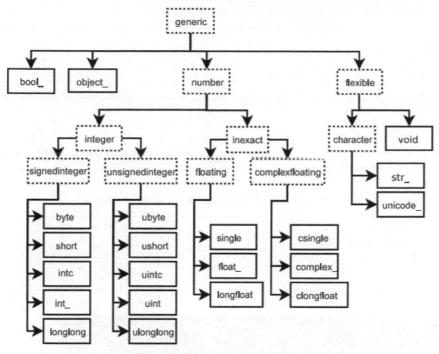

Figure 6: Objects Hierarchy in Python

```
import numpy as np
l = [1.0, 2.0, 3.0]
a = array(l)
print(a)
print(a.shape)
print(a.dtype)
```

Running the example prints the contents of the ndarray, the shape, which is a one-dimensional array with 3 elements, and the data type, which is a 64-bit floating point.

Empty Function

The empty () function will create a new array of the specified shape [2].

The argument to the function is an array or tuple that specifies the length of each dimension of the array to create. The values or content of the created array will be random and will need to be assigned before use.

```
# create empty array
from numpy import empty
a = empty([3,3])
print(a)
```

Running the example prints the content of the empty array. Your specific array contents will vary.

Zeros

The zeros () function will create a new array of the specified size with the contents filled with zero values.

The argument to the function is an array or tuple that specifies the length of each dimension of the array to create

The example below creates a 3×5 zero two-dimensional array.

```
# create zero array
from numpy import zeros
a = zeros([3,5])
print(a)
```

Ones

The ones () function will create a new array of the specified size with the contents filled with one values. The argument to the function is an array or tuple that specifies the length of each dimension of the array to create [2].

The example below creates a 5-element one-dimensional array (see figure).

```
# create one array
from numpy import ones
a = ones([5])
print(a)
```

Combining Arrays

NumPy provides many functions to create new arrays from existing arrays. Let's look at two of the most popular functions you may need or encounter [2].

- Vertical Stack
- Horizontal Stack

Vertical Stack

Given two or more existing arrays, you can stack them vertically using the vstack() function.

For example, given two one-dimensional arrays, you can create a new two-dimensional array with two rows by vertically stacking them.

This is demonstrated in the example below.

```
     # vstack
3    from numpy import array
4    from numpy import vstack
5    a1 = array([1,2,3])
6    print(a1)
7    a2 = array([4,5,6])
8    print(a2)
9    a3 = vstack((a1, a2))
10   print(a3)
     print(a3.shape)
```

Running the example first prints the two separately defined one-dimensional arrays. The arrays are vertically stacked resulting in a new 2×3 array, the contents and shape of which are printed.

Horizontal Stack

Given two or more existing arrays, you can stack them horizontally using the hstack() function (see Table 4 and Figure 7). For example, given two one-dimensional arrays, you can create a new one-dimensional array or one row with the columns of the first and second arrays concatenated.

This is demonstrated in the example below.

Table 4: HStack

```
# hstack
from numpy import array
from numpy import hstack
a1 = array([1,2,3])
print(a1)
a2 = array([4,5,6])
print(a2)
a3 = hstack((a1, a2))
print(a3)
print(a3.shape)
```

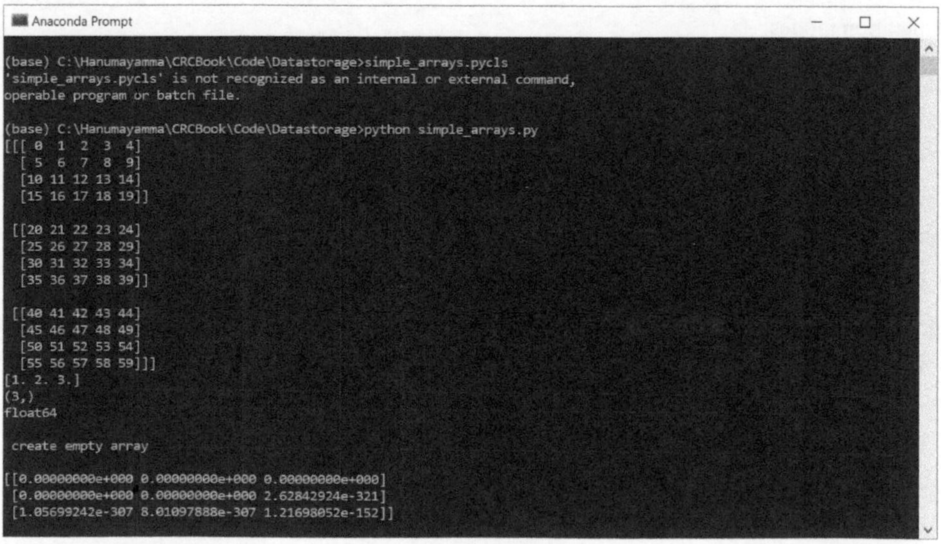

Figure 7: HStack output

Running the example first prints the two separately defined one-dimensional arrays. The arrays are then horizontally stacked resulting in a new one-dimensional array with 6 elements, the contents and shape of which are printed.

C Style vs. Fortran Style Allocation

Files in C

Embedded and SCADA based devices use Input/output (IO) operations that are driven from C device drivers. In nutshell, the IO operations are directly tied to how C based IO.

In this section, I would like to introduce very high-level C IO File operations. In the subsequent sections, the IO that is performed on EPROM, Android and iOS use basics of C style IO functions [3].

 C File operations File IO operations are performed in Embedded Storage software such as File IO and embedded data base SQLite.

File Operations

In C the following file operations are performed:

1. Creating file
2. Opening an existing file
3. Reading and Writing into the file
4. Closing the file

When working with files, you need to declare a pointer of type file. This declaration is needed for communication between the file and program.

FILE *fptr;

The above line defines a pointer to a file.

Opening a file—for creation or edit

Opening a file is performed by calling fopen—defined stdio.h

ptr = fopen("fileopen","mode")

File Operation modes:

File Mode	Meaning of Mode	During Inexistence of file
r	Open for reading.	If the file does not exist, fopen() returns NULL.
rb	Open for reading in binary mode.	If the file does not exist, fopen() returns NULL.
w	Open for writing.	If the file exists, its contents are overwritten. If the file does not exist, it will be created.
wb	Open for writing in binary mode.	If the file exists, its contents are overwritten. If the file does not exist, it will be created.
a	Open for append, i.e., Data is added to end of file.	If the file does not exists, it will be created.

File Mode	Meaning of Mode	During Inexistence of file
ab	Open for append in binary mode, i.e., Data is added to end of file.	If the file does not exists, it will be created.
r+	Open for both reading and writing.	If the file does not exist, fopen() returns NULL.
rb+	Open for both reading and writing in binary mode.	If the file does not exist, fopen() returns NULL.
w+	Open for both reading and writing.	If the file exists, its contents are overwritten. If the file does not exist, it will be created.
wb+	Open for both reading and writing in binary mode.	If the file exists, its contents are overwritten. If the file does not exist, it will be created.
a+	Open for both reading and appending.	If the file does not exists, it will be created.
ab+	Open for both reading and appending in binary mode.	If the file does not exists, it will be created.

Closing a File

The file should be closed after reading/writing. Closing a file is performed using library function fclose().

fclose(fptr); //fptr is the file pointer associated with file to be closed.

Reading a File

For reading and writing to a text file, we use the functions fprintf() and fscanf(). They are just the file versions of printf() and scanf(). The only difference is that, fprint and fscanf expects a pointer to the structure FILE.

In the code below, the scanf is loading a number of sensors in the variable.

```
printf("Enter number of sensors: ");
scanf("%d", &sensor_nums);
FILE *fptr;
      fptr = (fopen("C:\\Temp\\sensors.txt", "w"));
      if (fptr == NULL)
      {
            printf("Error!");
            exit(1);
      }
```

In the code above, the fptr points to a file on C:\Temp\Sensors.txt

The code below gets the number of sensors from the User and takes the sensor name and sensor readings and writes it to the file. To suppress security warnings in Microsoft Visual Studio environment, please go to Project Properties and Code Generation and Turn off Security Check (see Figure 8 and Table 5).

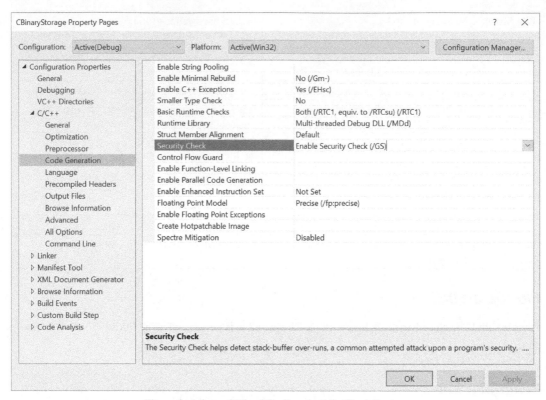

Figure 8: Microsoft Visual Studio—Security Check Feature

Table 5: File Storage

```
#include<stdio.h>
#define _CRT_SECURE_NO_WARNINGS
#include <stdio.h>
int main()
{
        char sensor_name[50];
        int readings, i, sensor_nums;
        printf("Enter number of sensors: ");
        scanf("%d", &sensor_nums);
        FILE *fptr;
        fptr = (fopen("C:\\Temp\\sensors.txt", "w"));
        if (fptr == NULL)
        {
                printf("Error!");
                exit(1);
        }
        for (i = 0; i < sensor_nums; ++i)
        {
                printf("For sensor%d\nEnter name: ", i + 1);
                scanf("%s", sensor_name);
                printf("Enter readings: ");
                scanf("%d", &readings);
                fprintf(fptr, "\nName: %s \nMarks=%d \n", sensor_name,
readings);
        }
        fclose(fptr);
        return 0;
}
```

Output

The code above generates the following output (see Figure 9):

```
Command Prompt

C:\Hanumayamma\CRCBook\Code\CFileStorage\CFileStorage\Debug>CFileStorage.exe
Enter number of sensors: 5
For sensor1
Enter name: water_fremont
Enter readings: 33
For sensor2
Enter name: water_sjc
Enter readings: 44
For sensor3
Enter name: water_sfo
Enter readings: 454
For sensor4
Enter name: water_la
Enter readings: 33
For sensor5
Enter name: water_lake_tahoe
Enter readings: 22

C:\Hanumayamma\CRCBook\Code\CFileStorage\CFileStorage\Debug>
```

Figure 9: File Storage Console Output

File Output

The file output is written to file (see Figure 10)

```
sensors.txt - Notepad
File Edit Format View Help

Name: water_fremont
Marks=33

Name: water_sjc
Marks=44

Name: water_sfo
Marks=454

Name: water_la
Marks=33

Name: water_lake_tahoe
Marks=22
```

Figure 10: File Output

Reading and writing to a binary file

Binary files are very similar to arrays of structures, except the structures are in a disk file rather than in an array in memory. Because the structures in a binary file are on disk, you can create very large collections of them (limited only by your available disk space). They are also permanent and always available. The only disadvantage is the slowness that comes from disk access time.

 Sterilization is the process of storing memory objects on to file and reloading them. C File operations make it possible.

Binary files[3] also usually have faster read and write times than text files, because a binary image of the record is stored directly from memory to disk (or vice versa). In a text file, everything has to be converted back and forth to text, and this takes time.

C supports the file-of-structures concept very cleanly. Once you open the file you can read a structure, write a structure, or seek to any structure in the file. This file concept supports the concept of a file pointer. When the file is opened, the pointer points to record 0 (the first record in the file). Any read operation reads the currently pointed-to structure and moves the pointer down one structure. Any write operation writes to the currently pointed-to structure and moves the pointer down one structure. Seek moves the pointer to the requested record.

Keep in mind that C thinks of everything in the disk file as blocks of bytes read from disk into memory or read from memory onto disk. C uses a file pointer, but it can point to any byte location in the file.

Functions fread() and fwrite() are used for reading from and writing to a file on the disk respectively, in case of binary files.

Writing to a Binary File

To write into a binary file, you need to use the function fwrite(). The functions takes four arguments: Address of data to be written in disk, Size of data to be written in disk, number of such type of data and pointer to the file where you want to write.

fwrite(address_data,size_data,numbers_data,pointer_to_file);

Sensor Diag is a simple three integer structure.

struct sensor_diag

{

      int x, y, z;

};

Open file for binary operations:

```
/* create the file of 10 records */
f = fopen("sensor_diag", "w");
```

Write 10 values into the structure and close the file.

```
for (i = 1;i <= 10; i++)
{
        sd_record.x = i;
        fwrite(&sd_record, sizeof(struct sensor_diag), 1, f);
}
fclose(f);
```

---

[3]   Binary files - https://computer.howstuffworks.com/c39.htm

FSeek to read records

```
f = fopen("sensor_diag", "r");
if (!f)
        return 1;
for (i = 9; i >= 0; i--)
{
        fseek(f, sizeof(struct sensor_diag)*i, SEEK_SET);
        fread(&sd_record, sizeof(struct sensor_diag), 1, f);
        printf("%d\n", sd_record.x);
}
fclose(f);
printf("\n");
```

Read records in alternate manner (see Figure 11):

```
/* use fseek to read every other record */
f = fopen("sensor_diag", "r");
if (!f)
        return 1;
fseek(f, 0, SEEK_SET);
for (i = 0;i<5; i++)
{
        fread(&sd_record, sizeof(struct sensor_diag), 1, f);
        printf("%d\n", sd_record.x);
        fseek(f, sizeof(struct sensor_diag), SEEK_CUR);
}
fclose(f);
printf("\n");
```

Code: The full code write to Binary file is listed as part of Table 6

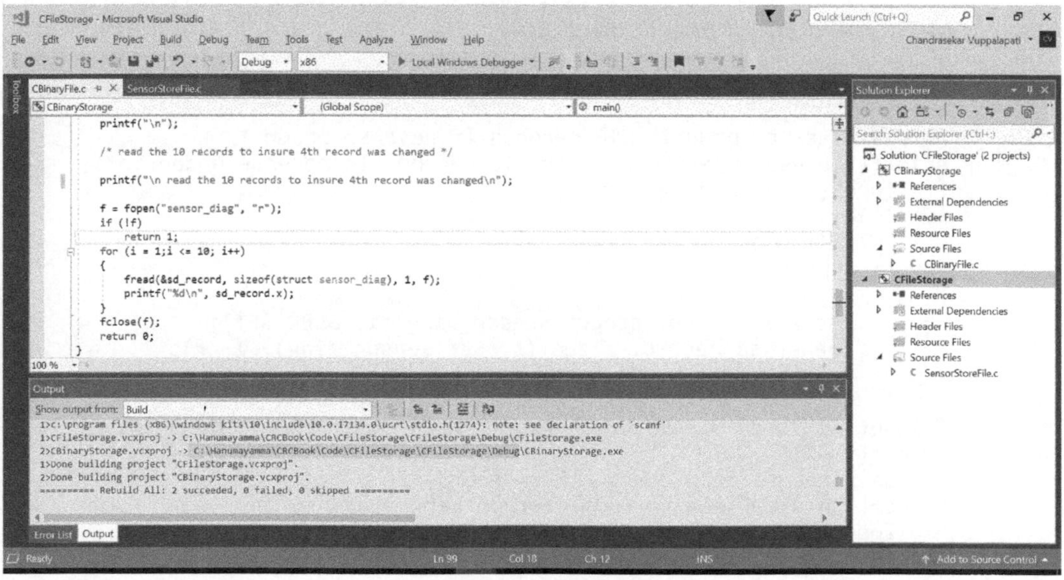

Figure 11: Visual Studio IDE

Table 6: Binary File Write Full Code

```c
#include <stdio.h>
#include <stdlib.h>
#include <stdio.h>
/* Sensor records */
struct sensor_diag
{
        int x, y, z;
};
/* writes and then reads 10 arbitrary records
from the file "sensor_diag". */
int main()
{
        int i, j;
        FILE *f;
        struct sensor_diag sd_record;
        /* create the file of 10 records */
        printf("\n create the file of 10 records \n");
        f = fopen("sensor_diag", "w");
        if (!f)
                return 1;
        for (i = 1;i <= 10; i++)
        {
                sd_record.x = i;
                fwrite(&sd_record, sizeof(struct sensor_diag), 1, f);
        }
        fclose(f);
        printf("\n read the 10 records \n");
        /* read the 10 records */
        f = fopen("sensor_diag", "r");
        if (!f)
                return 1;
        for (i = 1;i <= 10; i++)
        {
                fread(&sd_record, sizeof(struct sensor_diag), 1, f);
                printf("%d\n", sd_record.x);
        }
        fclose(f);
        printf("\n");
        /* use fseek to read the 10 records in reverse order */
        printf("\n use fseek to read the 10 records in reverse order \n");
        f = fopen("sensor_diag", "r");
        if (!f)
                return 1;
        for (i = 9; i >= 0; i--)
        {
                fseek(f, sizeof(struct sensor_diag)*i, SEEK_SET);
                fread(&sd_record, sizeof(struct sensor_diag), 1, f);
                printf("%d\n", sd_record.x);
        }
        fclose(f);
        printf("\n");
        /* use fseek to read every other record */
        printf("\n use fseek to read every other record \n");
        f = fopen("sensor_diag", "r");
        if (!f)
                return 1;
        fseek(f, 0, SEEK_SET);
        for (i = 0;i<5; i++)
```

Table 6 contd. ...

...Table 6 contd.

```
        {
                fread(&sd_record, sizeof(struct sensor_diag), 1, f);
                printf("%d\n", sd_record.x);
                fseek(f, sizeof(struct sensor_diag), SEEK_CUR);
        }
        fclose(f);
        printf("\n");
        /* use fseek to read 4th record, change it, and write it back */
        printf("\n use fseek to read 4th record, change it, and write it
back\n");
        f = fopen("sensor_diag", "r+");
        if (!f)
                return 1;
        fseek(f, sizeof(struct sensor_diag) * 3, SEEK_SET);
        fread(&sd_record, sizeof(struct sensor_diag), 1, f);
        sd_record.x = 100;
        fseek(f, sizeof(struct sensor_diag) * 3, SEEK_SET);
        fwrite(&sd_record, sizeof(struct sensor_diag), 1, f);
        fclose(f);
        printf("\n");
        /* read the 10 records to insure 4th record was changed */
        printf("\n read the 10 records to insure 4th record was changed\n");
        f = fopen("sensor_diag", "r");
        if (!f)
                return 1;
        for (i = 1;i <= 10; i++)
        {
                fread(&sd_record, sizeof(struct sensor_diag), 1, f);
                printf("%d\n", sd_record.x);
        }
        fclose(f);
        return 0;
}
```

The program above generates the following output (see Figure 12)

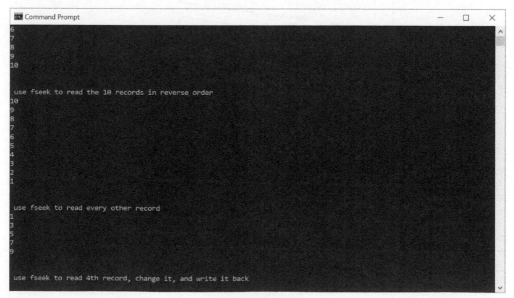

Figure 12: File Output

EPROM Data Storage

The embedded system On-chip ISP (In-System Programing) Flash allows the program memory to be reprogrammed In-System through an SPI (Serial Peripheral Interface) by a conventional nonvolatile memory programmer, or by an On-chip Boot program running on the AVR core. The Boot program can use any interface to download the application program in the Application Flash memory. The application with Kalman filter is developed using C Compilers.

The bootloader[4] is the little program that runs when you turn the embedded system on or press the reset button. Its main function is to load the application into the FLASH. The bootloader is what enables you to program the Arduino using just the USB cable.

Entering the Boot Loader takes place by a jump or call from the application program. This may be initiated by a trigger such as a command received via USART, or SPI interface.

The embedded system Program Flash memory space is divided in two sections, the Boot Program section and the Application Program section. The application Program section contains instructions for starting the connectivity (BLE), EPROM access and other services [4].

EEPROM provides nonvolatile data storage support. 0–255 record counters will be stored in EEPROM location zero (see Figure 13) [4]. 256–336 record counter with maximum of 2 weeks at 24 records/day will be stored in EEPROM location one. Records will start at EEPROM location two. Additionally, the data arrays store on EEPROM and retrieve command over Bluetooth Low Energy (BLE).

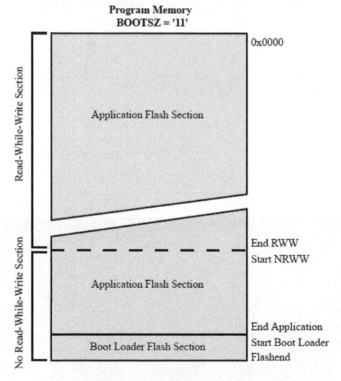

Figure 13: EPROM Program Memory

---

4   Bootloader   -   http://arduinoinfo.mywikis.net/w/images/3/35/Atmel-42735-8-bit-AVR-Microcontroller-ATmega328-328P_Datasheet.pdf

Hanumayamma Dairy IoT Sensor Data Storage for the IoT

BACKGROUND OF THE INVENTION

Problem Solved: Hardware Sensors, such as temperature, humidity, accelerometer and gyroscope, capture real-time environmental and location specific details. The frequency capture rate for these sensors varies. For instance, temperature sensors need to probe on an hourly basis in order to assess the changing gradient (increase or decrease) of location temperature. Other sensors, an accelerometer sensor for example, needs to be probed every minute so that accurate movement can be captured.

There are two design considerations that need to be made in order to capture and organize these sensor values: (1) Performance vs. Power requirements and (2) Storage space needed to capture these sensor values.

Power vs. Performance of the sensor is very important as it directly influences the longevity of the sensor's battery. The more active the sensors are, the more power is consumed and the shorter their battery life. This is particularly important for the sensors that are deployed in geographically distant places.

The storage requirement grows considerably if the sensors capture high-frequency sensor values. Another important factor on storage need is how frequently the data captured can be copied to a remote server over the Internet so that the sensor has to hold only a minimum amount of data. The capture, storage and transfer cycle influence battery power requirements and impact the overall functioning of the sensor. For power conserving purposes, the Internet transfer can be pre-configured (every three hours cycle or mid-night data transfer). The only design consideration that needs to assessed is how to inform if there are critical notifications such as sudden drastic change in temperature or high moment due to external factors.

Finally, the storage needs increase considerably if there is no consistent connectivity for uploading the captured sensor values. The sensor storage system needs to be optimal in order to tackle these requirements.

DETAILED DESCRIPTION OF THE INVENTION

The design of highly fault tolerant and consistent storage mechanism is required to tackle varying hardware sensor captures frequencies under low connectivity and how power efficient battery life cycle needs.

Creating a file system that can hold captured sensor values in a circular manner can help to overcome performance and power demands. The claimed invention differs from what currently exists in the following ways. In traditional file systems, as the new data is captured, the data appended to the end of existing file system and the footprint of the file system grows as more data are captured. With the increase in footprint, more space is required, and this in-turn puts the pressure on the battery life. Moreover, in order to retrieve the desired portion, the entire file needs to be processed, influencing more duty cycles on the operating system and, hence, increased power needs.

In the proposed innovation, efficient file system is designed to tackle storage, performance, retrieval and battery life requirements. The innovation is unique as it is developed with minimal software footprint on the embedded hardware sensors.

The file system at its core consists of in-memory software and underlying file storage, mirroring the data captured. The underlying file storage organizes the data on a record basis with pre-defined data placeholders. For instance, the first field of the record points to date and timestamp of capture, second field environmental temperature, third field humidity, this followed by contact temperature, contact humidity, and accelerometer X position, y position and z position values. All these fields are delimited by commas. In addition to data records, the software system consists of two-file pointers, current and recent, and a number of newly captured records. The file pointer is a simple address that the pointer holds for pointing purposes. As the new records are added, the current file pointer value changes to the latest memory position. Similarly, the Last retrieved file pointer stores the most recent retrieval of records by external connectivity mechanisms—Bluetooth or the Internet transfer.

The uniqueness of this innovation is that it maintains the file storage and pointer mechanism to pre-defined circular file system slots rather than expanding the footprint unfathomable manner. In addition, the innovation can produce:

a) Mobile Software Application that reprieves file system data on low powered connectivity such as Bluetooth.
b) Application Programmable Interface (APIs): The software exposes APIs to connect with external modules.

The Version of The Invention Discussed Here Includes (see Figure 14):
- In-memory dairy data cube
- Stream Collector
- Dairy Cattle Sensor Data Event Queue
- Stream Indexer
- Stream Injector
- Rule Processor
- Data Cube Archiver

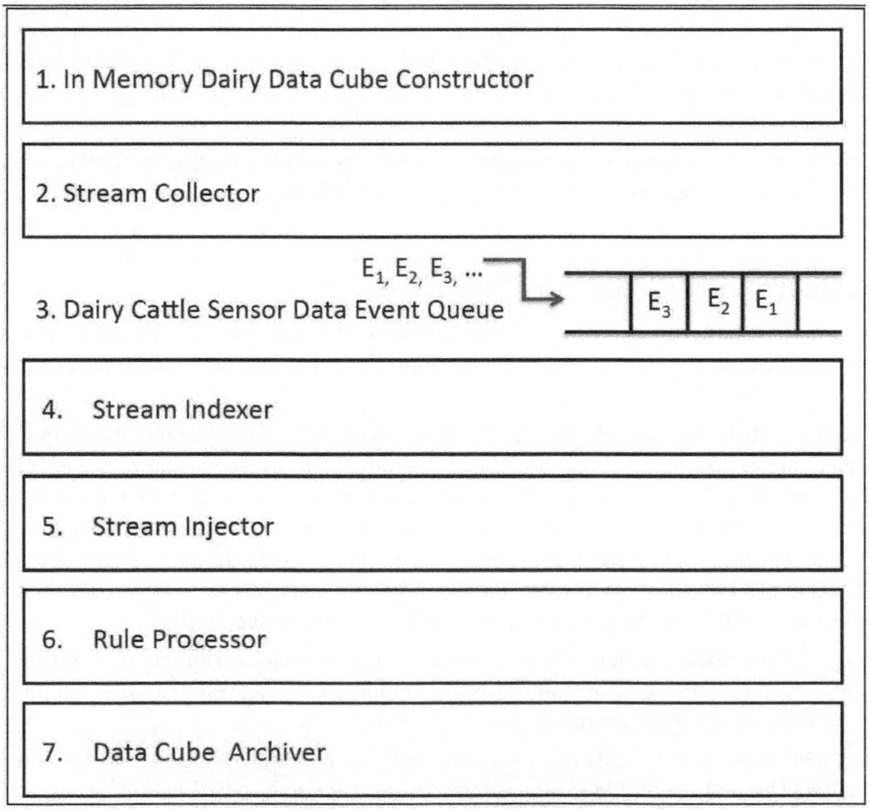

Figure 14: Data Cube Architecture

Relationship Between the Components (see Figure 15):

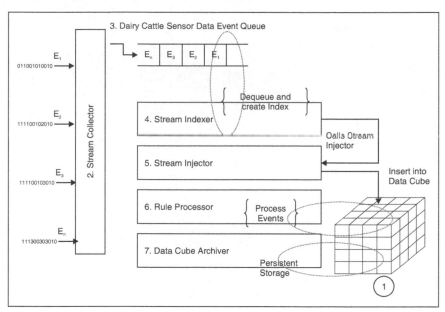

Figure 15: Data Cube Flow

Dairy Data Cube (1) is constructed using a software collections framework that can store the streams of data emitted from sensors in real-time. The stream collector (2) collects the sensor data through the Internet protocols or through the blue tooth low energy (BLE) communication protocols. Once the streams are emitted from the sensors, the communication protocol handler invokes the stream collector and passes the real-time sensor data in an asynchronous manner. The asynchronous invocation is required to un-block the events that are generated at Sensor level. Once the event data are received, the stream collector queues the data into the Dairy Cattle Sensor Data Event queue (3) and relinquishes the control back to the sensor communication handler. The Stream Indexer (4) dequeues the event data and creates a new hash index and passes the hash index and event sensor queue data to the Stream injector (5). The stream injector stores the new event data into a data cube cell (1). The cell position is determined based on the date and time of the sensor event. The completion of the Stream injector process ensures the proper stacking of an event queue in the dairy data cube (1). The Rule processor (6) is a background thread that periodically checks the availability of sensor event data in the dairy cube (1). Once available, the rule processor processes sensor data to validate any rules violations by application of stream sliding window algorithms. Once the rule processor (6) completes the rule evaluation, the data cube archiver (7) stores the elapsed time window dairy events (3) to persistent data storage. The persistent data storage could include a network server or local data storage (see Figure 16).

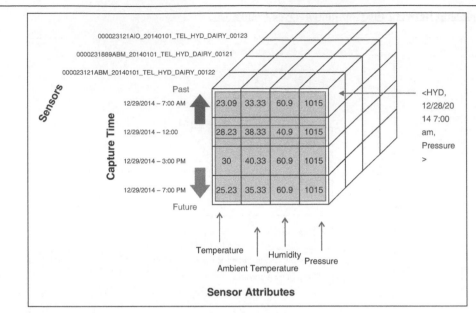

Figure 16: Data Cube

<u>How the Invention Works</u>:

As new streams of sensor data are generated, the software collects the sensor events and creates unique a data hash index in order to store new sensor values into the dairy sensor data event cube. The data event cube efficiently organizes the stream data by processing the events in the cube by the application of a sliding window algorithm and then persisting the events that were processed by the sliding algorithm. So, at any point, the dairy data cube holds the event data that is within the sliding range, i.e., current date & time, n events before current date and time and n events after current date and time. The value of "n" is fine-tuned based on actual hardware of the mobile device.

Setup real-time clock event handler that collects data on an hourly basis:

```
int tickEvent = t.every(3600000, HourlySensorReadAndStore); // hourly data grab 3600000 is an hour
```

The data is collected in the following order:

- Get Current Memory Pointer
- Read Sensor Values from the Store (see Table 7)
- Clear variables.

Table 7: Hourly Sensor Values

```
void HourlySensorReadAndStore ()
{
 GetCurrentRecordCount (); // get global count of existing valid stored records
 ReadSensorsAndStore(); // get all 5 sensor current readings and store in EEprom datalog
 ClearCowActivity (); // clear the running count of cow activity measurement to start again
 IncrementCurrentRecordCount (); // increment the total records now stored in EEprom datalog (on
EEprom)
 }
```

Get Current Record Count

The pointer to the record is stored in two register variables: EEPROMread(0) and EEPROMread(1)—see Table 8.

Table 8: Get Current Record Cound

```
// Returns record counter into int from two stored bytes.
void GetCurrentRecordCount () {
    totalRecordCount0 = EEPROMread(0); //get lower byte of current stored record count.
    totalRecordCount1 = EEPROMread(1); //get upper byte of current stored record count.
  if ((totalRecordCount0) < 253) //check if low order byte counter not full yet
  {
  currentCountBothBytes = (totalRecordCount0); //low order byte not full yet
  }
  else if ((totalRecordCount1) < 82)
  {
  currentCountBothBytes = (totalRecordCount0 + totalRecordCount1);
  }
}
```

EPROM Read

The following (see Table 9) code provides access to EPROM Read functionality:

Table 9: EPROM Read

```
//Reads out of EEprom and send Part One of the records over ble
//Part One is date time, amb temp, amb humidity, cow temp, cow humidity.
//This is the maximum data transfer possible with this ble arrangement apparently.
//Note that baseRecordByteAddrCount is set to zero on power up, first sw load, cmd "R", and "0" cmd.
void SendRecordPartOne() {
 GetCurrentRecordCount (); //0 to 335 integer count of records in eeprom. At 336 it restarts at 0.
 if ((currentCountBothBytes) > 0)// if no records yet, just exit. Otherwise process just one set.
 {
  //read section and send section
 EEPROMread((baseRecordByteAddrCount),(uint8_t*)timeDateData,int(sizeof(timeDateData)));
 // Send it over bluetooth.
 while (!lib_aci_send_data(PIPE_UART_OVER_BTLE_UART_TX_TX, (uint8_t*)timeDateData, sizeof
(timeDateData)))
 {
 for(int i=0; i<50; i++)
 aci_loop();
 delay (1);
 //SerialMonitorInterface.println(F("TX dropped!"));
 }
 nextRecordByteAddrCount = baseRecordByteAddrCount + (sizeof (timeDateData));
 EEPROMread((nextRecordByteAddrCount),(uint8_t*)ambientTemp, int(sizeof (ambientTemp)));
 while (!lib_aci_send_data(PIPE_UART_OVER_BTLE_UART_TX_TX, (uint8_t*)ambientTemp, sizeof
(ambientTemp)))
```

Table 9 contd. ...

Table 9 contd. ...

```
{
for(int i=0; i<50; i++)
aci_loop();
delay (1);
// SerialMonitorInterface.println(F("TX dropped!"));
}
nextRecordByteAddrCount2 = nextRecordByteAddrCount + (sizeof (ambientTemp));
EEPROMread((nextRecordByteAddrCount2),(uint8_t*)ambientHumd, int(sizeof (ambientHumd)));
while (!lib_aci_send_data(PIPE_UART_OVER_BTLE_UART_TX_TX, (uint8_t*)ambientHumd,
sizeof(ambientHumd)))
{
for(int i=0; i<50; i++)
aci_loop();
delay (1);
// SerialMonitorInterface.println(F("TX dropped!"));
}
nextRecordByteAddrCount3 = nextRecordByteAddrCount2 + (sizeof (ambientTemp));
EEPROMread((nextRecordByteAddrCount3),(uint8_t*)cowTemp, int(sizeof (cowTemp)));
while (!lib_aci_send_data(PIPE_UART_OVER_BTLE_UART_TX_TX, (uint8_t*)cowTemp,
sizeof(cowTemp)))
{
for(int i=0; i<50; i++)
aci_loop();
delay (1);
// SerialMonitorInterface.println(F("TX dropped!"));
}

nextRecordByteAddrCount4 = nextRecordByteAddrCount3 + (sizeof (cowTemp));
EEPROMread((nextRecordByteAddrCount4),(uint8_t*)cowHumd, int(sizeof (cowHumd)));
while (!lib_aci_send_data(PIPE_UART_OVER_BTLE_UART_TX_TX, (uint8_t*)cowHumd,
sizeof(cowHumd)))
{
for(int i=0; i<50; i++)
aci_loop();
delay (1);
//SerialMonitorInterface.println(F("TX dropped!"));
}
nextRecordByteAddrCount5 = nextRecordByteAddrCount4 + (sizeof (cowHumd));
}
}
```

Read Sensor and Store into EPROM

The following code (see Table 10) provides read & write EPROM Sensor data:

Table 10: Read Sensor and Store

```
void ReadSensorsAndStore () {
    String timeDateString;
    RTC.readTime(); // update RTC library's buffers from chip
    timeDateString += (int(RTC.getMonths()));
    timeDateString += "/";
    timeDateString += (int(RTC.getDays()));
    timeDateString += "/";
    timeDateString += (RTC.getYears());
    timeDateString += " ";
    timeDateString += (int(RTC.getHours()));
    timeDateString += ":";
    timeDateString += (int(RTC.getMinutes()));
    timeDateString.toCharArray(timeDateData, sizeof (timeDateData)); // copy into timeDateData buffer

    // Get ambient temp into array
    floatVal= (htu.readTemperature());
    dtostrf(floatVal, 3, 2, ambientTemp); //4 is mininum width, 4 is precision; float value is copied onto buff
ambientTemp

    // Get ambient humidity into array
    floatVal= (htu.readHumidity());
    dtostrf(floatVal, 3, 2, ambientHumd); //4 is mininum width, 4 is precision; float value is copied onto buff
ambientHumd

    // Get cow temp into array
    floatVal= (bme.readTemperature());
    dtostrf(floatVal, 3, 2, cowTemp); //4 is mininum width, 4 is precision; float value is copied onto buff
cowTemp

    // Get cow humidity into array
    floatVal= (bme.readHumidity());
    dtostrf(floatVal, 3, 2, cowHumd); //4 is mininum width, 4 is precision; float value is copied onto buff
cowHumd

    // Get cow activity into array, first convert integer count to string
    String str = String(activityCount);
    // Length (5 digits with one extra character for the null terminator), cowActivity is 6 bytes.
    str.toCharArray(cowActivity, sizeof (cowActivity)); // copy into cowActivity buffer

    //store sensor data as record by Writing to EEProm section. Have to use separate "W"address records
    //since this is part of a timed interrupt routine and can be called during other operations.
    EEPROMwrite((baseRecordByteAddrCountW),(uint8_t*) timeDateData,sizeof(timeDateData));

    nextRecordByteAddrCountW = baseRecordByteAddrCountW + (sizeof (timeDateData));
    EEPROMwrite ((nextRecordByteAddrCountW),(uint8_t*) ambientTemp,sizeof(ambientTemp));

    nextRecordByteAddrCount2W = nextRecordByteAddrCountW + (sizeof (ambientTemp));
    EEPROMwrite ((nextRecordByteAddrCount2W),(uint8_t*) ambientHumd,sizeof(ambientHumd));

    nextRecordByteAddrCount3W = nextRecordByteAddrCount2W + (sizeof (ambientTemp));
    EEPROMwrite ((nextRecordByteAddrCount3W),(uint8_t*) cowTemp,sizeof(cowTemp));
```

Table 10 contd. ...

...Table 10 contd.

```
    nextRecordByteAddrCount4W = nextRecordByteAddrCount3W + (sizeof (cowTemp));
     EEPROMwrite ((nextRecordByteAddrCount4W),(uint8_t*) cowHumd,sizeof(cowHumd));

    nextRecordByteAddrCount5W = nextRecordByteAddrCount4W + (sizeof (cowHumd));
    EEPROMwrite ((nextRecordByteAddrCount5W),(uint8_t*) cowActivity,sizeof(cowActivity));

    //last item in this record now written, so set baseRecordAddressCount to
    //point to location of start of next record

    baseRecordByteAddrCountW = nextRecordByteAddrCount5W + (sizeof (cowActivity));
    }
```

EPROM Read and Writes

```
byte EEPROMread(unsigned long addr){
 uint8_t val=255;
 uint8_t I2Caddr=EEPROM_ADDR;
 if(addr>0x0000FFFF){
   I2Caddr|=0x04;
 }
 Wire.beginTransmission(I2Caddr);
 Wire.write(addr>>8);
 Wire.write(addr);
 Wire.endTransmission();
 Wire.requestFrom(I2Caddr,(uint8_t)1);
 while(Wire.available()){
  val=Wire.read();
 }

 return val;
}

byte EEPROMwrite(unsigned long addr, byte val){
  uint8_t I2Caddr=EEPROM_ADDR;
  if(addr>0x0000FFFF){
   I2Caddr|=0x04;
  }
 Wire.beginTransmission(I2Caddr);
 Wire.write(addr>>8);
 Wire.write(addr);
 Wire.write(val);
 Wire.endTransmission();
 Wire.beginTransmission(I2Caddr);
 unsigned long timeout=millis();
 while(Wire.endTransmission() && millis()-timeout<10){
  Wire.beginTransmission(I2Caddr);
 }
}
```

Android Data Storage

Android provides several data storages for the embedded systems. Android storage includes:

- Network based storage
- Shared Preferences
- Internal Storage
- SQLite
- File Storage

In the context of IoT, especially, with constrained devices, the need to store on the Edge device is an essential part of the data resilience and provides a mechanism to upload to the central store upon the network being established.

 | Based on the need of the applications, the Edge application design can use file storage to embedded relational database systems (RDBMS).

Shared Preference File Storage

Android provides many ways of storing the data of an application. One method is called Shared Preferences. Shared Preferences allows you to save and retrieve data in the form of key, value pair.
The following APIs[5] provide access to shared preferences:

```
public abstract SharedPreferences getSharedPreferences (String name,
 int mode)
```

Retrieve and hold the contents of the preferences file 'name', returning a SharedPreferences through which you can retrieve and modify its values. Only one instance of the SharedPreferences object is returned to any callers for the same name, meaning they will see each other's edits as soon as they are made. This method is thead-safe [5].

Shared Preference App Manifest File

The App does not require any special or permissions (please see Table 11 and Figure 17).

Table 11: App Manifest file

```
<?xml version="1.0" encoding="utf-8"?>
<manifest xmlns:android="http://schemas.android.com/apk/res/android"
  package="com.secondassig.androiddatastorage"
  android:versionCode="1"
  android:versionName="1.0" >

 <uses-sdk
  android:minSdkVersion="8"
  android:targetSdkVersion="18" />

 <application
  android:allowBackup="true"
  android:icon="@drawable/ic_launcher"
```

Table 11 contd. ...

---

[5] Get Shared Preference - https://developer.android.com/reference/android/content/Context#getSharedPreferences(java.lang.String,%20int)

...Table 11 contd.

```
      android:label="@string/app_name"
      android:theme="@style/AppTheme" >
   <activity
      android:name="com.secondassig.androiddatastorage.MainActivity"
      android:label="@string/app_name" >
      <intent-filter>
        <action android:name="android.intent.action.MAIN" />
        <category android:name="android.intent.category.LAUNCHER" />
      </intent-filter>
   </activity>
   <activity
      android:name="com.secondassig.androiddatastorage.SetPreferencesActivity"
      android:label="@string/title_activity_set_preferences"
      android:parentActivityName="com.secondassig.androiddatastorage.
        MainActivity" >
   <meta-data
      android:name="android.support.PARENT_ACTIVITY"
      android:value="com.secondassig.androiddatastorage.MainActivity" />
   </activity>
   <activity
      android:name="com.secondassig.androiddatastorage.SQLiteActivity"
      android:label="@string/title_activity_sqlite"
      android:parentActivityName="com.secondassig.androiddatastorage.
        MainActivity" >
   <meta-data
      android:name="android.support.PARENT_ACTIVITY"
      android:value="com.secondassig.androiddatastorage.MainActivity" />
   </activity>
   </application>

</manifest>
```

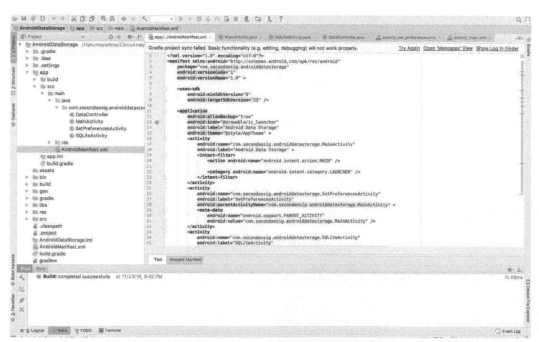

Figure 17: Android Manifest File

App IDE

Main Activity

The purpose of the main activity is to handle User interactions:

- Shared Preferences
- SQLite Data Storage

When the User clicks on the Share Preference button, the app calls openPreference() (see Table 12).

Table 12: Data Storage Android Code

```java
package com.secondassig.androiddatastorage;

import java.io.BufferedReader;
import java.io.InputStream;
import java.io.InputStreamReader;

import android.os.Bundle;
import android.app.Activity;
import android.content.Intent;
import android.view.Menu;
import android.view.View;
import android.widget.EditText;
import android.widget.TextView;

public class MainActivity extends Activity {
    @Override
    protected void onCreate(Bundle savedInstanceState) {
        super.onCreate(savedInstanceState);
        setContentView(R.layout.activity_main);

    }
    @Override
    public boolean onCreateOptionsMenu(Menu menu) {
        // Inflate the menu; this adds items to the action bar if it is present.
        getMenuInflater().inflate(R.menu.main, menu);
        return true;
    }
    @Override
    protected void onResume()
    {
        super.onResume();
        try
        {
         InputStream in=openFileInput(SetPreferencesActivity.STORE_PREFERENCES);
         if(in!=null)
         {
            InputStreamReader tmp=new InputStreamReader(in);
            BufferedReader reader=new BufferedReader(tmp);
            String str;
            StringBuilder buf=new StringBuilder();
            while((str=reader.readLine())!=null)
            {
               buf.append(str +"\n");
            }
            in.close();
            TextView savedPref=(TextView)findViewById(R.id.saved_data);
            savedPref.setText(buf.toString());
        }
    }
}
```

Table 12 contd. ...

...Table 12 contd.

```
  catch(Exception e)
  {
    e.printStackTrace();
  }
}

public void openPreference(View view)
{
  Intent intent=new Intent(this,SetPreferencesActivity.class);
  startActivity(intent);
}

public void saveInDatabase(View view)
{
  Intent intent =new Intent(this,SQLiteActivity.class);
  startActivity(intent);
}

public void exitApp(View view)
{
  finish();
  System.exit(0);
}
}
```

Clicking on the Preference button, creates an intent to invoke Set Preference Activity.

Set Preference Activity

Set Preference Activity initiates the Dialog to enter the preference values. This is achieved by calling OnCreate on the Activity.

The OnCreate creates a reference to System Shared Reference Object:

SharedPreferences sharedPrefs = PreferenceManager.*getDefaultSharedPreferences*(**this**);

In the App, the User can store preferences: Book name, Book Author, and Description. The entered values are stored to the preference by calling the Save button.

The Save button extracts the values from UI:

```
//save the preferences (if not null) in a file.
 EditText name_text=(EditText)findViewById(R.id.bookname_value);
 String name=name_text.getText().toString();
 EditText author_text=(EditText)findViewById(R.id.bookauthor_value);
 String author=author_text.getText().toString();
 EditText des_text=(EditText)findViewById(R.id.description_value);
 String description=des_text.getText().toString();
```

The save to shared preference is performed:

```
SharedPreferences sharedPreferences=PreferenceManager.getDefaultSharedPreferences(this);
 Editor editor=sharedPreferences.edit();
 editor.putInt("COUNTER", counter);
 editor.putInt("HELLOCMPE277", counter);
 editor.commit();
```

OutputStreamWriter out=**new** OutputStreamWriter(openFileOutput(***STORE_
PREFERENCES,MODE_APPEND***)));
String message="**\nSaved Preference** "+**counter**+", "+s.format(**new** Date());
out.write(message);
out.close();

First, it creates an editor object that is an edit of SharePreferences.

Second, values are inserted into the objects by calling puts.

Finally, the write results into saving the values into Shared preference (please see Table 13).

Table 13: Shared Preference App Code

```
package com.secondassig.androiddatastorage;

import java.io.FileNotFoundException;
import java.io.IOException;
import java.io.OutputStreamWriter;
import java.text.SimpleDateFormat;
import java.util.Date;

import android.os.Bundle;
import android.app.Activity;
import android.view.Menu;
import android.view.MenuItem;
import android.view.View;
import android.widget.EditText;
import android.support.v4.app.NavUtils;
import android.annotation.TargetApi;
import android.content.Context;
import android.content.Intent;
import android.content.SharedPreferences;
import android.content.SharedPreferences.Editor;
import android.os.Build;
import android.preference.PreferenceManager;

public class SetPreferencesActivity extends Activity {

  public final static String STORE_PREFERENCES="storePrefFinal.txt";
  //private static int counter;
  public int counter=0;
  private SimpleDateFormat s=new SimpleDateFormat("MM/dd/yyyy-hh:mm a");

  @Override
  protected void onCreate(Bundle savedInstanceState) {
    super.onCreate(savedInstanceState);
    setContentView(R.layout.activity_set_preferences);
    // Show the Up button in the action bar.
    setupActionBar();

    SharedPreferences sharedPrefs = PreferenceManager.getDefaultSharedPreferences(this);
    counter=sharedPrefs.getInt("COUNTER", 0);

  }
  @Override
  public void onResume()
  {
    super.onResume();
```

Table 13 contd. ...

...Table 13 contd.

```java
    SharedPreferences sharedPrefs = PreferenceManager.getDefaultSharedPreferences(this);
    counter=sharedPrefs.getInt("COUNTER", 0);
 }
 /**
 * Set up the {@link android.app.ActionBar}, if the API is available.
 */
@TargetApi(Build.VERSION_CODES.HONEYCOMB)
private void setupActionBar() {
   if (Build.VERSION.SDK_INT >= Build.VERSION_CODES.HONEYCOMB) {
      getActionBar().setDisplayHomeAsUpEnabled(true);
   }
 }
@Override
public boolean onCreateOptionsMenu(Menu menu) {
   // Inflate the menu; this adds items to the action bar if it is present.
   getMenuInflater().inflate(R.menu.set_preferences, menu);
   return true;
}
@Override
public boolean onOptionsItemSelected(MenuItem item) {
   switch (item.getItemId()) {
   case android.R.id.home:
   // This ID represents the Home or Up button. In the case of this
   // activity, the Up button is shown. Use NavUtils to allow users
   // to navigate up one level in the application structure. For
   // more details, see the Navigation pattern on Android Design:
   //
   // http://developer.android.com/design/patterns/navigation.html#up-vs-back
   //
    NavUtils.navigateUpFromSameTask(this);
    return true;
   }
   return super.onOptionsItemSelected(item);
 }
public void onSave(View view)
 {
   //save the preferences (if not null) in a file.
   EditText name_text=(EditText)findViewById(R.id.bookname_value);
   String name=name_text.getText().toString();
   EditText author_text=(EditText)findViewById(R.id.bookauthor_value);
   String author=author_text.getText().toString();
   EditText des_text=(EditText)findViewById(R.id.description_value);
   String description=des_text.getText().toString();
   if(name!=null && author!=null && description!=null)
   {
     try
     {
       counter+=1;

       SharedPreferences sharedPreferences=PreferenceManager.getDefaultSharedPreferences(this);
       Editor editor=sharedPreferences.edit();
```

Table 13 contd. ...

...Table 13 contd.

```
        editor.putInt("COUNTER", counter);
        editor.putInt("HELLOCMPE277", counter);
       editor.commit();

     OutputStreamWriter out=new OutputStreamWriter(openFileOutput(STORE_
PREFERENCES,MODE_APPEND));
     String message="\nSaved Preference "+counter+", "+s.format(new Date());
     out.write(message);
     out.close();
   }
  catch (FileNotFoundException e)
  {
    e.printStackTrace();
  }
    catch (IOException e)
  {
    e.printStackTrace();
  }

 }
//Retrieve the contents of the file in the OnResume() in MainActivity

//Go back to the main screen
Intent intent = new Intent(this,MainActivity.class);
startActivity(intent);

}

public void onCancel(View view)
{
   Intent intent = new Intent(this,MainActivity.class);
   startActivity(intent);
}

}
```

The saved preference is updated to the UI. The Shared Preferences file is saved on the device storage. Using Android Device Monitor, we can inspect the stored data (see Figure 18).

The device monitor provides the system view from the development point of view. On the actual device, the shared preference file is still stored at the same relative position on the App's directory (see Figure 19).

The location is:

APP Location\Data\Files\StrorePref.txt

The Contents of file: The above application creates the following Preferences file (see Figure 20).

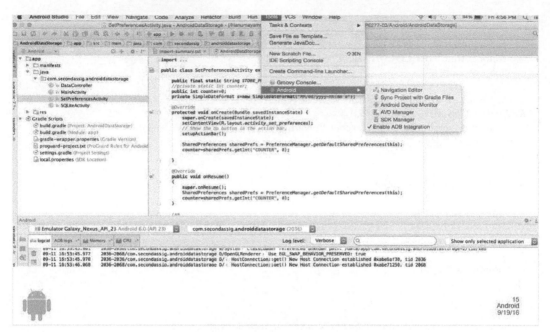

Figure 18: Android Device Monitor

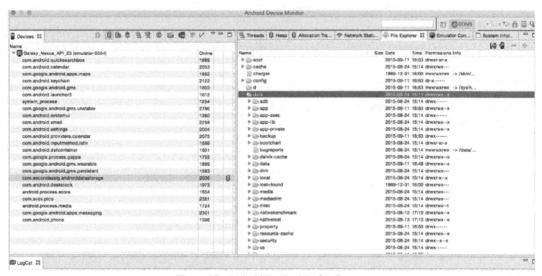

Figure 19: Android Device Monitor Process

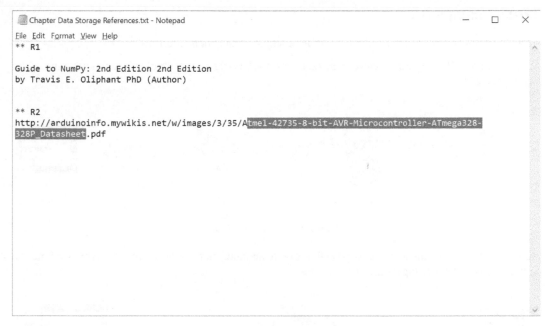

Figure 20: Shared Preferences Content

SQLite

SQLite is a self-contained, high-reliability, embedded, full-featured, public-domain, SQL database engine. The relational databases, generally, have a server process that provides callable interfaces to host or client application over the network (TCP/IP) and does not allow writing to the database directly. SQLite architecture is different from that of the traditional relational databases in that SQLite allows the process to write to SQLite directly on the file.

Classic Serverless vs. Neo-Serverless

Classic Serverless: The database engine runs as in-proc server, i.e., within the same process, thread, and address space as the application. There is no message passing or network activity.

Neo-Serverless: The database engine runs in an out of process server, i.e., a separate namespace from the application, probably on a separate machine, but the database is provided as a turn-key service by the hosting provider, requires no management or administration by the application owners, and is so easy to use that the developers can think of the database as being serverless even if it really does use a server.

SQLite is an example of a classic serverless database engine. To compare classic serverless to neo-serverless, please consider Azure Cosmos DB and Amazon S3. Both are Neo-serverless databases[6] [6].

---

[6] Serverless vs. Neo-serverless - https://www.sqlite.org/serverless.html

Azure Cosmos DB

Azure Cosmos DB provides three invocation interactions to the applications [6].

Azure Cosmos DB: Create an event-driven azure cosmos DB trigger that invokes a function on the Azure (see Figure 21).

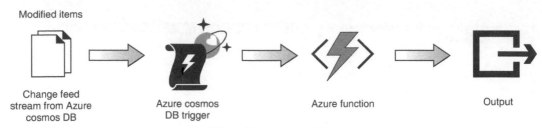

Figure 21: Azure Functions

Input binding (Azure function to Cosmos DB): Create an input function at Azure that invokes a function on the Cosmos DB (see Figure 22) [6].

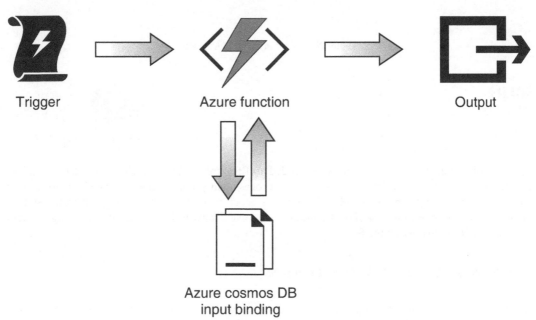

Figure 22: Azure Database Binding

Bind a function (see Figure 22) to an Azure Cosmos DB container using an *output binding*. Output bindings write data to a container when a function completes (see Figure 23) [7].

 Serverless embedded databases are essential ingredients to make edge level data storage.

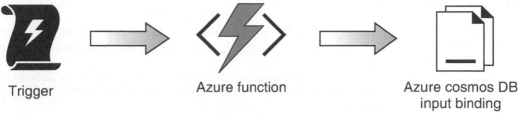

Trigger

Azure function

Azure cosmos DB input binding

Figure 23: Azure Trigger Functions

SQLite Transactional

A transactional database is one in which all changes and queries appear to be Atomic, Consistent, Isolated, and Durable (ACID). SQLite implements serializable transactions that are atomic, consistent, isolated, and durable, even if the transaction is interrupted by a program crash, an operating system crash, or a power failure to the computer.[7]

Embedded requirements: Size of SQLite Library

The embedded systems need to be low-footprint; especially, for the constrained devices support. Compiling with GCC and -Os results in a binary that is slightly less than 500 KB in size. (Update 2018-07-07: Due to the addition of new features such as UPSERT and window functions, the library footprint is now slightly larger than 500 KB.) The best build option is GCC 7 (see Figure 24).

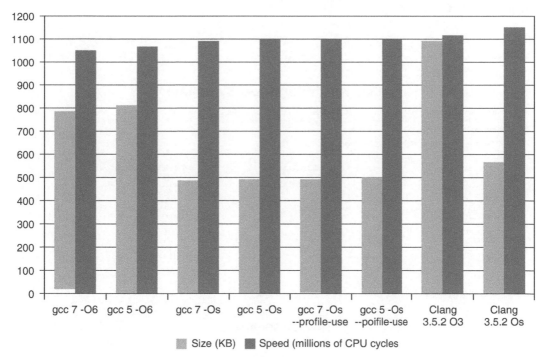

Figure 24: SQL Build

---

7   SQLite Transactional - https://www.sqlite.org/transactional.html

SQLite & IoT

SQLite is a high-reliability storage solution. The high-reliability of SQLite is proven in practice. SQLite has been used without problems in billions of smart-phones, IoT devices, and desktop applications, around the world, and for over a decade.[8]

	SQLite is well suited for IoT and embedded devices for the following reasons: • 500 kb foot-print • Highly reliable DO-178B[9] • Supports huge number of transactions

The maximum number of bytes in a string or BLOB in SQLite is defined by the preprocessor macro SQLITE_MAX_LENGTH. The default value of this macro is 1 billion (1 thousand million or 1,000,000,000). You can raise or lower this value at compile-time using a command-line option like this:

-DSQLITE_MAX_LENGTH=123456789[10]

Some benefits of using SQLite for local storage[11]

SQLite is light-weight and self-contained. It's a code library without any other dependencies. There's nothing to configure.

- There's no database server. The client and the server run in the same process.

- SQLite is in the public domain, so you can freely use and distribute it with your app.

- SQLite works across platforms and architectures.

Supported SQL Commands (please see Figure 25):[12]

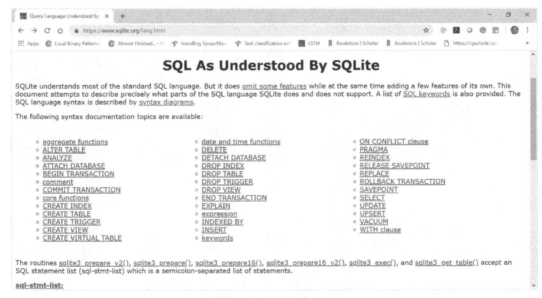

Figure 25: SQL References

---

8   SQLite & Reliability - https://www.sqlite.org/hirely.html

9   https://www.sqlite.org/hirely.html

10   Limits - https://sqlite.org/limits.html

11   Some benefits of using SQLite for local storage - https://docs.microsoft.com/en-us/windows/uwp/data-access/sqlite-databases

12   SQL Commands - https://www.sqlite.org/lang.html

SQLite and Embedded C App

SQLite Database can be build and distributed as part of an embedded application. The foot-print of SQLite data is less than 500 KB. To build SQLite, Eclipse Mars IDE with MinGW GCC (see Figure 26 and Figure 27).

Step 1: Create C Project
Step 2: Enter the Project name and select the compiler

Figure 26: Eclipse C Project

Figure 27: Ming GCC

Add sqlite3.c and sqlite3.h in your project. Create a file iDispenser.c and add it in the project. This is the program where we will create a database, create a table and insert data in the database. Step 3: Create a main driver in C to access SQLite functionalities (see Figure 28).

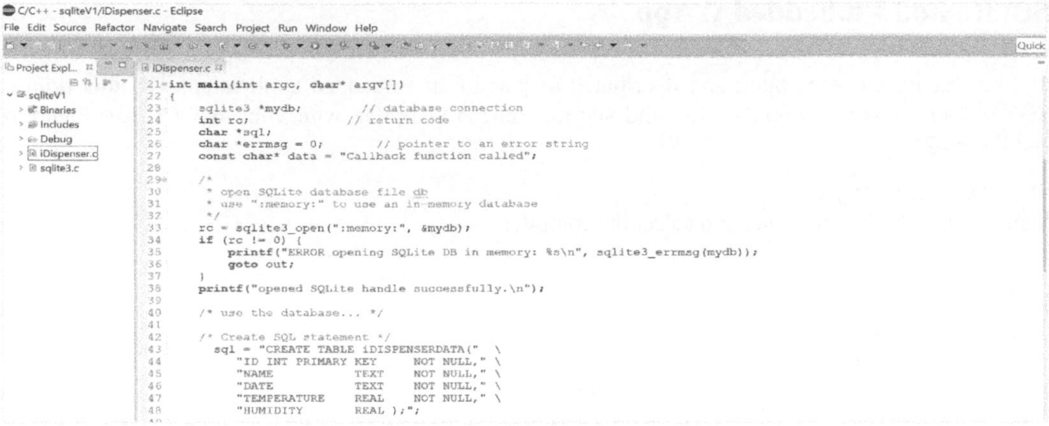

Figure 28: SQL C Program

Data insert successful (see Figure 29).

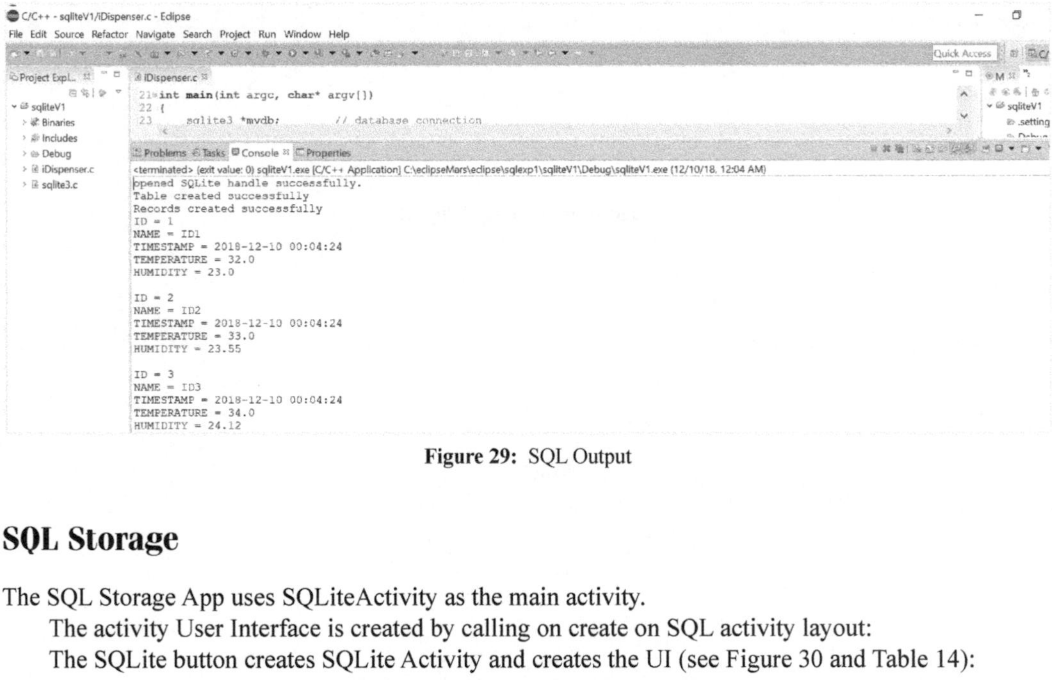

Figure 29: SQL Output

SQL Storage

The SQL Storage App uses SQLiteActivity as the main activity.

The activity User Interface is created by calling on create on SQL activity layout:

The SQLite button creates SQLite Activity and creates the UI (see Figure 30 and Table 14):

Figure 30: SQL App

Table 14: Android Data Storage SQL Code

```
package com.secondassig.androiddatastorage;

import java.io.OutputStreamWriter;
import java.text.SimpleDateFormat;
import java.util.Date;

import android.os.Bundle;
import android.app.Activity;
import android.view.Menu;
import android.view.MenuItem;
import android.view.View;
import android.widget.EditText;
import android.widget.Toast;
import android.support.v4.app.NavUtils;
import android.annotation.TargetApi;
import android.content.Context;
import android.content.Intent;
import android.content.SharedPreferences;
import android.content.SharedPreferences.Editor;
import android.os.Build;
import android.preference.PreferenceManager;
```

Table 14 contd. ...

...Table 14 contd.

```
public class SQLiteActivity extends Activity {

    public int counter=0;
    private SimpleDateFormat s=new SimpleDateFormat("MM/dd/yyyy-hh:mm a");

    @Override
    protected void onCreate(Bundle savedInstanceState) {
        super.onCreate(savedInstanceState);
        setContentView(R.layout.activity_sqlite);
        // Show the Up button in the action bar.
        setupActionBar();

        SharedPreferences sharedPrefs = PreferenceManager.getDefaultSharedPreferences(this);
        counter=sharedPrefs.getInt("SQL_COUNTER", 0);
    }

    @Override
    public void onResume()
    {
        super.onResume();
        SharedPreferences sharedPrefs = PreferenceManager.getDefaultSharedPreferences(this);
        counter=sharedPrefs.getInt("SQL_COUNTER", 0);
    }
    /**
     * Set up the {@link android.app.ActionBar}, if the API is available.
     */
    @TargetApi(Build.VERSION_CODES.HONEYCOMB)
    private void setupActionBar() {
        if (Build.VERSION.SDK_INT >= Build.VERSION_CODES.HONEYCOMB) {
            getActionBar().setDisplayHomeAsUpEnabled(true);
        }
    }

    @Override
    public boolean onCreateOptionsMenu(Menu menu) {
        // Inflate the menu; this adds items to the action bar if it is present.
        getMenuInflater().inflate(R.menu.sqlite, menu);
        return true;
    }

    @Override
    public boolean onOptionsItemSelected(MenuItem item) {
        switch (item.getItemId()) {
        case android.R.id.home:
        // This ID represents the Home or Up button. In the case of this
        // activity, the Up button is shown. Use NavUtils to allow users
        // to navigate up one level in the application structure. For
        // more details, see the Navigation pattern on Android Design:
        //
        // http://developer.android.com/design/patterns/navigation.html#up-vs-back
        //
        NavUtils.navigateUpFromSameTask(this);
        return true;
        }
        return super.onOptionsItemSelected(item);
    }
```

Table 14 contd. ...

...Table 14 contd.

```java
public void saveMessage(View view)
{
   EditText editText=(EditText)findViewById(R.id.msg);
   String message=editText.getText().toString();
   DataController dataController=new DataController(getBaseContext());
   dataController.open();
   long retValue= dataController.insert(message);
   dataController.close();
   if(retValue!=-1)
   {
     Context context = getApplicationContext();
     CharSequence text=getString(R.string.save_success_msg);
      int duration=Toast.LENGTH_LONG;
     Toast.makeText(context, text, duration).show();

      try
      {
       counter+=1;

       SharedPreferences sharedPreferences=PreferenceManager.getDefaultSharedPreferences(this);
       Editor editor=sharedPreferences.edit();
       editor.putInt("SQL_COUNTER", counter);
       editor.commit();

       OutputStreamWriter out=new OutputStreamWriter(openFileOutput(SetPreferencesActivity.STORE_
PREFERENCES,MODE_APPEND));
       out.write("\nSQLite "+counter+", "+s.format(new Date()));
       out.close();
   }
   catch(Exception e)
   {
      e.printStackTrace();
   }
 }

 Intent intent=new Intent(this,MainActivity.class);
 startActivity(intent);

}

 public void cancelMessage(View view)
 {
    Intent intent = new Intent(this,MainActivity.class);
    startActivity(intent);
 }

}
```

Data Controller

The Data Controller class connects to SQLite and performs SQL operations:
   The main SQLite libraries used are:

```java
import android.content.ContentValues;
import android.content.Context;
import android.database.Cursor;
```

```
import android.database.sqlite.SQLiteDatabase;
import android.database.sqlite.SQLiteException;
import android.database.sqlite.SQLiteOpenHelper;
```

SQLlite Tables and Database

The goal of the app is to store user entered text in the database. The DATABASE_NAME and TABLE_ NAME is defined as follows:

public static final String *MESSAGE*="**Message**";
public static final String *TABLE_NAME*="**Msg_Table**";
public static final String *DATABASE_NAME*="**Assignment2.db**";
public static final int *DATABASE_VERSION*=4;
public static final String *TABLE_CREATE*="**create table Msg_Table (Message text not null);**";

SQL Commands

The Data Controller class connects to SQLite and performs SQL operations (see Table 15).

Table 15: Database Helper Code

```
DataBaseHelper dbHelper;
 Context context;
 SQLiteDatabase db;

 public DataController(Context context)
 {
 this.context=context;
 dbHelper=new DataBaseHelper(context);
 }

 public DataController open()
 {
 db=dbHelper.getWritableDatabase();
 return this;
 }

 public void close()
 {
 dbHelper.close();
 }

 public long insert(String message)
 {
 ContentValues content=new ContentValues();
 content.put(MESSAGE, message);
 return db.insertOrThrow(TABLE_NAME, null, content);
 }
 public Cursor retrieve()
 {
 return db.query(TABLE_NAME, new String[]{MESSAGE}, null, null, null, null, null);
 }

 private static class DataBaseHelper extends SQLiteOpenHelper
 {
```

Table 15 contd. ...

...Table 15 contd.

```
public DataBaseHelper(Context context)
{
super(context, DATABASE_NAME, null, DATABASE_VERSION);
}

@Override
public void onCreate(SQLiteDatabase db) {
// TODO Auto-generated method stub
try
{
db.execSQL(TABLE_CREATE);
}
catch(SQLiteException e)
{
e.printStackTrace();
}
}

@Override
public void onUpgrade(SQLiteDatabase db, int oldVersion, int newVersion) {
// TODO Auto-generated method stub
db.execSQL("DROP TABLE IF EXISTS Msg_Table");
onCreate(db);
}
 }
```

iOS Data Storage

An iPhone or iPad user can install multiple applications on a single device. The iOS platform is responsible for ensuring that these applications do not interfere with each other, both in terms of memory usage and data storage. This is achieved by using application Sandbox[13] behavior. iOS platform (see Figure 31) the files under application folders either TMP or DOCUMENTS directory.

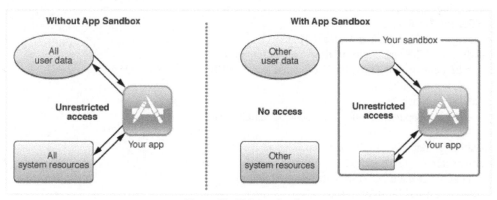

Figure 31: iOS App Sandbox

---

[13] Apple Sandbox - https://developer.apple.com/library/archive/documentation/Security/Conceptual/AppSandboxDesign Guide/AboutAppSandbox/AboutAppSandbox.html

File Storage in iOS

The Foundation Framework provides three classes that are indispensable when it comes to working with files and directories:

- *NSFileManager*—The NSFileManager class can be used to perform basic file and directory operations, such as creating, moving, reading and writing files and reading and setting file attributes. In addition, this class provides methods for, amongst other tasks, identifying the current working directory, changing to a new directory, creating directories and listing the contents of a directory.
- *NSFileHandle*—The NSFileHandle class is provided for performing lower level operations on files, such as seeking to a specific position in a file and reading and writing a file's contents by a specified number of byte chunks and appending data to an existing file.
- *NSData*—The NSData class provides a useful storage buffer into which the contents of a file may be read, or from which dynamically stored data may be written to a file.

File Store App

The File Store App stores the user entered text into a file on the iOS system.

To write data to the file system, first, we need to get a reference to the file system.

```
filemgr = [NSFileManager defaultManager];
```

The default Manager gives the reference to the file manager. As iOS Sandbox principles dictate, one must get a reference to the Document directory.

```
// Build the path to the data file
dataFile = [docsDir stringByAppendingPathComponent:
@"datafile.dat"];
```

The Data File above points to App path in the document directory and points to datafile.dat

Next, when the user clicks on the save button, having entered the text, call document file to save it.

```
NSFileManager *filemgr;
NSData *databuffer;
NSString *dataFile;
NSString *docsDir;
NSArray *dirPaths;
filemgr = [NSFileManager defaultManager];
dirPaths = NSSearchPathForDirectoriesInDomains(
    NSDocumentDirectory, NSUserDomainMask, YES);
docsDir = dirPaths[0];
dataFile = [docsDir
    stringByAppendingPathComponent: @"datafile.dat"];

NSLog(@"dataFile %@", dataFile)
databuffer = [_textBox.text
    dataUsingEncoding: NSASCIIStringEncoding];
[filemgr createFileAtPath: dataFile
    contents: databuffer attributes:nil];
```

The contents are saved in the file by first extracting text entered in the text box into NSData object. NSData object functions as a buffer.

Next, call createFileAtPath and sending the data buffer and full qualified path.

The complete code (see Table 16):

Table 16: iOS File Example Code

```
//
// FileExampleViewController.m
// FileExample
//
// Created by Chandrasekar Vuppalapati – 11/24/2018
// Copyright (c) 2018 Hanumayamma. All rights reserved.
//
#import "FileExampleViewController.h"
@interface FileExampleViewController ()
@end
@implementation FileExampleViewController
- (void)viewDidLoad
{
    [super viewDidLoad];
        NSFileManager *filemgr;
        NSString *dataFile;
        NSString *docsDir;
        NSArray *dirPaths;

        filemgr = [NSFileManager defaultManager];

        // Identify the documents directory
        dirPaths = NSSearchPathForDirectoriesInDomains(
            NSDocumentDirectory, NSUserDomainMask, YES);
         docsDir = dirPaths[0];

      NSLog(@"\ndocsDir %@\n", docsDir);


        // Build the path to the data file
        dataFile = [docsDir stringByAppendingPathComponent:
           @"datafile.dat"];
      NSLog(@"\ndataFile %@\n", dataFile);

        // Check if the file already exists
        if ([filemgr fileExistsAtPath: dataFile])
        {
            // Read file contents and display in textBox
            NSData *databuffer;
            databuffer = [filemgr contentsAtPath: dataFile];
            NSString *datastring = [[NSString alloc]
              initWithData: databuffer
              encoding:NSASCIIStringEncoding];
              _textBox.text = datastring;
        }
 }
- (void)didReceiveMemoryWarning
{
    [super didReceiveMemoryWarning];
    // Dispose of any resources that can be recreated.
}
```

Table 16 contd. ...

...Table 16 contd.

```
- (IBAction)saveText:(id)sender {
    NSFileManager *filemgr;
    NSData *databuffer;
    NSString *dataFile;
    NSString *docsDir;
    NSArray *dirPaths;
    filemgr = [NSFileManager defaultManager];
    dirPaths = NSSearchPathForDirectoriesInDomains(
        NSDocumentDirectory, NSUserDomainMask, YES);
    docsDir = dirPaths[0];
    dataFile = [docsDir
        stringByAppendingPathComponent: @"datafile.dat"];

  NSLog(@"dataFile %@", dataFile);

    databuffer = [_textBox.text
        dataUsingEncoding: NSASCIIStringEncoding];
    [filemgr createFileAtPath: dataFile
        contents: databuffer attributes:nil];
}
@end
```

The Path is located under (see Figure 32):
The file location is as shown above (see Figure 33).

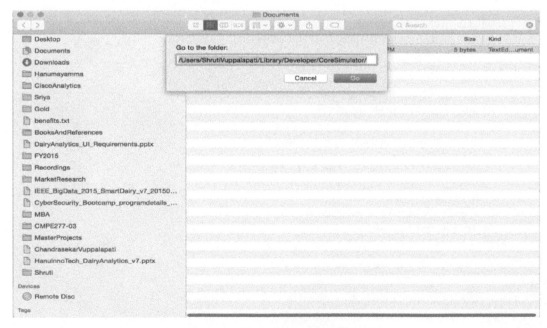

Figure 32: File Location on the Device

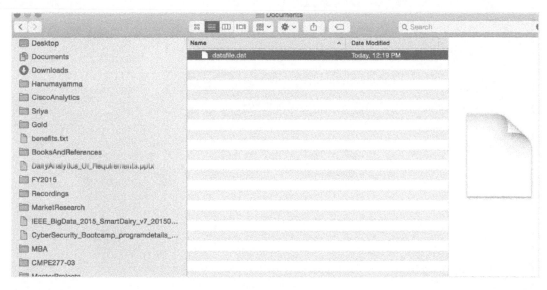

Figure 33: File on Device

iOS Simulator path structure:

```
/Users/<user name>/Library/Application Support
/iPhone Simulator/<sdk version>/Applications/<app id>/Documents
```

iOS Physical Device path:

```
/var/mobile/Applications/<app id>/Documents
```

SQLite Storage App in iOS

 | MacOS X is shipped with SQLite pre-installed, including an interactive environment for issuing SQL commands from within a Terminal window.

iOS App needs to add reference to libsqlite3.0dylib in order to enable SQL commands in iOS App (see Figure 34).

The Database app has simple screen: enter User Name, address, and Phone click save should save the values into SQL Database.

Tensors as Storage

The Database View Controller Header File (please see Table 17)

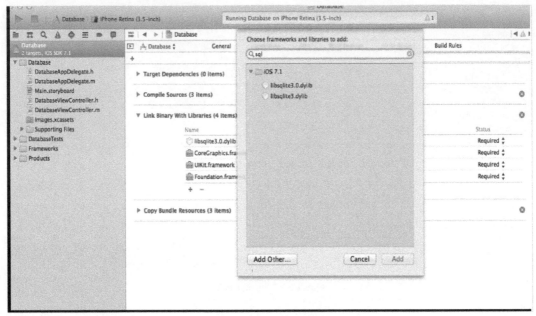

Figure 34: SQLite Settings

Table 17: Database View Controller Header File

```
//
// DatabaseViewController.h
// Database
//
// Created by Chandrasekar Vuppalapati
// Copyright (c) 2018 Hanumayamma. All rights reserved.
//
#import <UIKit/UIKit.h>
#import <sqlite3.h>
@interface DatabaseViewController : UIViewController
@property (strong, nonatomic) NSString *databasePath;
@property (nonatomic) sqlite3 *contactDB;
@property (weak, nonatomic) IBOutlet UITextField *name;
@property (weak, nonatomic) IBOutlet UITextField *address;
@property (weak, nonatomic) IBOutlet UITextField *phone;
@property (weak, nonatomic) IBOutlet UILabel *status;
- (IBAction)saveData:(id)sender;
- (IBAction)findContact:(id)sender;
@end
```

The Database View Controller header file creates Interface Builder Property and action outlets.

@property (weak, nonatomic) IBOutlet UITextField *name;

The values from the Screens are stored to the database by calling save method.
The code checks

(a) Database file is open by calling sqlite3_open command.

(b) If open is successful, the code inserts data into DB, by calling SQL INSERT

```
sqlite3_stmt *statement;
    const char *dbpath = [_databasePath UTF8String];
    if (sqlite3_open(dbpath, &_contactDB) == SQLITE_OK)
    {
        NSString *insertSQL = [NSString stringWithFormat:
        @"INSERT INTO CONTACTS (name, address, phone) VALUES (\"%@\", \"%@\", \"%@\")",
        _name.text, _address.text, _phone.text];
        const char *insert_stmt = [insertSQL UTF8String];
        sqlite3_prepare_v2(_contactDB, insert_stmt,
        -1, &statement, NULL);
        if (sqlite3_step(statement) == SQLITE_DONE)
        {
            _status.text = @"Contact added";
            _name.text = @"";
            _address.text = @"";
            _phone.text = @"";
        } else {
        _status.text = @"Failed to add contact";
        }
        sqlite3_finalize(statement);
        sqlite3_close(_contactDB);
    }
```

The full code: Please see Table 18:

Table 18: Database Full Code

```
//
// DatabaseViewController.m
// Database
//
// Created by Neil Smyth on 9/18/13.
// Copyright (c) 2013 Neil Smyth. All rights reserved.
//
#import "DatabaseViewController.h"
@interface DatabaseViewController ()
@end
@implementation DatabaseViewController
- (void)viewDidLoad
{
    [super viewDidLoad];
    NSString *docsDir;
    NSArray *dirPaths;
    // Get the documents directory
    dirPaths = NSSearchPathForDirectoriesInDomains(
    NSDocumentDirectory, NSUserDomainMask, YES);
    docsDir = dirPaths[0];
    // Build the path to the database file
    _databasePath = [[NSString alloc]
    initWithString: [docsDir stringByAppendingPathComponent:
    @"contacts.db"]];
```

Table 18 contd. ...

...Table 18 contd.

```objc
  NSLog(@"\nDBPath %@\n", docsDir);


  NSFileManager *filemgr = [NSFileManager defaultManager];
  if ([filemgr fileExistsAtPath: _databasePath ] == NO)
  {
   const char *dbpath = [_databasePath UTF8String];
   if (sqlite3_open(dbpath, &_contactDB) == SQLITE_OK)
   {
     char *errMsg;
     const char *sql_stmt =
     "CREATE TABLE IF NOT EXISTS CONTACTS (ID INTEGER PRIMARY KEY AUTOINCRE-
MENT, NAME TEXT, ADDRESS TEXT, PHONE TEXT)";
     if (sqlite3_exec(_contactDB, sql_stmt, NULL, NULL, &errMsg) != SQLITE_OK)
     {
        _status.text = @"Failed to create table";
     }
     sqlite3_close(_contactDB);
   } else {
        _status.text = @"Failed to open/create database";
   }
  }
}
- (void)didReceiveMemoryWarning
{
  [super didReceiveMemoryWarning];
  // Dispose of any resources that can be recreated.
}
- (IBAction)saveData:(id)sender {
  sqlite3_stmt *statement;
  const char *dbpath = [_databasePath UTF8String];
  if (sqlite3_open(dbpath, &_contactDB) == SQLITE_OK)
  {
    NSString *insertSQL = [NSString stringWithFormat:
    @"INSERT INTO CONTACTS (name, address, phone) VALUES (\"%@\", \"%@\", \"%@\")",
    _name.text, _address.text, _phone.text];
    const char *insert_stmt = [insertSQL UTF8String];
    sqlite3_prepare_v2(_contactDB, insert_stmt,
      -1, &statement, NULL);
    if (sqlite3_step(statement) == SQLITE_DONE)
    {
      _status.text = @"Contact added";
      _name.text = @"";
      _address.text = @"";
      _phone.text = @"";
    } else {
      _status.text = @"Failed to add contact";
    }
    sqlite3_finalize(statement);
    sqlite3_close(_contactDB);
  }
}
- (IBAction)findContact:(id)sender {
  const char *dbpath = [_databasePath UTF8String];
```

Table 18 contd. ...

...Table 18 contd.

```
sqlite3_stmt *statement;
if (sqlite3_open(dbpath, &_contactDB) == SQLITE_OK)
{
  NSString *querySQL = [NSString stringWithFormat:
    @"SELECT address, phone FROM contacts WHERE name=\"%@\"",
    _name.text];
   const char *query_stmt = [querySQL UTF8String];
   if (sqlite3_prepare_v2(_contactDB,
     query_stmt, -1, &statement, NULL) == SQLITE_OK)
   {
     if (sqlite3_step(statement) == SQLITE_ROW)
     {
       NSString *addressField = [[NSString alloc]
       initWithUTF8String:
       (const char *) sqlite3_column_text(
        statement, 0)];
      _address.text = addressField;
     NSString *phoneField = [[NSString alloc]
       initWithUTF8String:(const char *)
       sqlite3_column_text(statement, 1)];
      _phone.text = phoneField;
      _status.text = @"Match found";
   } else {
       _status.text = @"Match not found";
      _address.text = @"";
      _phone.text = @"";
   }
   sqlite3_finalize(statement);
 }
 sqlite3_close(_contactDB);
 }
}
@end
```

The SQL App enables the User to insert and query SQL data (see figure SQL Query).

Chapter Summary

After reading the chapter:

- Should able to define two, three dimensional arrays in C
- Should able to create data storages in android and ios platforms
- Should be able to create embedded database in SQLite
- Should be able to create in memory arrays

References

1. Jason Brownlee, A Gentle Introduction to N-Dimensional Arrays in Python with NumPy, January 31, 2018, https://machinelearningmastery.com/gentle-introduction-n-dimensional-arrays-python-numpy/, Access date: June 2018
2. Travis E. Oliphant, Guide to NumPy: 2nd Edition 2nd Edition, ISBN-13: 978-1517300074, http://web.mit.edu/dvp/Public/numpybook.pdf
3. Marshall Brain, The Basic of C Programming, 2016, https://computer.howstuffworks.com/c39.htm , June 2018
4. Atmel Corporation, 8-bit AVR Microcontroller, Nov 2016, http://arduinoinfo.mywikis.net/w/images/3/35/Atmel-42735-8-bit-AVR-Microcontroller-ATmega328-328P_Datasheet.pdf, Access date: June 2018

5. Google Android, Context, March 2016, https://developer.android.com/reference/android/content/Context#get SharedPreferences, Access date: June 2018
6. Google Android, Context, March 2016, SQLite is Serverless, https://www.sqlite.org/hirely.html, Access date: June 2018
7. Apple Corporation, App Sandbox, 2016-09-13, https://developer.apple.com/library/archive/documentation/Security/Conceptual/AppSandboxDesignGuide/AboutAppSandbox/AboutAppSandbox.html, Access date: June 2018

CHAPTER 7

Machine Learning at the Edge

This Chapter Covers
- Machine Learning at the Edge
- Supervised Learning that includes Decision Trees
- Naïve-Bayes Theorem
- Clustering Regulation
- Artificial Neural Networks

IEEE defined embedded systems are the devices that are used to Control, Monitor or Assist the Operation of Equipment Machinery or Plants. As we can see, development of embedded systems has seen a significant increase over the past few years, their application changing our lives enormously. From the traditional embedded systems to the advanced systems that we use today, one severe issue we are facing is how to effectively and efficiently deal with the massive amounts of data collected from sensors and inter-device communication.

To process this data, we need machine learning techniques that can scale at the embedded or constrained IoT devices. Machine Learning is a trending technology which uses data to draw meaningful insights, like prediction, classification and clustering. Machine Learning is widely used in most sectors these days for various problem-solving mechanisms. Embedded Systems typically include microcontrollers, CPU cycles, memory constraints and cost and consumption of power. One trending topic is the utilization of machine learning techniques on data coming from sensors in a microcontroller environment and analyzing it.

Advancement in Sensors and reduction in cost in micro controllers has led to emerging technologies where companies use sensors attached to embedded systems like MCU board and plummeted sensor based artificial intelligence and autonomous based vehicles. Machine learning uses sophisticated tools to optimize signal issues and can aid in speeding up the process for embeddable code. It drives developers to work on the code functionality rather than the statistics and mathematical logical behind it. This is a big pain point for most developers in the real world.

The most important advantage of ML is that it eludes traditional models by detecting fault or anomaly in a system. This makes ML efficient and best usage of data to overcome variation. Machine learning models can learn independently with raw data and directly detect anomaly without any human intervention. This plays an important role in embedded systems.

Machine learning (see Figure 1) denotes the application of artificial intelligence (AI) to provide systems with the ability to automatically learn and improve from experience without being explicitly programmed to do so. Machine Learning started from the computer, but the emerging trend suggests that machine learning has fostered a new generation of applications and devices viz. mobile phones and embedded systems that can sense, perceive, learn from, and respond to their users and environment, transforming businesses like

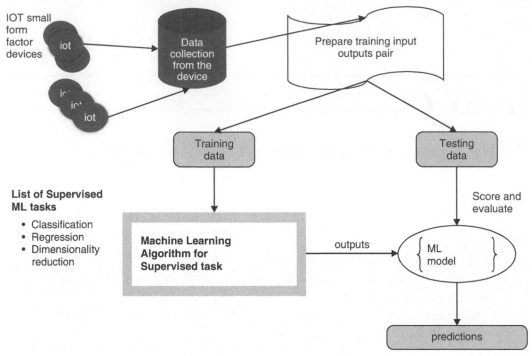

Figure 1: Machine Learning at the Edge

never before. The modern mobile devices show a high productive capacity level that is enough to perform appropriate tasks to the same degree as traditional computers do. Among the machine learning techniques, Supervised techniques are particularly important for designing mobile apps, as they require guided examples to understand human needs and activities. Some key machine learning applications for mobile devices include Sensor based activity recognition, Mobile text categorization, Malware detection on mobile devices, Language understanding, etc. Mobile apps are designed for convenient, quick and easy access by humans. The guided learning processes of humans can be used to incorporate their reasoning in the apps.

The most common ML model techniques used in mobile and farm factors are listed below–

1. K-Nearest Neighbor
2. Case Based Reasoning
3. Artificial Neural Network
4. Support vector Machine
5. Classification and Regression Trees
6. Random Forest
7. Reinforcement Learning
8. Fuzzy Logic

Case Study: Intelligent Dispenser for Long Tail Venues

In today's competitive business environment, creating memorable experiences and emotional connections with consumers is critical in winning customer spending and long term brand loyalty. Brands want their customers to be in pleasing, subliminally scented environments. Even a few micro particles of scent can do a lot of marketing's heavy lifting, from improving consumer perceptions of quality to increasing the number of visits. Hence, high roller venues, such as Trump Towers and Caesars Palace of the World, use digitally connected intelligent scent disperser systems that deliver seamless olfactory experiences in order to connect with consumers on a deeper, emotional and personal level.

The challenges, however, for venues with limited digitally connected infrastructure and deficient intelligent systems, are a lack of engagement with patrons at a personal and emotional level, this leading to missed recurring business opportunities and poor long-term brand loyalty.

The Long Tail:[1] Issue at stake

The internet has changed the dynamics of the venue and hospitality marketplace (see Figure 2). Since the development of the internet, all the attributes of the hospitality & venue industry have deeply and continuously evolved: Usage of connected networks, usage of consumer digital data & mining of consumer preferences, exclusive recommendations gleaned from Web 2.0 & 3.0 data sources, Internet Of Things (IoT), wearable data sources and new business models reshape the industry. This process has even been sped up recently by the emergence of disruptive innovations, from the evolutions of web & mobile technologies to the Artificial Intelligence (AI) platforms. The long tail [1] is used in this paper (Case Study), as a multitude of a small, independent venue operators with limited digital connected infrastructure and deficient intelligent systems can grasp and synthesize this complexity.

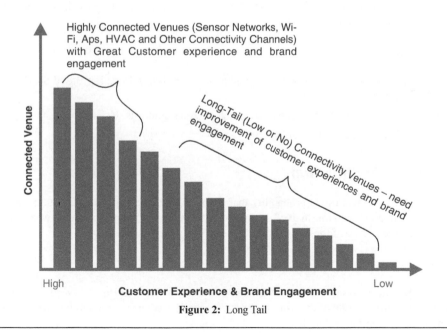

Figure 2: Long Tail

---

[1]   Long Tail in Tourism Industry - https://halshs.archives-ouvertes.fr/halshs-01248302/document

Supervised Learning Techniques useful for small form Factor Devices

K-Nearest Neighbor

This method performs categorization of a new input instance based on the classification of pre-recorded instances which have the most representatives within the nearest neighbors of the point, using a simple majority vote of the nearest neighbors of each point. It is a popular algorithm for pattern recognition.

Applications

User Activity Recognition

Analyze raw data from mobile device accelerometer sensor, apply KNN to help classify user activity. Application finds out which template is closer to the current activity vector by measuring distance of each point of our target pattern with all points of each template. The development of the proposed activity recognition classifier proceeds with a supervised learning algorithm. Supervised techniques rely on labeled data. An application is developed to execute on the phone in order to collect the data. The application allows the user to set the sampling rate, start and stop the data collection, and label the activity being performed with an elegant graphical user interface. Data are collected every 10 milliseconds. The user is granted the freedom to set the orientation of the mobile phone at will. The user can hold the phone in hand, or put it in a pocket or in a handbag. User activities include staying, walking, running, ascending stairs, and descending stairs.

Travel Guide

Using location dependent systems in Android Mobile Phones to provide to provide tourism information to the mobile users on the move using GPS.

Other App Types

- Monitoring applications over a wide geographical region. Using Location Based Service to retrieve the information on the nearest K nodes.
- Mobile dating apps—identify friends that match personal interests. Use KNN to match patterns.
- Recommendation systems, face detection.

Naive-Bayes Classification

This method infers the conditional probability that a new instance belongs to a particular class based on applying Bayes' theorem with the "naive" assumption of conditional independence between every pair of features for the given value of the class variable.

Application

- Text categorization—Naive Bayes classification model works with Bag-of-Word feature extraction which avoids the position of the word in the document. It predicts the probability that a given feature set belongs to a particular label.
- Malware Detection—Using android application file which contain dataset and attributes, get permission to access the attribute in the application, xml file contains actual object, transfer text file into integer value. The malware detection is based on probability rule, which predicts the value and permission access for different applications. It contains a data set in terms of matrix which produce output in the form of integer value, which shows the result of detected malware. It is found that malware can be present in SMS, MMS, Bluetooth, Videos, or social networking, etc.

- Spam filtering
- Sentiment Analysis
- Fraud detection

Support Vector Machines

This method aims to create an optimal hyperplane in a multidimensional space that can maximize the margin between the closest data points and the boundary in order to accurately divide a population of data points into two groups.

Applications

Automatic task classification—can be implemented for mobile devices by using the support vector machine algorithm and crowdsourcing.

- The training data set obtained via crowdsourcing using the Amazon Mechanical Turk platform.

Personal assistant services—A user's voice command is first transmitted to a server, which is running the speech-to-text converter and the task classifier. The received speech at the server is converted to text, and the converted text is given to the task classifier, a core part of the overall system. The task classifier classifies the given text into the most probable task, and the predicted task is transmitted back to the user's mobile device. Then, the task launcher of the personal assistant system performs the predicted task accordingly on behalf of the user.

AdaBoost

Adaptive Boosting is an ensemble algorithm used to train a boosted classifier by training learners on the filtered output of other learners. For instance, it can boost the performance of decision trees on binary classification problems.

Applications

- For boosting the performance of a Mobile phone crowdsourcing parking app. It combines multiple trust prediction approaches with a real-time data fusion method in order to create the final map.

Artificial neural network (ANN)

ANN are crude models which are based on the human brain's network model.

Applications

- For the identification of ADL (Activities of Daily living)—The mobile application captures the data in 5 seconds slots every 5 minutes, where the frequency of the data acquisition is around 10 ms. For the definition of the experiments, 25 individuals aged between 16 and 60 years old were selected. The dataset is composed by Standard Deviation of the raw signal, Average of the raw signal, Variance of the of the raw signal, and Median of the raw signal, extracted from the accelerometer and the magnetometer sensors
- Computer vision, speech recognition, machine translation, social network filtering and medical diagnosis.

Gaussian Process

A Gaussian process in a machine learning algorithm uses lazy learning. It measures the similarity between points in order to predict the value for an unseen point from training data.

Applications

- Build a context aware application—When several users are in a room and each of them carries a mobile device, we can find the matching relationship between the users and the mobile devices by advanced inertial sensors of the phone.
- Prediction of spatiotemporal fields by the use of mobile sensor networks.
- Nonparametric Gaussian process regression (or kriging in geo-statistics) has been widely used as a nonlinear regression technique in order to estimate and predict geostatistical data.
- Disease tracking
- Weather forecasting

Decision Tree

Tree data structure is formed by recursively splitting points from a set of input values, based on a cost function, to predict an output value. The information gain is calculated at every node, and the node with the lowest entropy is chosen as the split node.

Applications

- To get the real-life impression of the most common user behavior and infer presence status—Mobile Data Challenge (MDC) data set was collected from around 200 individuals over more than a year. The logs contain information related to GPS, WiFi, Bluetooth and accelerometer traces, as well as call and SMS logs, multimedia and application usage. All the features are turned to nominal. GSM cell identifier is also used as a nominal feature after determining all the possible values during the preprocessing. For leaf nodes, only the nodes with class labels were used. Classification error is used as fitness function.

Linear Regression

Tree data structure is formed by recursively splitting points from a set of input values, based on a cost function, in order to predict an output value.

Clustering

Clustering is a data mining algorithm used to divide data into groups or segments based on some similarity criteria. Real time embedded Systems include devices which monitor and observe the physical world with sensors and provide real time control or reaction. There is a wide range of such systems, from a stand-alone small devices to global networked real time embedded systems. Some examples are Temperature Sensors, Traffic control sensors, electric power grid controls, IoT devices, etc. All these devices generate streaming time series data or multivariate time series data. Clustering of the time series data may include the similar behavior of attributes in a given time window, similar time series pattern, etc. However, Clustering with such a large amount of data (time series data is generated over time and is huge in comparison to static data) is challenging, since multivariate time series data cannot be scanned multiple times, unlike regular static data.

Clustering Techniques

In the following, we will discuss some clustering techniques for time series data as per below:

1. Partitioning methods (k-means, k-medoid)
2. Hierarchical methods (AGNES, DIANA)
3. Incremental Clustering (COBWEB, BIRCH)
4. Online Divisive Agglomerative Clustering
5. Density based methods (DBSCAN)
6. Grid based methods (STING)
7. Model based methods (Expectation Maximization)

Partitioning methods—Also called Portioning methods. Partitioning methods tend to minimize the cost function while maintaining the optimality. The Dataset is divided into k-partitions, such that the object distance among the cluster is minimum, whereas inter-cluster distance is maximized. Major approaches in partitioning methods are: k-means and k-medoid.

Hierarchical methods—There are three features that give hierarchical clustering the advantage over Partitioning methods—(1) No need to pre-define the number of clusters k. (2) No assumptions on data distribution, (3) Representation in pair-wise dissimilarity matrix. In hierarchical methods, split can be done based on K-means technique at each hierarchy.

IANA (Divisive ANAlysis) is a hierarchical clustering technique that uses a top-down approach—from top (large cluster) to bottom (several clusters). The top cluster contains all the data. At each hierarchy, the cluster with the largest diameter is selected and divided into new clusters until all the clusters have only one piece of data.

Incremental Clustering—To overcome the limitation of time, memory and sample size for multivariate time series data, over which multiple passes are not achievable, an incremental clustering system was developed for compact representation, quickly and incrementally processing new data. Several methods introduced for this purpose:

- COBWEB—conceptual, incremental, hierarchical algorithm for hill-climbing search on classifiers.
- BIRCH—builds CF-tree by agglomerative procedure which is a hierarchical structure.
- Single Pass K-means—buffer to keep all points in compressed form.
- CURE: each group is represented by a constant number of points, arbitrary shape of groups can be found out.
- STREAM—minimizes the sum-of-squares quality measure.
- CluStream—online method extends CF vectors of BIRCH, offline method performs global clustering.
- One problem with these models is how to determine the minimum number of observations that are necessary to achieve a converged model.

Online Divisive Agglomerative Clustering (ODAC)—This method of clustering is a time series whole clustering system which constructs a tree-shaped hierarchical structure using top-down strategy. Leaves of the tree are the final clusters with a set of variables at each leaf, the Union of which gives the complete set of variables and intersection gives an empty set. Cluster diameters are monitored at each level in order to minimize the intra-cluster dissimilarity after each split. For each cluster, the system finds the diameter of the cluster based on two variables, if the given condition is met on this cluster, it is split into two new clusters for each chosen variable called pivot variable for that cluster. All remaining variables on the old cluster are assigned a new cluster with closest pivot. A test is performed on this model to see whether the diameters of children leaves approached parent's diameter, if yes, then the previous structure needs to be re-aggregated. This agglomerative phase which starts new statistics on selected node is used to detect concept drift.

Density based methods (DBSCAN)—Density based methods find data groups or clusters based on the density instead of centers unlike the hierarchical or partitioning methods. DBSCAN is an example of this method.

Grid based methods—An object is divided into n number of cells and then clustering is performed on them.

STING (Statistical Information Grid) is an example of this method. It decomposes spatial area into rectangular cells.

Model based methods—model based methods try to fit the data into a predefined mathematical model. Expectation maximization is a n example of this method.

These days there are many IOT devices prompting data scientists to use clustering algorithms on real time data because of the massive amounts of data an IOT device can provide. One popular unsupervised clustering partitioning algorithm that is K means clustering. This clustering algorithm partitions a data set into k number of clusters by comparing how similar a data point is to the centroid value for each cluster; however, the drawbacks to this clustering method are that it cannot be applied to nominal attributes/ categorical values and it cannot handle outliers. The k is a value that is determined by the developer, based on his or her understanding of the data. A common example where k-means is used with an IOT device is fitness trackers. The data collected from these devices can be used to determine which workouts are similar and put them into the same cluster. STREAM and CluStream are extensions to the K-means algorithms that are widely used today for partitioning clustering algorithms.

Hierarchical clustering algorithms are also commonly used with real time data. The main goal behind hierarchical clustering is to form tree clusters by decomposing the data. It creates a binary-tree, also known as a dendrogram, and can be built both ways, i.e., bottom up or top down. This tree, once created, can be modified by a user by changing the number of levels, without having to rerun the clustering algorithm. One drawback, however, is that once a split in the algorithm is made, it cannot be reverted.

Balanced Iterative Reducing and Clustering using Hierarchies (BIRCH) is a clustering algorithm for real time data that is being proposed to improve the hierarchical clustering. It clusters multidimensional data dynamically and incrementally. This algorithm was designed with the thought that not every data point is important, so not everything should be stored in memory, therefore, BIRCH proposes better usage of the system's available resources, such as memory.

G-Stream is a clustering algorithm based on the GNG algorithm. Growing Neural Gas algorithm is where a graph is created from data and then clustered to find similar structure. It is an unsupervised clustering algorithm mostly used with topological map. G-stream is an industry used GNG algorithm and is often used for the visualization purpose of the clustering algorithm.

D-Stream clustering is a commonly used Grid based clustering algorithm. Unlike K-means/partitioning clustering grid-based clustering, for example, D-Stream can handle outliers well. D-stream is a grid-based stream method, which means the clustering is density based. There are two parts to this algorithm; first, the input data is mapped on a grid and second, the density is calculated for the grids and then clustered based on the density. D-Stream clustering doesn't require the user to specify the number of clusters, instead the algorithm dynamically determines the number of clusters.

After reading the chapter, you should be able to

• Understand and apply Machine Learning Models at the Edge
• Apply clustering

References

1. Christian Longhi, Sylvie Rochhia, Long tails in the tourism industry: towards knowledge intensive service suppliers, 2015, https://halshs.archives-ouvertes.fr/halshs-01248302/document, Access Date: July 2018

CHAPTER **8**

Edge Analytics

This Chapter Covers
- Machine Learning at the Edge
- Design of Decision Trees, Implementation, and Deployment into embedded device
- Kalman Filter Design, Implementation, and Deployment into an IoT device
- K means clustering Design, Implementation
- Sliding Window Design Implementation
- Clustering Design and Implementation
- Neural Network for Sound Classification
- Design and Implementation of Fuzzy Logic and Re-enforcement Learning

In just ten years, the Internet of Things (IoT) has gone from science fiction, to reality, to exceeding our wildest expectations. By 2020 there will be nearly 26 billion connected devices,[1] generating more bits of data than there are stars in the known universe.[2] Companies need a way to process and analyze this data faster and smarter across the network in order to achieve distinct business value [1].

The upside for businesses that can harness more of their data is undeniable. For example, McKinsey estimates the potential economic impact of IoT to be as much as $11.1T per year by 2025 [2]. Factories are estimated to feel an estimated impact of up to $3.7T, while worksites in sectors such as mining, oil and gas and construction could see an impact on the scale of $930B[3] [3].

However, the massive amount of data being generated is both the promise of IoT and its most critical pain point. Not only do businesses need solutions that can help them tap more of their IoT data, but they also need a way to separate high-value data from the noise. In light of these IoT challenges, organizations increasingly recognize the advantages of leaving certain data right where it is captured—at the edge—and analyzing it there. The result of this is the ability to take instant action and affect immediate control over the connected things.[4]

> Wikibon IoT Project [A] estimates that data processing costs are reduced by 30% when the edge is 200 miles away from the network and by 60% when the edge is 100 miles away.[4]

---

[1] Gartner Says the Internet of Things Installed Base Will Grow to 26 Billion Units By 2020 http://www.gartner.com/newsroom/id/2636073

[2] Data Growth, Business Opportunities, and the IT Imperatives - https://www.emc.com/leadership/digital-universe/2014iview/executive-summary.htm

[3] Unlocking the potential of the internet of things - http://www.mckinsey.com/business-functions/digital-mckinsey/our-insights/the-internet-of-things-the-value-of-digitizing-the-physical-world

[4] The Vital Role of Edge Computing in the Internet of Things - https://wikibon.com/the-vital-role-of-edge-computing-in-the-internet-of-things/

Edge Analytics help organizations harness the power of the data they generate remotely by delivering analytics and machine learning at the point of data collection. In this chapter, some of the Edge level analytics are introduced: Supervised, Unsupervised, and reinforcement models.

Case Study: Smart City—Intelligent Dispenser

Intelligent Dispenser (iDispenser)[5] works by applying precision of decision, learned through analyzing & applying data science, with real-time sensing of sensors (see Figure 1), adaptively builds data models through historical and Venue analytics, and improves the overall performance through learning. The iDispenser communicates through the Internet of Things (IoT) stack. It uses Bluetooth Low Energy (BLE) protocol and Wi-Fi communication modules to perform intelligent dispensing [4].

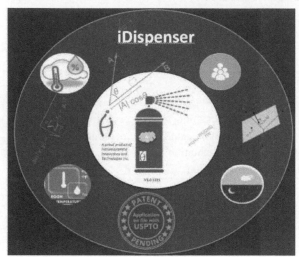

Figure 1: iDispenser

Sensor Modules

The custom hardware, built on top of ARM processor, collects environmental and location specific influential attributes (as predicted by our Data science model) and evaluates the dispensing both at IoT Edge and Cloud compute level. The sensor module captures real-time venue temperature and humidity characteristics (see Figure 2). Additionally, it calculates the number of people in the venue. Finally, it correlates venue characteristics with location geo weather.

Microcontroller

SJ One Board - **LPC1758 ARM Cortex M3**, 512K ROM, 64K RAM. Requires power supply of 5V.

Operating System:

Real-time operating system - **FreeRTOS**. The RTOS kernel requires 5 to 10K ROM. The scheduler consumes 236 bytes and the task / thread consumes 64 bytes.

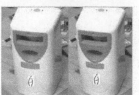

Mechanical Enclosure & Battery:
Product Dimensions: Custom Made – samples include:

- 6.1 x 5.5 x 7.4 inches; 2.6 pounds
- 7.5 x 6.5 x 14 inches; 4.1 pounds

Waterproof & flexible rotator design
Number of USB 2.0 Ports: 1
1.5V AAA Alkaline Batteries: 2
Voltage: 240 volts

Figure 2: iDispenser Technical Specification

5　Hanumayamma iDispenser - http://hanuinnotech.com/images/VenueAnalytics_ProductBrochure_v7_20160801_Final.pdf

Machine Learning & Algorithms

Algorithms at the edge level, based on business requirements, can include supervised, unsupervised, and reinforcement models. The application of an algorithm is based on the business outcome or the pain-point we're solving for the customer.

Decision Trees

A decision tree algorithm is used to classify the attributes and select the outcome of the class attribute. To construct a decision tree, both class attribute and item attributes are required. A decision tree is a tree like structure wherein the intermediate nodes represent attributes of the data, the leaf nodes represent the outcome of the data and the branches hold the attribute value. Decision trees are widely used in the classification process because no domain knowledge is needed to construct the decision tree. Following Figure 3 shows simple decision trees.

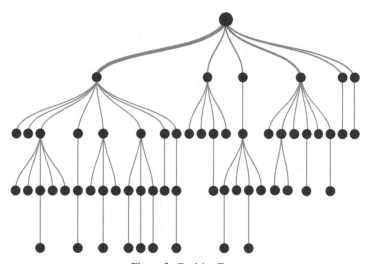

Figure 3: Decision Tree

The primary step in the decision tree algorithm is to identify the root node for the given set of data. Multiple methods exist to decide the root node of the decision tree. Information gain and Gini impurity are the primary methods used to identify the root node. The root node plays an important role in deciding which side of decision tree the data falls into. Like every classification method, decision trees are also constructed using the training data and tested with the test data.

Information Gain: Information gain is used to identify the root node and the branch nodes in the decision tree. Information gain is calculated using entropy and information. The purpose of the information gain calculation is to identify the node that has the least randomness or impurity. Entropy is calculated using the following formula

$$Info\ (D) = -\sum_{i=1}^{m} pi\ \log_2\ (pi) \tag{1}$$

Information of the attribute is calculated using the following formula:

$$Info_A\ (D) = \sum_{j=1}^{v} \frac{|Dj|}{|D|}\ x\ info\ (Dj) \tag{2}$$

The information gain of an attribute is the difference between entropy and the information of that attribute. The attribute with the highest information gain is the root node, and the next level nodes are identified using the next highest information gain attributes.

Edge Level Temperature Collector and Adaptive Decision Tree Coefficients

The combination of other supervised learning algorithms, such as decision trees and Neural networks, make for a very adaptive edge. For instance, we have collected the following temperature trend parameters and applied a fine-tuned Edge coefficient in real-time that accurately or proactively considers not only number of people but also the environmental predictive factors.

We ran a supervised data collection and collected one week's worth of the following data:

Outlook	Temp (°F)	Humidity (%)	Windy	Adaptive Coefficient
Sunny	75	70	True	Low Value
Sunny	80	90	True	High Value
Sunny	85	85	False	High Value
Sunny	72	95	False	High Value
Sunny	69	70	False	Low Value
overcast	72	90	true	Low Value
overcast	83	78	False	Low Value
overcast	64	65	True	Low Value
overcast	81	75	False	Low Value
Rain	71	80	True	High Value
Rain	65	70	True	High Value
Rain	75	80	False	Low Value
Rain	68	80	False	Low Value
Rain	70	96	False	Low Value

Model development (on Paper)

For the given dataset, the class label "Adaptive Coefficient" has two distinct outcomes (High Value, Low Value). Therefore, there are two distinct classes (m = 2)

C1 = Low Value   C2 = High Value

9 tuples of class **Low Value**

5 tuples of class **High Value**

Information Gain (D) = $-\Sigma$ (Probability of 'Low Value * Log$_2$ (Probability of 'Low Value) + Probability of 'High Value * Log$_2$ (Probability of High Value))

Information Gain (D) = $-\Sigma$ ((5/14) * Log2 (5/14) + (9/14) * Log2 (9/14))

Information gain (D) = 0.9402 bits

Expected information required for each attribute:

1. For Outlook attribute

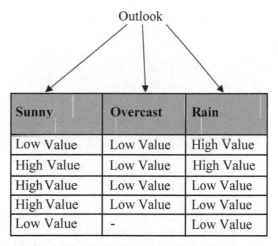

Sunny	Overcast	Rain
Low Value	Low Value	High Value
High Value	Low Value	High Value
High Value	Low Value	Low Value
High Value	Low Value	Low Value
Low Value	-	Low Value

2. For Temperature attribute

For continuous valued attributes, there will be many possible split points.

- Evaluate information gain for every possible split point for the attribute
- Choose the best split point
- Information gain for the best split point becomes information gain for that attribute

Sort the values in the increasing order

64	65	68	69	70	71	72	72	75	75	80	81	83	85
Low	High	Low	Low	Low	High	High	Low	Low	Low	High	Low	Low	High

Consider a split at the value 71.5

Example: Temperature < 71.5 & Temperature >= 71.5

Temperature

<71.5	>=71.5
Low Value	High Value
High Value	Low Value
Low Value	Low Value
Low Value	Low Value
Low Value	High Value
High Value	Low Value
-	Low Value
	High Value

Information Gain (Temperature):

Info Temperature(D) = 8/14 *(−(5/8 * Log2(5/8)) − (3/8 * Log2(3/8))) + 6/14 * (−(4/6 * Log2(4/6)) − (2/6 * Log2(2/6)))

= 0.9388 bits

Gain (A) = 0.9402 – 0.9388

Gain (A) = **0.0014 bits**

1. For attribute Humidity
Consider split at value 79.5

Humidity

<79.5	>=79.5
Low Value	High Value
Low Value	High Value
Low Value	High Value
Low Value	Low Value
Low Value	High Value
High Value	Low Value
	Low Value
	Low Value

Information Gain (Humidity):

Info Humidity(D) = 10/14 *(–(6/10 * Log2(6/10)) – (4/10 * Log2(4/10))) + 4/14 * (–(3/4 * Log2(3/4)) – (1/4 * Log2(1/4)))
= 0.8963

Gain (A) = 0.9402 – 0.8963

Gain (A) = **0.0439 bits**

For attribute **wind**

Wind

True	False
Low	High
Low	High
High	Low
High	Low
Low	Low
Low	Play
-	Low

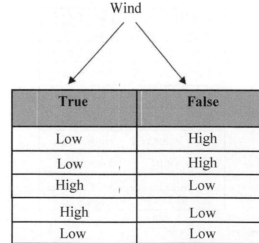

Information Gain (Windy)

Info Wind(D) = 8/14 *(–(6/8 * Log2(6/8)) – (2/8 * Log2(2/8))) + 6/14 * (–(3/6 * Log2(3/6)) – (3/6 * Log2(3/6)))

= 0.8921

Gain (A) = 0.9402 – 0.8921

Gain (A) = **0.0479 bits**

Therefore, Outlook is selected as root node based on Gain values
Once the root node is known, the next step is to select which attributes become internal nodes (see Figure 4).

Info(D) = $- {}^25 \log 2({}^25) - {}^35 \log 2({}^35) = 0.971$ bits

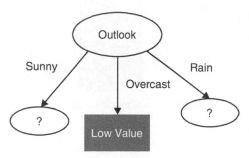

Figure 4: Outlook node

Once the root node is known, the step is to select with attributes become internal nodes (see Figure 5).

Info (D) = $- {}^25 \log 2({}^25) - {}^35 \log 2({}^35) = 0.971$ bits

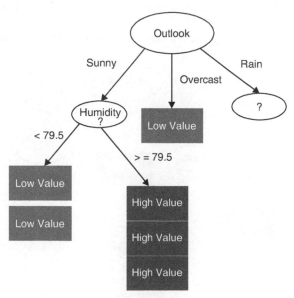

Figure 5: Humidity Node

Infowindy(D) = 25 x (−12 log2(12) + −12 log2(12)) + 35 x (−32 log2(23) + −31 log2(13)) 0.4 + 0.551 0.951 bits (see Figure 6)

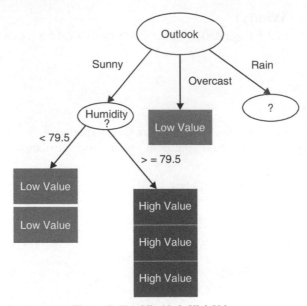

Figure 6: Humidity Node High Value

Temperature

InfoTemperature(D) = 0.6490 bits (based on Information Gain calculation Eq. 2)

Gain(Humidity) = Info(D) – InfoHumidity(D) = 0.971 – 0 = 0.971 bits (see Figure 6)

Gain(Windy) = Info(D) – Infowindy(D) = 0.971 – 0.951 = 0.02 bits

Gain(Temperature) = Info(D) – InfoTemperature(D) = 0.971 – 0.6490 = 0.322 bits

Therefore, Humidity is selected as internal node (see Figure 7 and see Figure 8).

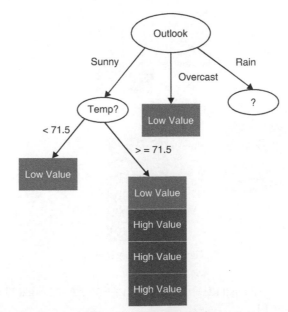

Figure 7: Temperature Node

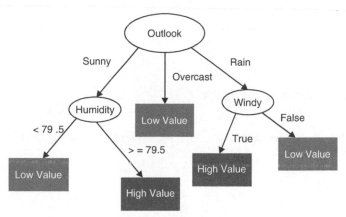

Figure 8: Decision Tree

Model Development (Python)

For the given dataset, the following python code generates the same decision tree (please see Table 1 and Figure 9).

Table 1: Decision Tree Python Code

```
# -*- coding: utf-8 -*-
"""
Created on Sat Dec 8 18 : 33 : 27 2018
@author: cvuppalapati
"""
from sklearn.tree import DecisionTreeClassifier, export_graphviz
from subprocess import call
import numpy as np
"""
Outlook Temp(oF)      Humidity(%)    Windy    Adaptive Coefficient
Sunny   75      70    True   Low Value
Sunny   80      90    True   High Value
Sunny   85      85    False  High Value
Sunny   72      95    False  High Value
Sunny   69      70    False  Low Value
overcast        72    90     true   Low Value
overcast        83    78     False  Low Value
overcast        64    65     True   Low Value
overcast        81    75     False  Low Value
Rain    71      80    True   High Value
Rain    65      70    True   High Value
Rain    75      80    False  Low Value
Rain    68      80    False  Low Value
Rain    70      96    False  Low Value
Codification
Outlook Attributes
Tuples :
Sunny 1
Overcast 2
Rain 3
```

Table 1 contd. ...

...Table 1 contd.

```
Windy Attribute
Tuples
True 1
False 0
Class Variable - adaptive coeffcent
Low Value 0
High Value 1
OutlookTemp(oF)        Humidity(%)    Windy   Adaptive Coefficient
1       75      70      1                      0
1       80      90      1              1
1       85      85      0              1
1       72      95      0              1
1       69      70      0              0
2       72      90      1              0
2       83      78      0              0
2       64      65      1              0
2       81      75      0              0
3       71      80      1              1
3       65      70      1                      1
3       75      80      0                      0
3       68      80      0                      0
3       70      96      0                      0
"""""
X = np.array([[1, 75, 70, 1],
        [1, 80, 90, 1],
        [1, 85, 85, 0],
        [1, 72, 95, 0],
        [1, 69, 70, 0],
        [2, 72, 90, 1],
        [2, 83, 78, 0],
        [2, 64, 65, 1],
        [2, 81, 75, 0],
        [3, 71, 80, 1],
        [3, 65, 70, 1],
        [3, 75, 80, 0],
        [3, 68, 80, 0],
        [3, 70, 96, 0]])
        y = ['Low Value', 'High Value', 'High Value', 'High Value', 'Low
Value', 'Low Value', 'Low Value', 'Low Value', 'Low Value', 'High Value',
'High Value', 'Low Value', 'Low Value', 'Low Value']
        dtree = DecisionTreeClassifier()
        dtree.fit(X, y)
        dot_data = export_graphviz(dtree,
                out_file = 'edgedecisiontree.dot',
                feature_names = ['Outlook', 'Temperature', 'Humidity','Windy'],
                class_names = ['Low Value', 'High Value'],
                filled = True,
                rounded = True)
        call(['dot', '-Tpng', 'edgedecisiontree.dot', '-o', 'edgedecisiontree.
png'])
```

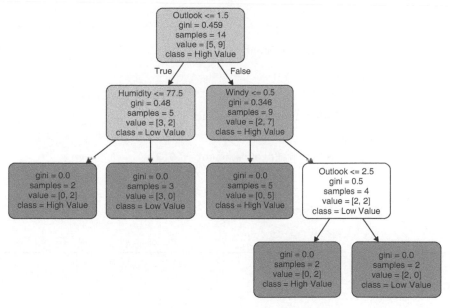

Figure 9: Decision Tree (python model)

Decision Rule

The above code is converted into decision rules within the edge processor. The values for the Humidity, Windy, and Temperature are used to dynamically compute the adaptive edge coefficient (see Figure 9 and Table 2).

Table 2: Edge Processor Decision Rules

```
/*
Codification
Outlook Attributes
Tuples:
Sunny 1
Overcast 2
Rain 3
Windy Attribute
Tuples
True 1
False 0

*/
double computeAdaptiveEdgeCoefficent(int Outlook, bool Windy)
{
        If(Outlook == 1)
        {
                if (Humidity < 79.5)
                {
                        adaptiveEdgeCoefficent = 0.3
                }
                else {
                        adaptiveEdgeCoefficent = 0.8
                }
        } Else if (Outlook == 2)
```

Table 2 contd. ...

...Table 1 contd.

```
    {
            adaptiveEdgeCoefficent = 0.3
    }
    else if (outlook == 3) {
            if (Windy == True)
            {
                    adaptiveEdgeCoefficent = 0.8
            }
            else {
                    adaptiveEdgeCoefficent = 0.3
            }
    }
    else {
            // Do nothing
    }
}
```

Sliding Window

Sliding-window (see Figure 10) analysis is a common way to analyze and identify trends in the stream data. The term "sliding-window analysis" refers to a common pattern analysis of real-time and continuous data and uses rolling counts of incoming data to examine trending, temperature, and humidity.

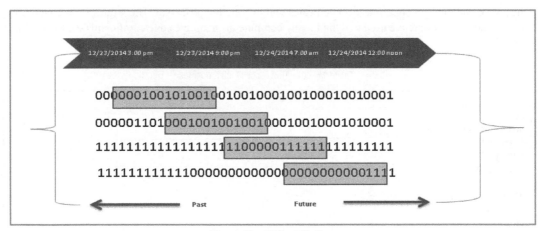

Figure 10: Sliding Window

Sliding window Algorithm (see Figure 11) is implemented to monitor the changes in temperature. The sensor data is retrieved after every n second. So, the sliding window algorithm's input data is also getting generated after every n second with the fixed window size, say k. The window starts from position 1st and keeps shifting right by one element as shown in Figure 11. The below Figure 11 shows t0, t1, t2 as the window with (k = 5) temperatures after each after 1st, 2nd and 3rd shift.

The main objective of this algorithm is to monitor the k temperatures. If it's found to be increasing consecutively in all k readings, then it will generate an alert and notification, to take necessary actions (see Figure 11).

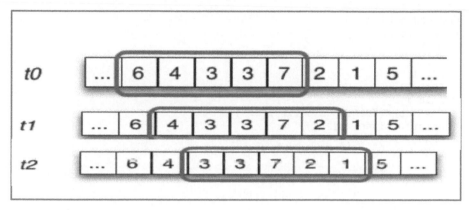

Figure 11: Sliding Window with time intervals

Sliding Window—C Code

The following C Code generates Sliding window and transfers the data by Window Bucket by Bucket.

Sample Data: SensorData.txt

> Test message hello world. This is a sample sliding window in C.

For Window Size: 25, the parsing is performed as follows:

For Window Size greater than length of the text in the file, a single Window transfer is sufficient (see Figure 12).

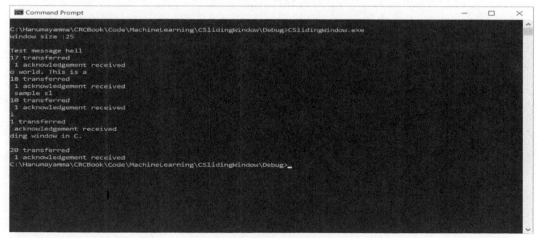

Figure 12: Sliding Window CUI

```
#include<stdio.h>
#include<conio.h>
void main()
{
        int i, m, n, j, w, l;
        char c;
        FILE*f;
        f = fopen("sensordata.txt", "r");
        printf("window size");
        scanf("%d", &n);
        m = n;
        if (f == NULL)
        {
                printf("File is not open");
                return 0;
        }
        while (!feof(f))
        {

                i = rand() % n + 1;
                j = i;
                l = i;

                if (m>i)
                {
                        m = m - i;
                        if (m>0)
                        {
                                printf("\n");
                                while (i>0 & !feof(f))
                                {
                                        c = getc(f);
                                        printf("%c", c);
                                        i--;
                                }
                                printf("\n%d transferred", j);
                                if (j>3)
                                        printf("\n 1 acknowledgement received");
                                else
                                        printf("\n acknowledgement received", j +
1);
                        }
                }
                m = m + j - 1;
        }
}
```

Sliding Window in Python

Smoothing is a technique applied to time series in order to remove the fine-grained variation between time steps. The goal of smoothing is to remove noise and better expose the signal of the underlying causal processes. Moving averages are a simple and common type of smoothing used in time series analysis and time series forecasting. Calculating a moving average involves creating a new series where the values are comprised of the average of raw observations in the original time series. A moving average requires that you specify a window size called the window width. This defines the number of raw observations used to calculate the moving average value. The "moving" part in the moving average refers to the fact that the window defined by the window width is slid along the time series in order to calculate the average values in the new series.

There are two main types of moving average that are used: Centered and Trailing Moving Average. The moving average value can also be used directly to make predictions. It is a naive model and assumes that the trend and seasonality components of the time series have already been removed or adjusted for. The moving average model for predictions can easily be used in a walk-forward manner. As new observations are made available (e.g., daily), the model can be updated, and a prediction made for the next day. We can implement this manually in Python (see Table 3). Below is an example of the moving average model used in a walk-forward manner.

Table 3: Sliding window in Python

```
# -*- coding: utf-8 -*-
""""
Created on Sat Dec 8 19:33:31 2018
@author: cvuppalapati
""""
from pandas import Series
from numpy import mean
from sklearn.metrics import mean_squared_error
from matplotlib import pyplot
series = Series.from_csv('daily-total-female-births.csv', header=0)
# prepare situation
X = series.values
window = 3
history = [X[i] for i in range(window)]
test = [X[i] for i in range(window, len(X))]
predictions = list()
# walk forward over time steps in test
for t in range(len(test)):
        length = len(history)
        yhat = mean([history[i] for i in range(length-window,length)])
        obs = test[t]
        predictions.append(yhat)
        history.append(obs)
        print('predicted=%f, expected=%f' % (yhat, obs))
error = mean_squared_error(test, predictions)
print('Test MSE: %.3f' % error)
# plot
pyplot.plot(test)
pyplot.plot(predictions, color='red')
pyplot.show()
# zoom plot
pyplot.plot(test[0:100])
pyplot.plot(predictions[0:100], color='red')
pyplot.show()
```

The Data (see Figure 13) for the sliding window is from

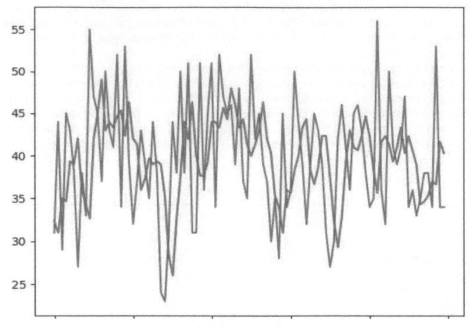

Figure 13: Data

Predicted: See the console output of predicted values (see Figure 14)

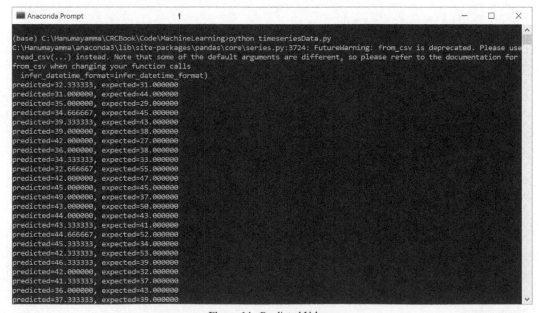

Figure 14: Predicted Values

Model Equation—Regression Analysis

In statistics, linear regression is an approach for modeling the relationship between a scalar dependent variable y and one or more explanatory variables denoted X. The case of one explanatory variable is called simple linear regression. When the outcome, or class, is numeric, and all the attributes are numeric, linear regression is a natural technique to consider. This is a staple method in statistics. The idea is to express the class as a linear combination of the attributes, with predetermined weights:

$$x = w_0 + w_1a_1 + w_2a_2 + \ldots + w_ka_k$$

where x is the class; a1, a2,…, ak are the attribute values; and w0, w1,…, wk are weights

The weights are calculated from the training data. Here, the notation gets a little heavy, because we need a way of expressing the attribute values for each training instance. The first instance will have a class, say x(1), and attribute values, $a_1^{(1)}$, $a_2^{(1)}$,…, $a_k^{(1)}$, where the superscript denotes that it is the first example. Moreover, it is notationally convenient to assume an extra attribute a0, whose value is always 1.

Crowdedness to Temperature Modeling (Edge State Model)

We have developed two edge analytics models: (1) Crowdedness to temperature and (2) Temperature to Humidity correlation. Both models have used case level applicability to the Smart City edge analytics. Finally, we have developed the Kalman filter and deployed the machine learning models in test hardware.

Dataset

We have downloaded Kaggle dataset (see figure Kaggle Dataset) to get crowdedness to temperature machine learning model [10].

Dataset: https://www.kaggle.com/nsrose7224/crowdedness-at-the-campus-gym

The dataset consists of 26,000 people counts (about every 10 minutes) over the last year. In addition, extra information was gathered, including weather and semester-specific information that might affect how crowded it is. The label is the number of people, which is predicted, given some subset of the features.

Label:

1. Number of people

Features:

- date (string; datetime of data)
- timestamp (int; number of seconds since beginning of day)
- day_of_week (int; 0 [monday] - 6 [sunday])
- is_weekend (int; 0 or 1) [boolean, if 1, it's either saturday or sunday, otherwise 0]
- is_holiday (int; 0 or 1) [boolean, if 1 it's a federal holiday, 0 otherwise]
- temperature (float; degrees fahrenheit)
- is_start_of_semester (int; 0 or 1) [boolean, if 1 it's the beginning of a school semester, 0 otherwise]
- month (int; 1 [jan] - 12 [dec])
- hour (int; 0 - 23) (see Figure 15)

Model Development

To develop a model equation, we have used the Excel regression data model packet (see Figure 16).

Our goal of the model is to predict the influencing factors that affect the room temperature. The influencing parameters include: Number of people, day of the week, holiday and other dataset parameters (see Figure 17).

number_people	date	timestamp	day_of_week	is_weekend	is_holiday	temperature	is_start_of_semester	is_during_semester	month	hour
37	2015-08-14 17:00:11-07:00	61211	4	0	0	71.76	0	0	8	17
45	2015-08-14 17:20:14-07:00	62414	4	0	0	71.76	0	0	8	17
40	2015-08-14 17:30:15-07:00	63015	4	0	0	71.76	0	0	8	17
44	2015-08-14 17:40:16-07:00	63616	4	0	0	71.76	0	0	8	17
45	2015-08-14 17:50:17-07:00	64217	4	0	0	71.76	0	0	8	17
46	2015-08-14 18:00:18-07:00	64818	4	0	0	72.15	0	0	8	18
43	2015-08-14 18:20:08-07:00	66008	4	0	0	72.15	0	0	8	18
53	2015-08-14 18:30:09-07:00	66609	4	0	0	72.15	0	0	8	18
54	2015-08-14 18:40:14-07:00	67214	4	0	0	72.15	0	0	8	18
43	2015-08-14 18:50:15-07:00	67815	4	0	0	72.15	0	0	8	18
39	2015-08-14 19:00:16-07:00	68416	4	0	0	69.97	0	0	8	19
38	2015-08-14 19:20:07-07:00	69607	4	0	0	69.97	0	0	8	19
45	2015-08-14 19:30:08-07:00	70208	4	0	0	69.97	0	0	8	19
41	2015-08-14 19:40:14-07:00	70814	4	0	0	69.97	0	0	8	19
36	2015-08-14 19:50:16-07:00	71416	4	0	0	69.97	0	0	8	19
42	2015-08-14 20:00:17-07:00	72017	4	0	0	68.8	0	0	8	20
35	2015-08-14 20:21:14-07:00	73274	4	0	0	68.8	0	0	8	20
36	2015-08-14 20:31:14-07:00	73874	4	0	0	68.8	0	0	8	20
48	2015-08-14 20:41:15-07:00	74475	4	0	0	68.8	0	0	8	20
40	2015-08-14 20:51:17-07:00	75077	4	0	0	68.8	0	0	8	20
49	2015-08-14 21:01:18-07:00	75678	4	0	0	68.04	0	0	8	21
37	2015-08-14 21:20:06-07:00	76806	4	0	0	68.04	0	0	8	21
48	2015-08-14 21:30:07-07:00	77407	4	0	0	68.04	0	0	8	21
45	2015-08-14 21:40:08-07:00	78008	4	0	0	68.04	0	0	8	21
45	2015-08-14 21:50:09-07:00	78609	4	0	0	68.04	0	0	8	21
38	2015-08-14 22:00:10-07:00	79210	4	0	0	67.55	0	0	8	22
41	2015-08-14 22:20:07-07:00	80407	4	0	0	67.55	0	0	8	22
41	2015-08-14 22:30:08-07:00	81008	4	0	0	67.55	0	0	8	22
37	2015-08-14 22:40:09-07:00	81609	4	0	0	67.55	0	0	8	22

Figure 15: RKaggle Excel Data [10]

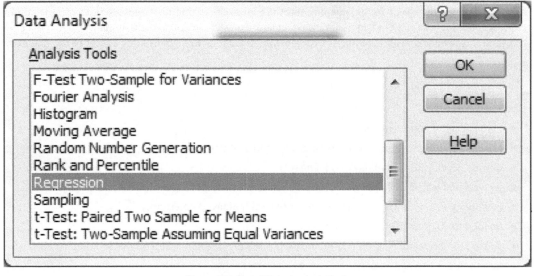

Figure 16: Excel Data Analysis Pack

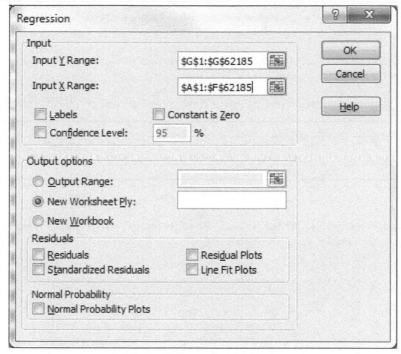

Figure 17: Regression Parameters

Class Variable: Temperature

Model generation: Running data analysis pack with regression parameters creates mode (see Figure 18)

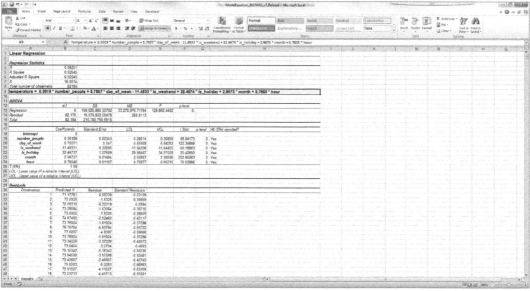

Figure 18: Model Output

Model Validation

F-Test

The F-test for linear regression tests whether any of the independent variables in a multiple linear regression model are significant (see Figure 19).

ANOVA					
	d.f.	SS	MS	F	p-level
Regression	6.	199,625,860.30702	33,270,976.71784	128,652.4462	0.
Residual	62,178.	16,079,933.58478	258.6113		
Total	62,184.	215,705,793.8918			

Figure 19: ANOVA

Since, F value of regression (128652.652) falls in the reject region (> F-Critical), the Null Hypothesis is rejected. That is, the model is valid (see Figure 20).

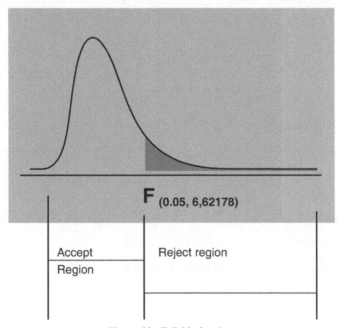

Figure 20: F-Critical region

T-Test

The T-Tests are used to conduct hypothesis tests on the regression coefficients obtained in simple linear regression. The goal is to find out the most influencing parameters in the regression model equation (see Figure 21).

T Critical = TINV (0.05, 62178) = 1.960002138

All the above values of the fall within the reject region of the t critical. Hence, all the above variables have the explanatory power in explaining the temperature (see Figure 22).

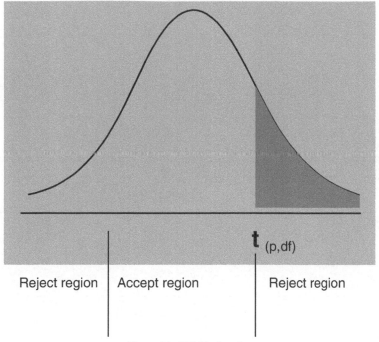

Reject region | **Accept region** | **Reject region**

Figure 21: T-Critical region

	Coefficients	Standard Error	LCL	UCL	t Stat	p-level	H0 (5%) rejected?
Intercept	0						
number_people	0.30186	0.00343	0.29514	0.30858	88.04173	0.	Yes
day_of_week	5.75071	0.047	5.65858	5.84283	122.34898	0.	Yes
is_weekend	-11.49331	0.22895	-11.94206	-11.04455	-50.19903	0.	Yes
is_holiday	32.46737	1.27689	29.96467	34.97008	25.42693	0.	Yes
month	2.96727	0.01464	2.93857	2.99596	202.65263	0.	Yes
hour	0.78046	0.01107	0.75877	0.80215	70.52986	0.	Yes
T (5%)	1.96						
LCL - Lower value of a reliable interval (LCL)							
UCL - Upper value of a reliable interval (UCL)							

Figure 22: Model Attributes

Data Assumptions

Residually are normally distributed: From the residuals plot, the data points are normally distributed (see Figure 23).

Assumptions 2: Linear - Residuals are independent

To confirm residuals are independent, we need to complete Durbin-Watson test.

A test that the residuals from a linear regression or multiple regression are independent.

Since most regression problems involving time series data exhibit positive autocorrelation, the hypotheses usually considered in the Durbin-Watson test are:

$H_0 : \rho = 0$

$H_1 : \rho > 0$

Durbin-Watson Numerator = 2020.722066

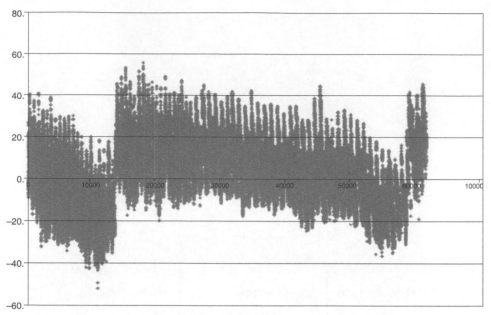

Figure 23: Data Samples

The test statistic is

$$d = \frac{\sum_{i=2}^{n}(e_i - e_{i-1})^2}{\sum_{i=1}^{n} e_i^2}$$

where $e_i = y_i - \hat{y}_i$ and y_i and $\hat{y}_i$ are, respectively, the observed and predicted values of the response variable for individual i. d becomes smaller as the serial correlations increase. Upper and lower critical values, d_U and d_L have been tabulated for different values of k (the number of explanatory variables) and n.

If $d < d_L$ reject $H_0 : \rho = 0$

If $d > d_U$ do not reject $H_0 : \rho = 0$

If $d_L < d < d_U$ test is inconclusive.

Figure 24: Statistical test

Durbin-Watson Denominator = 62183

Durbin-Watson statics Test (DW-Test) = 2020.722066/62183= 0.032496

Therefore, the test passes the DW-Test (see Figure 21).

Please note:

As a rough rule of thumb, if Durbin–Watson is less than 1.0, there may be cause for alarm. Small values of d indicate successive error terms are, on average, close in value to one another, or positively correlated. If d > 2, successive error terms are, on average, much more different in value from one another, i.e., negatively correlated. In regressions, this can imply an underestimation of the level of statistical significance.

Model Equation

From the dataset, we have derived model equation that will calculate the temperature value bases on parameters: Number people, day of the week, is_weekend, is_holiday, month and hour.

Temperature = 0.3019 * number_people + 5.7507 * day_of_week – 11.4933 * is_weekend + 32.4674 * is_holiday + 2.9673 * month + 0.7805 * hour

Model Equation & Independent Parameters Coefficients

As provided by the model equation, the independent parameter "number of people" influences to the state variable "temperature". What this means is, if a number of people suddenly entered into a room, keeping all other effect variables constant, the room temperature would increase by a factor of 0.3019.

In other words:

Change of Temperature given number of people in a room is (see Figure 25):

Temperature Change $_{Number\ of\ People}$ = 0.30198

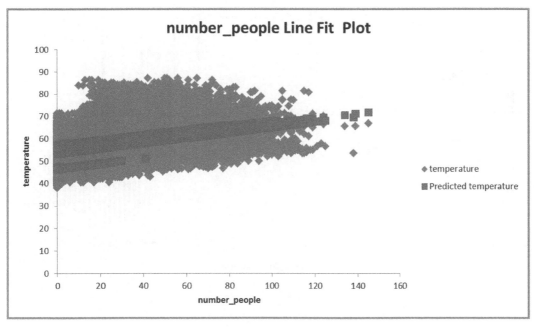

Figure 25: Predicated Values

Model Development in Python

Checking Linearity

Before you execute a linear regression model, it is advisable to validate that certain assumptions are met. As noted earlier, you may want to check that a linear relationship exists between the dependent variable and the independent variable/s. In our example, you may want to check that a linear relationship exists between:

- The Temperature (dependent variable) and the Number of People (independent variable);
- The Temperature (dependent variable) and the day_of_the_week (independent variable);
- The Temperature (dependent variable) and the is_weekend (independent variable);
- The Temperature (dependent variable) and the is_holiday (independent variable);
- The Temperature (dependent variable) and the month (independent variable);
- The Temperature (dependent variable) and the hour (independent variable);

Scatter Diagram

To perform a linearity check, a scatter plot diagram will help (see figure linearity).

As you can see Figure 26, a linear relationship exists in all cases.

The sklearn performs linear regression (see Table 4).

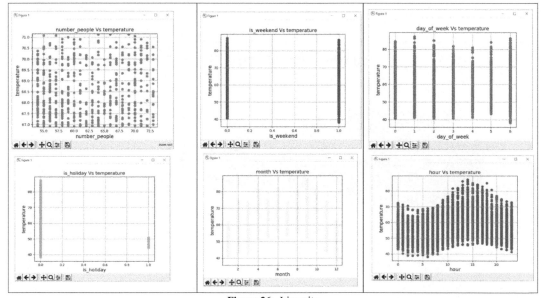

Figure 26: Linearity

Table 4: sklearn Linear Regression Python Code

```
Stock_Market = pd.read_csv(r'C:\Hanumayamma\CRCBook\Code\MachineLearning\Crowdedness_To_
Temperature_20170403.csv')
df = DataFrame(Stock_Market,columns=['number_people','day_of_week','is_weekend','is_holiday','mon
th','hour','temperature'])
X = df[['number_people','day_of_week','is_weekend','is_holiday','month','hour']] # here we have 5 vari-
ables for multiple regression. If you just want to use one variable for simple linear regression, then use X =
df['Interest_Rate'] for example.Alternatively, you may add additional variables within the brackets
Y = df['temperature']

# with sklearn
regr = linear_model.LinearRegression()
regr.fit(X, Y)

# prediction with sklearn
New_number_people = 48
New_day_of_week = 5
New_is_weekend= 0
New_is_holiday = 0
New_month= 9
New_hour=20
print ('Predicted Temperature: \n', regr.predict([[New_number_people ,New_day_of_week, New_is_
weekend,New_is_holiday,New_month,New_hour]]))

# with statsmodels
X = sm.add_constant(X) # adding a constant

model = sm.OLS(Y, X).fit()
predictions = model.predict(X)

print_model = model.summary()
print(print_model)
```

The output contains both model and sklearn output (see Table 5 and Figure 27):

Table 5: Model Output

```
Intercept:
53.97404514884691
Coefficients:
[ 0.11713168 0.01120827 1.2402437 -6.95555267 0.17810501 -0.04197969]
Predicted Temperature:
[60.41575833]
coef std err t P>|t| [0.025 0.975]
```

const	53.9740	0.083	646.981	0.000	53.811	54.138
number_people	0.1171	0.001	92.547	0.000	0.115	0.120
day_of_week	0.0112	0.019	0.587	0.557	−0.026	0.049
is_weekend	1.2402	0.085	14.650	0.000	1.074	1.406
is_holiday	−6.9556	0.463	−15.015	0.000	−7.863	−6.048
month	0.1781	0.007	26.171	0.000	0.165	0.191
hour	−0.0420	0.004	−10.049	0.000	−0.050	−0.034

Omnibus:	1796.583	Durbin-Watson:	0.025
Prob(Omnibus):	0.000	Jarque-Bera (JB):	3150.086
Skew:	0.248	Prob(JB):	0.00
Kurtosis:	3.984	Cond. No.	786.

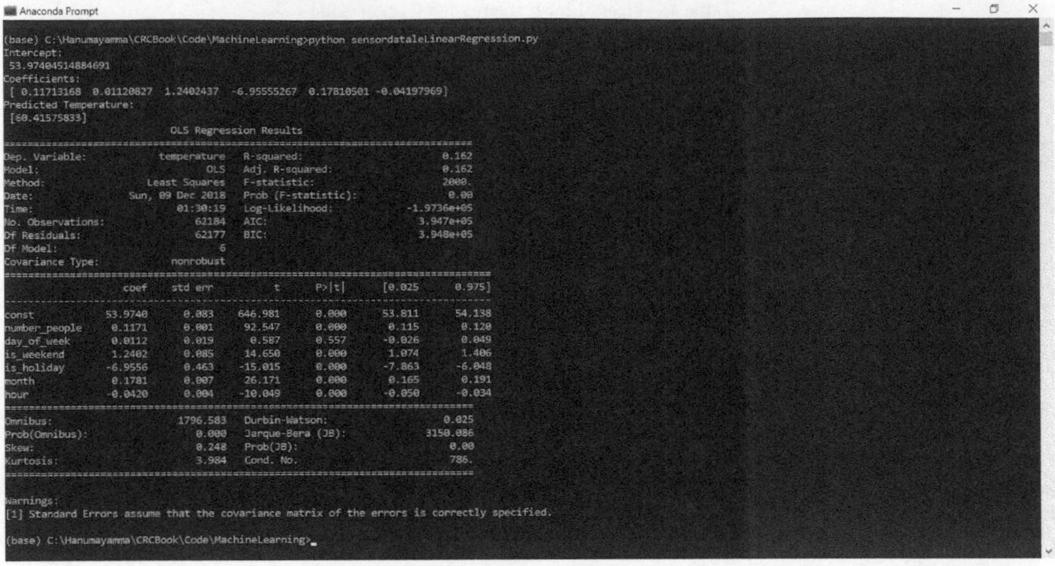

Figure 27: sklearn model output

Kalman Filter

The Kalman filter theory has been proved an efficient tool for correcting systematic forecast errors, combining observations with model forecasts. Applications of the Kalman filter theory for the correction of surface temperature forecasts are based either only on temperature information where other parameters, such as relative humidity and surface winds, are used. A complete description of the Kalman filter can be found in Kalman. For the convenience of the reader, some basic notions of the general Kalman filter theory are presented in the following paragraphs.

In the Kalman filter, the fundamental concept is the notion of the state. By this is meant, intuitively, some quantitative information (a set of numbers, a function, etc.) that is the least amount of data one has to know about the past behavior of the system in order to predict its future behavior. The dynamics are then described in terms of state transitions, i.e., one must specify how one state is transformed into another as time passes.

In order to use the Kalman filter to estimate the internal state of a process given only a sequence of noisy observations, one must model the process in accordance with the framework of the Kalman filter. This means specifying the following matrices:

- F_k, the state-transition model;
- H_k, the observation model;
- Q_k, the covariance of the process noise;
- R_k, the covariance of the observation noise; and
- Sometimes B_k, the control-input model, for each time-step, k, as described below

The Kalman filter model assumes the true state at time k is evolved from the state at $(k-1)$ according to

$$\mathbf{x}_k = \mathbf{F}_k \mathbf{x}_{k-1} + \mathbf{B}_k \mathbf{u}_k + \mathbf{w}_k$$

where

- F_k is the state transition model which is applied to the previous state $\mathbf{x}_{k-1}$;
- B_k is the control-input model which is applied to the control vector $\mathbf{u}_k$;
- $\mathbf{w}_k$ is the process noise which is assumed to be drawn from a zero mean

At time k an observation (or measurement) $\mathbf{z}_k$ of the true state $\mathbf{x}_k$ is made according to

$$\mathbf{z}_k = \mathbf{H}_k \mathbf{x}_k + \mathbf{v}_k$$

where

- $\mathbf{H}_k$ is the observation model which maps the true state space into the observed space and
- $\mathbf{v}_k$ is the observation noise which is assumed to be zero mean Gaussian

The goal of the Kalman filter is to generate next predicted values and compare the predicted value to Sensor computed value. The deviated values are notified (see Figure 28).

Time – k –1 → Historical value

Time = k → Current Value

Time = k+1 → Future value

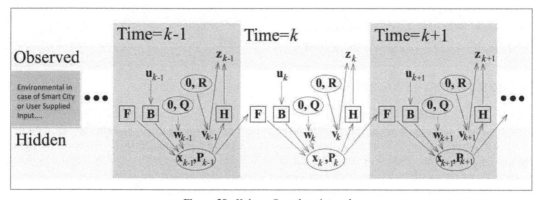

Figure 28: Kalman Over three intervals

Kalman Filter Block Diagram Representation

The Kalman filter forms an optimal state estimate for a state-space plant model in the presence of process and measurement noise. The optimality criterion for the steady-state Kalman filter is the minimization of the steady-state error covariance matrix P shown

$$\mathbf{P} = \lim_{t \to \infty} E \left(\{\hat{x} - x\} \{\hat{x} - x\}^T \right)$$

where E represents the expectation (or mean) of the parenthesized expression, $\hat{x}$ is the state estimate, and x is the true state vector. The term $\{\hat{x} - x\}$ is the state estimation error. The diagonal elements of **P** contain the variance (the square of the standard deviation) of the state estimation error. The off-diagonal terms represent the correlation between the errors of the state vector elements.

The integrator stands for n integrators such that the output of each is a state variable; F(t) indicates how the outputs of the integrators are fed back to the inputs of the integrators. Thus fij(t) is the coefficient with which the output of the jth integrator is fed back to the input of the ith integrator. It is not hard to relate this formalism to more conventional methods of linear system analysis (see Figure 29).

Where

- u(t) is an m-vector (m ≤ n) representing the inputs to the system

Kalman filter is a recursive data processing algorithm that generates the best measurement based on all previous measurements. It basically consists of two methods, prediction and correction. Thus, it is good fit to predict anomaly behavior in the system at the edge level. Another use of the Kalman filter is

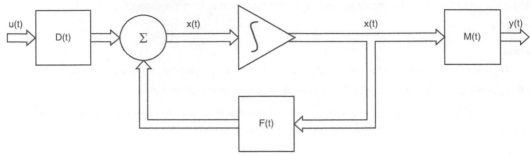

Figure 29: Kalman Model

automatic re-calibration of the failed sensor from the neighboring sensor. The reasons behind choosing the Kalman filter is as follows:

- It is light weight
- It works efficiently for real-time processing
- Removes Gaussian noise and gives good result

Kalman Filter for Smart City

Kalman filtering is useful for analyzing real-time inputs about a system that can exist in certain states. Typically, there is an underlying model of how the various components of the system interact with and affect each other. A Kalman filter processes the various inputs, attempts to identify the errors in the input, and predicts the current state. For example, a Kalman filter in our Smart Cities Ambiance Maintenance System can process various inputs, such as number of people, day of the week, month, hour and location, and update the estimate of the room temperature.

In our system, the model is:

temperature = 0.3019 * number_people + 5.7507 * day_of_week - 11.4933 * is_weekend + 32.4674 * is_holiday + 2.9673 * month + 0.7805 * hour

Please note: day_of_week, is_weekend, is_holiday and month parameters are computed by tapping into onboard embedded real-time clock. Since, these values are critical to the embedded system, we have not modeled filtering for these parameters as part of the Kalman Filter design.

The use of Kalman filters is motivated by: (a) their support for streaming analysis using only current input measurements (therefore making the solution more memory efficient), (b) they do not require matrix calculations (therefore, the solution is more computationally efficient), (c) the ease of the algorithm tuning process, and (d) their implementation simplicity.

The number of people can be retrieved either interfacing with Venue system or by adding data from strip sensor.

During the initialization process the parameters which need tuning are the process noise covariance q, the sensor measurement noise covariance r, the initial estimated error covariance p and an initial measurement x.[6]

---

[6]   Filters - http://commons.apache.org/proper/commons-math/userguide/filter.html

State Transition Matrix :: x = [0.3019 * NUMBER OF PEOPLE]

Control Input Matrix :: B = NULL;

Measurement Matrix:: H =

$$\begin{bmatrix} 0.3019 * \text{NUMBER OF PEOPLE}_1 \\ 0.3019 * \text{NUMBER OF PEOPLE}_2 \\ 0.3019 * \text{NUMBER OF PEOPLE}_3 \\ 0.3019 * \text{NUMBER OF PEOPLE}_4 \end{bmatrix}$$

Process Noise Covariance Matrix:: q = [0.25%] or [0.0025]

As per Si 7020 – A10 Sensor Data Sheet - the sensor shows an accuracy of 0.25%RH[7] [5].

Measurement Noise Covariance Matrix :: r = [0.005]

Kalman Filter implementation for Smart City (Number of people in avenue vs. Temperature)—see Table 6:

Table 6: Kalman Filter Code - Smart City

```
 -*- coding: utf-8 -*-
"""
Created on Wed Nov 28 18:37:06 2018
@author: cvuppalapati
"""
# Kalman filter example demo in Python
# A Python implementation of the example given in pages 11–15 of "An
# Introduction to the Kalman Filter" by Greg Welch and Gary Bishop,
# University of North Carolina at Chapel Hill, Department of Computer
# Science, TR 95-041,
# https://www.cs.unc.edu/~welch/media/pdf/kalman_intro.pdf

# by Andrew D. Straw
# https://scipy-cookbook.readthedocs.io/items/KalmanFiltering.html

import numpy as np
import matplotlib.pyplot as plt

plt.rcParams['figure.figsize'] = (10, 8)

# intial parameters
n_iter = 50
sz = (n_iter,) # size of array
x = 0.3019 # Temperature state Transition matrix 0.3019 * NUMBER OF PEOPLE z = np.random.
normal(x,0.1,size=sz) # observations (normal about x, sigma=0.1)

Q = 0.0025 # process variance

# allocate space for arrays
xhat=np.zeros(sz)          # a posteri estimate of x
P=np.zeros(sz)             # a posteri error estimate
xhatminus=np.zeros(sz)     # a priori estimate of x
Pminus=np.zeros(sz)        # a priori error estimate
K=np.zeros(sz)             # gain or blending factor

R = 0.1**2                 # estimate of measurement variance, change to see effect
# intial guesses
```
Table 6 contd. ...

---

[7]   Si7020 – A 10 http://www.mouser.com/ds/2/368/Si7020-272416.pdf

...Table 6 contd.

```
xhat[0] = 0.0
P[0] = 1.0
for k in range(1,n_iter):
    # time update
    xhatminus[k] = xhat[k-1]
    Pminus[k] = P[k-1]+Q

    # measurement update
    K[k] = Pminus[k]/( Pminus[k]+R )
    xhat[k] = xhatminus[k]+K[k]*(z[k]-xhatminus[k])
    P[k] = (1-K[k])*Pminus[k]
plt.figure()
plt.plot(z,'k+',label='noisy measurements')
plt.plot(xhat,'b-',label='a posteri estimate')
plt.axhline(x,color='g',label='truth value')
plt.legend()
plt.title('Estimate vs. iteration step', fontweight='bold')
plt.xlabel('Iteration')
plt.ylabel('Temperature')

plt.figure()
valid_iter = range(1,n_iter) # Pminus not valid at step 0
plt.plot(valid_iter,Pminus[valid_iter],label='a priori error estimate')
plt.title('Estimated $\it{\mathbf{a \ priori}}$ error vs. iteration step', fontweight='bold')
plt.xlabel('Iteration')
plt.ylabel('$(Temperature)^2$')
plt.setp(plt.gca(),'ylim',[0,.01])
plt.show()
```

The goal of the Kalman filter is to predict the deviations in the actual temperature vs. predicted values (see Figure 30).

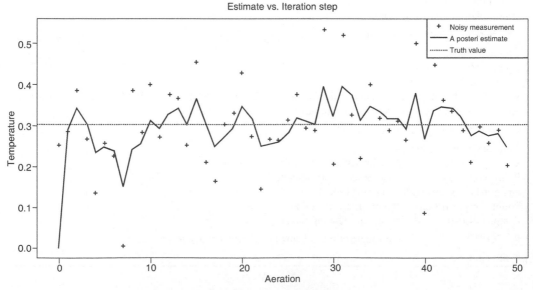

Figure 30: Predicted Values

K-Means Clustering

K-means clustering[8] is a type of unsupervised learning, which is used when you have unlabeled data (i.e., data without defined categories or groups). The goal of this algorithm is to find groups in the data, with the number of groups represented by the variable K. The algorithm works iteratively to assign each data point to one of K groups based on the features that are provided. Data points are clustered based on feature similarity. The results of the K-means clustering algorithm are [7]:

- The centroids of the K clusters, which can be used to label new data
- Labels for the training data (each data point is assigned to a single cluster)

Rather than defining groups before looking at the data, clustering allows you to find and analyze the groups that have formed organically. The "Choosing K" section below describes how the number of groups can be determined.

Each centroid of a cluster is a collection of feature values which define the resulting groups. Examining the centroid feature weights can be used to qualitatively interpret what kind of group each cluster represents.

This introduction to the K-means clustering algorithm covers:

- Common business cases where K-means is used
- The steps involved in running the algorithm

The K-means clustering algorithm is used to find groups which have not been explicitly labeled in the data. This can be used to confirm business assumptions about what types of groups exist or to identify unknown groups in complex data sets. Once the algorithm has been run and the groups are defined, any new data can be easily assigned to the correct group (see Figure 31).

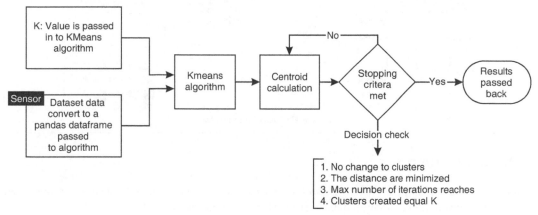

Figure 31: K-Means Sequence Flow

This is a versatile algorithm that can be used for any type of grouping. Some examples of Case Study are:

- Behavioral segmentation:
  - o Segment by purchase history
  - o Segment by activities on application, website, or platform
  - o Define personas based on interests
  - o Create profiles based on activity monitoring

---

[8]   K Means Clustering - https://www.datascience.com/blog/k-means-clustering

- Sorting sensor measurements:
  - ○ Detect activity types in motion sensors
  - ○ Group images
  - ○ Separate audio
  - ○ Identify groups in health monitoring

The algorithm works with the following steps:

1. Data assignment step
2. Centroid Calculation Step

Case Study: Sensor Signal and Data Interference & Machine Learning

In Cluster Sensor Deployments, for instance, the distance between Sensors and Signal Strengths, the edge device has to make the decision by taking the most appropriate cluster of sensors with high accuracy (see Figure 32). For example, in the below figure Dairy Cow necklaces are tied and the edge device, mobile, could get the data from all the sensors. To connect to appropriate Cow necklace, the K-means clustering is used:

Figure 32: Dairy Architecture

K Means Example

As a simple illustration of a k-means algorithm, consider the following data set consisting of the signal strength (Received Signal Strength Indicator) and Distance from seven sensors:

Sensors	RSSI	Distance
1	1.0	1.0
2	1.5	2.0
3	3.0	4.0
4	5.0	7.0
5	3.5	5.0
6	4.5	5.0
7	3.5	4.5

The goal at the Edge level is to pick the sensor value that is more accurate and weed out any signal or sensory interferes. Here, we can apply K means to separate the Sensor based on RSSI and distance parameters: Let the RSSI and distance parameters values of the two sensors furthest apart (using the Euclidean distance measure), define the initial cluster means, giving:

	Sensors	Mean Vector (Centroid)
Group 1	1	(1.0,1.0)
Group 2	4	(5.0, 7.0)

The remaining Sensors are now examined in sequence and allocated to the cluster to which they are closest, in terms of Euclidean distance to the cluster mean. The mean vector is recalculated each time a new member is added. This leads to the following series of steps:

	Cluster 1 (Closer Sensors)		Cluster 2 (Further Sensors)	
Step	Sensors	Mean Vector (Centroid)	Sensors	Mean Vector (Centroid)
1	1	(1.0, 1.0)	4	(5.0, 7.0)
2	1,2	(1.2, 1.5)	4	(5.0, 7.0)
3	1,2,3	(1.8,2.3)	4	(5.0, 7.0)
4	1,2,3	(1.8,2.3)	4,5	(4.2,6.0)
5	1,2,3	(1.8,2.3)	4,5,6	(4.3,5.7)
6	1,2,3	(1.8,2.3)	4,5,6,7	(4.1,5.4)

Now the initial partition has changed, and the two clusters at this stage having the following characteristics:

	Sensors	Mean Vector (Centroid)
Cluster 1	1,2,3	(1.8, 2.3)
Cluster 2	4,5,6,7	(4.1, 5.4)

But we cannot yet be sure that each Sensor has been assigned to the right cluster. So, we compare each Sensor distance to its own cluster mean and to that of the opposite cluster. And we find:

Distance between Sensor 1 & Mean Vector 1
$= SQRT ((1.8 - 1.0)^2 + (2.3 - 1)^2)$
Distance between Sensor 1 & Mean Vector 2
$= SQRT ((4.1 - 1.0)^2 + (5.4 - 1.0)^2)$

Sensor	Distance mean (Centroid) of Cluster 1	Distance mean (Centroid) of Cluster 2
1	1.5	5.4
2	0.4	4.3
3	2.1	1.8
4	5.7	1.8
5	3.2	0.7
6	3.8	0.6
7	2.8	1.1

Only Sensor 3 is nearer to the mean of the opposite cluster (Cluster 2) than its own (Cluster 1). In other words, each sensor's distance to its own cluster mean should be smaller than the distance to the other cluster's mean (which is not the case with Sensor 3). Thus, Sensor 3 is relocated to Cluster 2, resulting in the new partition:

	Sensor	Mean Vector (Centroid)
Cluster 1	1,2	(1.3, 1.5)
Cluster 2	3,4,5,6,7	(3.9,5.1)

Through K-Means, now, we can separate sensors (RSSI and Distance) in two clusters and take appropriate actions.

K Means Clustering – Python Code

The following code performs the K-Means clustering that we have performed in the section above:

Step 1: plot existing Sensor deployment (see Figure 33)

```
x1 = np.array([1.0, 1.5, 3.0, 5.0, 3.5, 4.5, 3.5])
x2 = np.array([1.0, 2.0, 4.0, 7.0, 5.0, 5.0, 4.5])
plt.plot()
plt.xlim([0, 10])
plt.ylim([0, 10])
plt.title('Dataset')
plt.scatter(x1, x2)
plt.show()
# create new plot and data
plt.plot()
X = np.array(list(zip(x1, x2))).reshape(len(x1), 2)
colors = ['b', 'g', 'r','y']
markers = ['o', 'v', 's','+']
```

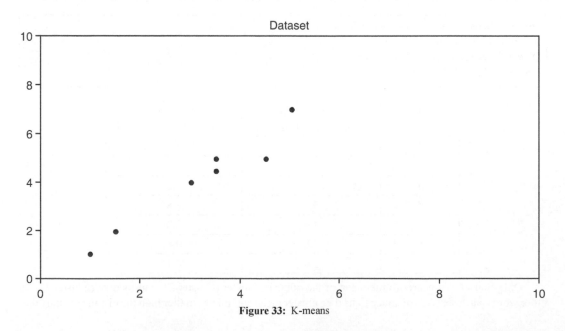

Figure 33: K-means

Step 2: perform K-means clustering

```
# KMeans algorithm
K = 4
kmeans_model = KMeans(n_clusters=K).fit(X)
plt.plot()
for i, l in enumerate(kmeans_model.labels_):
 plt.plot(x1[i], x2[i], color=colors[l], marker=markers[l],ls='None')
 plt.xlim([0, 10])
 plt.ylim([0, 10])
plt.show()
```

You can perform K with different values (k = 2,3, or 4).

Choose appropriate cluster:

K = 2

Cluster 1(Triangles) and Cluster 2(Circles) (see figure K=2 Cluster)

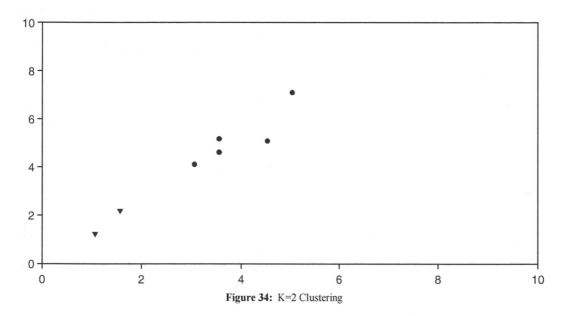

Figure 34: K=2 Clustering

K = 3
Cluster 1(Triangles), Cluster 2(Circles) and Cluster 3 (Triangles) (see K=3 Cluster)

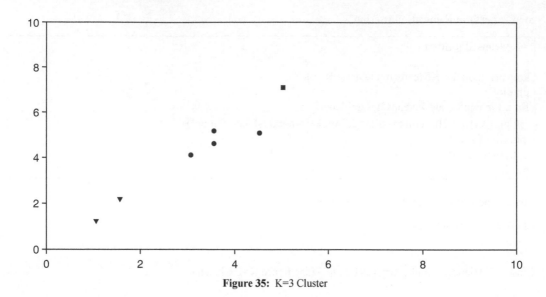

Figure 35: K=3 Cluster

K = 4

Cluster 1(Triangles), Cluster 2(Circles), Cluster 3 (Squares) and Cluster 4 (see figure K=4 cluster) (Plus Sign)

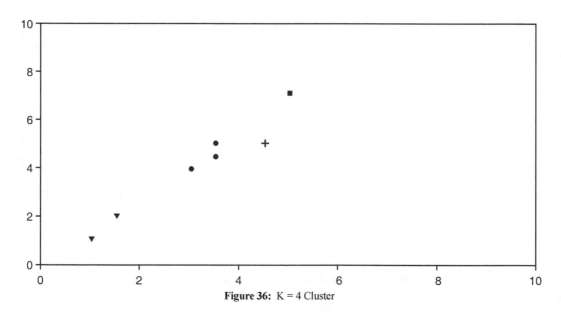

Figure 36: K = 4 Cluster

The complete source code:

```
# -*- coding: utf-8 -*-
"""
Created on Wed Oct 10 17:33:07 2018
@author: cvuppalapati
"""
# -*- coding: utf-8 -*-
"""
Created on Wed Oct 10 17:31:34 2018
@author: cvuppalapati
"""
# clustering dataset
from sklearn.cluster import KMeans
from sklearn import metrics
import numpy as np
import matplotlib.pyplot as plt
x1 = np.array([1.0, 1.5, 3.0, 5.0, 3.5, 4.5, 3.5])
x2 = np.array([1.0, 2.0, 4.0, 7.0, 5.0, 5.0, 4.5])
plt.plot()
plt.xlim([0, 10])
plt.ylim([0, 10])
plt.title('Dataset')
plt.scatter(x1, x2)
plt.show()
# create new plot and data
plt.plot()
X = np.array(list(zip(x1, x2))).reshape(len(x1), 2)
colors = ['b', 'g', 'r','y']
markers = ['o', 'v', 's','+']
# KMeans algorithm
K = 4
kmeans_model = KMeans(n_clusters=K).fit(X)
plt.plot()
for i, l in enumerate(kmeans_model.labels_):
 plt.plot(x1[i], x2[i], color=colors[l], marker=markers[l],ls='None')
 plt.xlim([0, 10])
 plt.ylim([0, 10])
plt.show()
```

Fuzzy Logic (FL)

Fuzzy Logic (FL) incorporates a simple rule-based approach, If X and Y Then Z, to solving a control problem. The FL model is empirically based, relying on an operator's experience rather than their technical knowhow of the system. For building an intelligent dispenser model, we have modeled a Venue scent diffusion system using predefined rules governing the behavior of venue activity and developed activity indicator. Hence, Fuzzy Logic is the most appropriate ML model.

Linguistic Variables

Fuzzy or Linguistic variables are central to FL and were introduced by Professor Lotfi Zadeh in 1973. Succinctly put, Linguistic variables are language artifacts not numbers.

The control inputs, namely hardware sensors, in intelligent dispenser model are Fuzzy variables. For instance, "BLE Devices", "Noise Level", "Proximity Indicator—received signal strength indicator (RSSI) Strength", "Microphone" and "Passive Infra Red (PIR) Sensor" are modeled as Fuzzy parameters. In addition, the error values of Fuzzy parameters could be modeled as Fuzzy Variable.

Another salient principle of FL is fuzzy variables (see Table 7) which are themselves adjectives that modify the variable (e.g., "large positive" error, "small positive" error "zero" error, "small negative" error, and "large negative" error). As a minimum, one could simply have "positive", "zero", and "negative" variables for each of the parameters. Additional ranges such as "very large burst of devices" and "very small noise thresholds" are added to handle the BURST response or very nonlinear behavior.

Table 7: Fuzzy Variables

"N"	=	Negative Error or error-dot level
"Z"	=	Zero Error or error-dot level
"P"	=	Positive Error or error-dot level
"D"	=	"Dispense" Output response
"-"	=	"No Change" or Current Output
"ND"	=	No Dispense

Control Process

The goal of the control process is to model/control dispense of Scent through the FL by detecting people activity in a venue (as Figure 37): Number of people in a room (number of BLE devices), Motion Sensor detection and Noise detected on the microphones

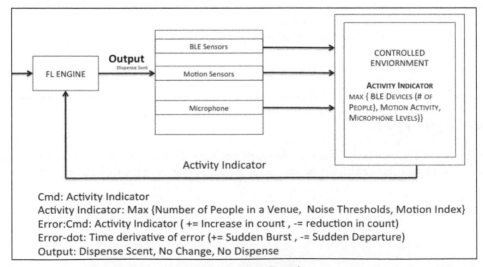

Figure 37: Fuzze Control

Definitions

INPUT# 1 System Status

- Error = Command-Feedback (Activity Indicator Eq. 2)
- P = Increased Activity Index (Increase of venue occupancy or Noise or Motion thresholds)
- Z = Muted or no Activity
- N = Decreased Activity Index (Decrease of venue occupancy or Noise or Motion thresholds)

INPUT# 2 System Status Error-dot = d (Error)/dt

- P = Getting Increased Activity Index
- Z = Not changing
- N = getting decreased activity index

OUTPUT Conclusion and System Response
Output:

- DISP = Dispense Scent
- = don't change
- NDISP = No Dispense

Matrix Rules

The fuzzy parameters of error (# BLE Devices with RSSI, PIR Sensor threshold and Microphone Noise Levels) and error-dot (rate-of-change-of-error) were modified by the adjectives "negative", "zero", and "positive" (Table 8).

In the Figure 37, the linguistic parameter "BLE Devices" indicates number of devices within a venue (negative value indicates people leaving the proximity of the iDispenser & positive implies entering into proximity of the iDispenser).

Table 8: Matrix Rules

1.	IF CMD_AI = N AND d(CMD_AI)/dt = N THEN OUTPUT = NDISP
2.	IF CMD_AI = Z AND d(CMD_AI)/dt = N THEN OUTPUT = NDISP
3.	IF CMD_AI = P AND d(CMD_AI)/dt = N THEN OUTPUT = DISP
4.	IF CMD_AI = N AND d(CMD_AI)/dt = Z THEN OUTPUT = NDISP
5.	IF CMD_AI = Z AND d(CMD_AI)/dt = Z THEN OUTPUT = NC
6.	IF CMD_AI = P AND d(CMD_AI)/dt = Z THEN OUTPUT = DISP
7.	IF CMD_AI = N AND d(CMD_AI)/dt = P THEN OUTPUT = DISP
8.	IF CMD_AI = Z AND d(CMD_AI)/dt = P THEN OUTPUT = DISP
9.	IF CMD_AI = P AND d(CMD_AI)/dt = P THEN OUTPUT = DISP

Matrix

The fuzzy parameters of error (Activity Indicator) and error-dot (rate-of-change-of-activity_Indicator) were modified by the adjectives "negative", "zero", and "positive". The following diagram provides a 3-by-3 matrix (Figure 38) input and output outcomes. The columns represent "negative error", "zero error", and "positive error" inputs from left to right. Please note: "DISP" results into dispensing of the scent. NDISP—no Dispensing required and NC implies no Change to the condition (see Figure 38)

The membership function is a graphical representation of the magnitude of participation of each control input (#BLE Devices, Microphone Thresholds and PIR Sensor Thresholds). It associates a weighting with # BLE Devices, Microphone and PIR Thresholds inputs, defines functional overlap between inputs (BLE

| | | Error (CMD_AI) | |
	N	Z	P
N	NDISP	DISP	DISP
Z	NDISP	NC	DISP
P	NDISP	NDISP	DISP

Error (d(CMD_AI) / dt)

Figure 38: Fuzzy Matrix

Devices to Noise Level on Microphone) (BLE Devices to PIR Thresholds) and (Microphone Noise Levels to PIR Thresholds), and ultimately determines an output response of dispenser (Scent diffusion or not).

Reinforcement Learning (RL)

RL is the learning process of an agent to act to maximize its rewards. In RL, learning is by trial-and-error to map situations to actions that maximize a numerical reward. The following figure has standard RL architecture (see Figure 39).

The agent is defined as a scent-dispensing algorithm, the action is the command to dispense or not dispense, the environment is the target object, in this case activity indicator that is formed in a closed venue, and the reward is performance improvement after applying the action. The goal of RL is to choose an action in response to a current stat, based on activity indicator, to maximize the reward, dispense scent or not. Generally, RL approaches learn estimates of the initialized Q Values Q(s,a), which maps all system states s to their best action a. We initialize all Q(s,a) and, during learning, choose an action a for state s based on $\in$ greedy policy to target platform. Then, we observe the new state s' and reward r and update Q-value of the last state-action pair Q(s,a), with respect to the observed outcome state (s') and reward r. However, in our case, we have used State-Action-Reward-State-Action (SARSA).

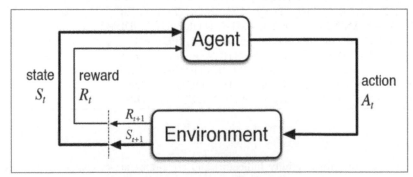

Figure 39: Reinforcement Learning

The Mode-Free vs. Mode Based

The Model stands for the simulation of the dynamics of the environment (dispense based on activity). That is, the model learns the transition probability T (s1/((s0,a))) from the pair of current state s0 and action as to the new state s1 (see Figure 40). In order to derive an action, an agent uses a policy that aims to increase the future rewards (indefinitely or long run environments). If the transition probability is successfully learned, the agent will know how likely to enter a specific state given the current state and action. The model-based approach becomes impractical if state space and action space grows (S * S * A).

Trial-and-error process enables model-free systems to update knowledge. Model free-based systems are more practical, as the interactions of environment and agent are unknown in many physical systems.

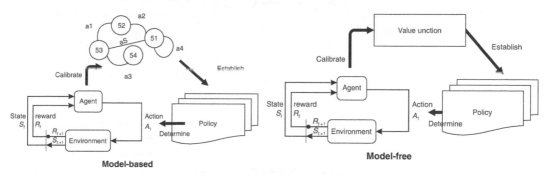

Figure 40: Reinforcement Learning

On-policy and Off-policy

An on-policy agent, our current dispenser in case, learns the value based on an action derived from the current policy, an off-policy agent learns based on the action a* obtained from another policy (a greedy policy in the case of Q-learning).

Neural Networks

Deep learning is a specific subfield of machine learning: A new take on learning representations from data that puts an emphasis on learning successive layers of increasingly meaningful representations. The "deep" in deep learning isn't a reference to any kind of deeper understanding achieved by the approach; rather, it stands for this idea of successive layers of representations. How many layers contribute to a model of the data is called the depth of the model. Other appropriate names for the field could have been layered representations learning and hierarchical representations learning. Modern deep learning often involves tens or even hundreds of successive layers of representations—and they're all learned automatically from exposure to training data. Meanwhile, other approaches to machine learning tend to focus on learning only one or two layers of representations of the data; hence, they're sometimes called shallow learning.

The specification of what a layer does to its input data is stored in the layer's weights, which are essentially a bunch of numbers. In technical terms, we'd say that the transformation implemented by a layer is parameterized by its weights. (Weights are also sometimes called the parameters of a layer.) In this context, learning means finding a set of values for the weights of all the layers in a network, such that the network will correctly map example inputs to their associated targets (see Figure 41).

The *loss function* takes the predictions of the network and the true target (what you wanted the network to output) and computes a distance score, capturing how well the network has done on this specific example.

The fundamental trick in deep learning is to use this score as a feedback signal to adjust the value of the weights a little, in a direction that will lower the loss score for the current example. This adjustment is the job of the *optimizer*, which implements what's called the ***Backpropagation*** algorithm: The central algorithm in deep learning (see Figure 42).

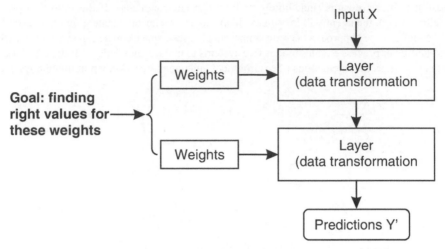

Figure 41: Neural Network Weights

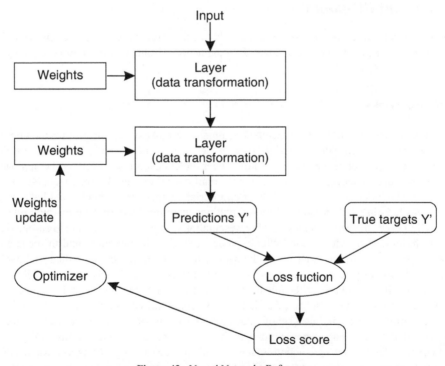

Figure 42: Neural Networks Reference

Voice Detection (Neural Networks)

To perform audio classification, we need to extract the most appropriate and informative acoustic parameters. Traditionally, pitch, jitter and shimmer have been used for this purpose. In recent years, the Mel Frequency Cepstral Coefficient (MFCC) has gained popularity as a successful parameter for audio classification. We are using MFCC to classify between healthy and pathological patient audio.

We initially started with Machine Learning models like the Random Forest classifier and the XGBoost classifier. The model performance was moderate, and the accuracy was around 50% to 60%.

The first one is a plain, fully connected neural network. This architecture is equivalent to stacking multiple perceptron (neurons) and averaging their predictions.

The second architecture is a Convolution Neural Network (CNN). Initially CNNs gained popularity for their state of the art performance on image dataset. In the recent years, research has shown that CNN performs well in Text and audio classification as well.

The third architecture is that of an Recurrent Neural Network (RNN). RNN is a neural network with memory that leverages sequential information and has been used in Natural Language Processing (NLP). It assumes the inputs are dependent on each other. Hence, RNN has produces state of art results for NLP tasks, such as predicting the next word in a sentence. Audio data, unlike Image data, is time dependent and, hence, research shows RNN is more relevant for audio classification too.

Data Samples

The Voice samples were obtained from a voice clinic in a tertiary teaching hospital (Far Eastern Memorial Hospital, FEMH), these included 50 normal voice samples and 150 samples of common voice disorders, including vocal nodules, polyps, and cysts (collectively referred to as Phono trauma); glottis neoplasm; unilateral vocal paralysis. Voice samples of a 3-second sustained vowel sound /a:/ were recorded at a comfortable level of loudness, with a microphone-to-mouth distance of approximately 15–20 cm, using a high-quality microphone (Model: SM58, SHURE, IL), with a digital amplifier (Model: X2u, SHURE) under a background noise level between 40 and 45 dBA. The sampling rate was 44,100 Hz with a 16-bit resolution, and data were saved in an uncompressed .wav format (see Figure 43).

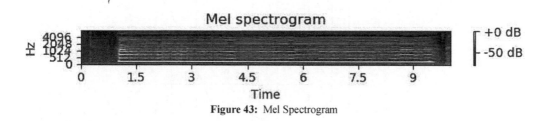

Figure 43: Mel Spectrogram

Voice Feature Extraction Techniques

Our aim is to identify the best features that can be applied to perform sound recognition. As indicated in [13], there are several frames to sounds features. Time domain features include: Zero Crossing Rate (ZCR) and Short Time Energy (STE). Spectral Features Include: Linear Predictive Coding (LPC) coefficients, Relative Spectral Predictive Linear Coding (RASTA PLP), Pitch, Sone, Spectral Flux (SF) and coefficients from basic time to frequency transforms (FFT, DFT, DWT, CWT and Constant Q-Transform). Cepstral domain features are: Mel Frequency Cepstral Coefficient (MFCC) and Bark Frequency Cepstral Coefficient (BFCC).

Table 9: Melspectrogram

```
sig, sample_rate = librosa.load('002.wav')
mel = librosa.feature.melspectrogram(y=sig, sr=sample_rate)
```

The above code (see Table 9) loads "002.wav" file and generates Mel spectrogram, Figure 44, by calling Librosa function.

MFCC

The mel-frequency cepstrum (MFC) [14] is the representation of short-term power spectrum of a sound that is derived from a linear cosine transform of a log power spectrum on a nonlinear of a mel scale of frequency. Please note mel scale is a scale of pitches with reference point set to 1000 Hz tone, 40 dB above listener's threshold, Figure 44, with a pitch of 1000 mels [15].

The MFCC is derived in the following order [16]:

Signal → Pre-emphasis → Hamming Window → Fast Fourier Transform → Log → cosine → Mel-frequency Cepstral coefficients → MFCC

We have considered Mel-Scale Spectrogram and MFCC as our audio features for our Neural network. Mel-Scale's goal is to represent the non-linear perception of sound by human. It is more discriminative in lower frequencies and less discriminative in higher frequencies. The Spectrogram as a result is highly correlated. When Discrete Cosine Transform (DCT) is applied on the Mel scaled features, we get the compressed representation of the Spectrogram by retaining the most prominent features (top 40 in our case) which are decorrelated [17]. Mel-Scale Spectrogram gives the entire information and bigger picture of the audio composition and MFCC highlights the top features to focus on.

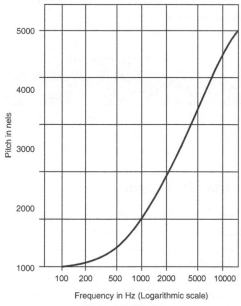

Figure 44: MFCC

Data Wrangling

We are wrangling our audio data into a 60×41×2 array. Here 60 is the number of MFCC, 41 is the number of windows over the audio signal and 2 is the number of channels. In order to have fixed size input for our varying size audio files, we perform windowing over the audio.

In our case, we can feed in a different feature for each channel. In our first channel, we feed in the MFCCs and in the second channel, we feed in the delta features (local estimate of the derivative of MFCC) [12]. Now the CNN can learn from not only MFCC but also their deltas. Librosa has been used to extract the MFCC and delta features. Keras has been used to implement the 5-layer CNN [18,19]. Apart from feeding only MFCC, we have also trained the neural network with both Mel-Scale Spectrogram and MFCC features. Another method employed to improve the accuracy is using the Mel-Scale Spectrogram and MFCC features as input to separate Neural networks and bagging of the multiple Neural networks, by taking a majority voting based on the class probability.

Metrics

We have used Accuracy score and ROC (Receiver Operating Characteristic) metric to evaluate our classifier. In healthcare applications, sensitivity is given higher importance. It defines the percentage of people the model can correctly predict as unhealthy. Hence, we have also used sensitivity and specificity as our metric.

Architecture

We studied and applied various Deep learning architectures: 5-layer network, 5-layer CNN and RNN.

A. *Neural Network with a 5-layer architecture*

We started with a plain Neural Network with a 5-layer architecture [20]. The tuning of the hyper parameters of the model improved the accuracy by 10%. Each parameter value was tuned with the following values.

- regularization_rate = 0.1, 0.01, 0.001
- activation = 'tanh', 'relu'
- Number of hidden nodes = 64, 40, 100

The values that gave stable and good performance where

- regularization_rate = 0.001
- activation = 'tanh'
- Number of hidden nodes = 64

The following Table 10 and Figure 45 displays the accuracies for various epoch values.

Table 10: Parameters & Accuracies

Parameters	Accuracy	ROC
Convolution_filter_size 1x1	48%	0.741
Convolution_filter_size 2x2	76%	0.939
Convolution_filter_size 3x3	82%	0.959
Max_Pooling_filter_size 2x2	84%	0.976
Learning rate = 0.01	69%	0.918
SGD_momentum = 0.9	90%	0.986
Adam optimizer	93%	0.994
Tanh for all layers	94%	0.994
Tanh for1st layer & relu for rest of layers	94%	0.996
Tanh for1st layer & LeakyReLU for rest of layers	93%	0.99

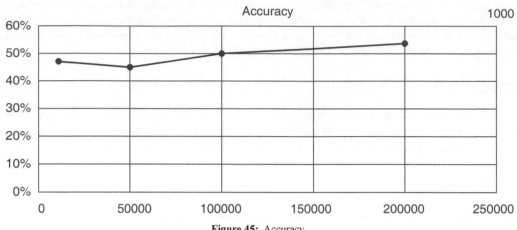

Figure 45: Accuracy

5-layer Convolution Neural Network (CNN)

We move to CNN for better accuracy. We have implemented a 5-layer CNN [12,16,21] to classify our audio dataset. This architecture has 3 convolutions + Max Pooling layers and one fully connected dense layer. We have used 2D convolution and 2D Max Pooling. The CNN was trained for 1000 epochs and has achieved an accuracy on test data of 93% and ROC of 0.99.

1) Parameters
   Below are the parameter values experimented:
- regularization_rate = 0.1, 0.01, 0.001
- learning rate = 0.1, 0.01, 0.001
- activation = 'tanh', 'relu', LeakyReLU, ELU, PReLU
- Number_nodes = 64, 40, 100
- Convolution filter size = 1,2,3
- Max Pooling filter size = 2x2, 4x2
- Optimizers: Adam, SGD
- SGD_momentum: between 0.5 to 0.9

   We tuned one parameter with the rest of them fixed. The optimized parameter values achieved are:
- Convolution_filter_size = 3x3
- pooling_filter_size = 2x2
- Convolution stride = 1x1
- pooling stride = 2x2
- learning rate = 0.001
- regularization rate = 0.001
- optimizer = Adam
- activation = LeakyReLU

   CNN (see Figure 46) has achieved 94% accuracy, when trained on only one feature—MFCC and achieved 93% accuracy, when trained on 2 features. It achieved a sensitivity of 97% and specificity of 94% on the test partition of training dataset and a sensitivity of 96% and specificity of 18% on the testing dataset. Below are the parameters tuned to achieve this accuracy (please see Figure 47).

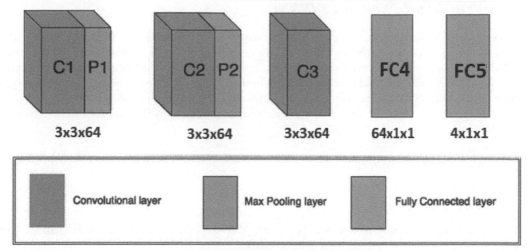

Figure 46: Neural Network Architecture

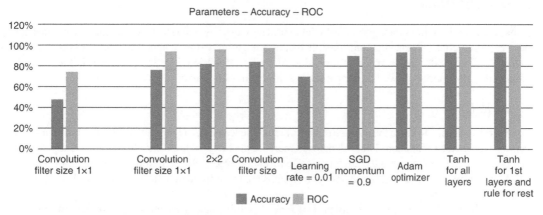

Figure 47: Parameters—Accuracy ROC

Parameter intuition

A larger convolution filter size gives a larger receptive window over the audio pattern. A smaller pooling filter size helps us to extract the right local features. A larger pooling filter size might miss the local features retrieved in the convolution step. A learning rate reduction of 0.001 affects the model's learning negatively.

Optimizers that aid in CNN learning are: Adam and SGD. However, for our data, Adam seems to push the accuracy higher. It is a general practice to use Relu as the activation function in the hidden layers and Tanh for the input layer. Leaky Relu solves the problem of Relu, where if Relu's output is consistently zero, then the gradient is zero and the error back propagated is multiplied by zero, meaning the Relu has died. As we can see, this again impacts the ROC positively (see Figure 47). Our final test accuracy achieved is 93%.

Recurrent Neural Networks (RNN)

RNNs are used in applications where the data has sequential information with respect to time. RNN's nodes not only process the current inputs but also have a function to store the weights of the last inputs. LSTM—long short term memory, is more popular for their performance in Natural language processing tasks. LSTM is similar to RNN in architecture, but uses a different function to record the previous input states. As a result, LSTM can capture longer term dependencies in a sequence than RNN.

The architecture used for LSTM, is a 4-layer network. We used the below parameters to tune our RNN to give an accuracy of 90%:

- dropout = 0.5
- optimizer = Adam
- learning rate = 0.001
- recurrent_dropout = 0.35

We achieved sensitivity of 8% and specificity: 97.7% on the RNN prediction evaluation. CNN performed better than RNN with respect to these metrics.

Loss Functions

A loss function is an important part of artificial neural networks, which is used to measure the inconsistency between predicted value and actual label. It is a non-negative value, where the robustness of the model increases along with the decrease of the value of the loss function. The loss function is the core of an empirical risk function as well as a significant component of a structural risk function.

Leaky ReLU

Leaky ReLU is used to solve the dying ReLU problem. The leaky ReLU function is nothing but an improved version of the ReLU function. The leak helps to increase the range of the ReLU function. Usually, the value is 0.01 or so. When it is not 0.01, then it is called Randomized ReLU. Therefore, the range of the Leaky ReLU is (-infinity to infinity). Leaky ReLU functions are monotonic in nature. Also, their derivatives also monotonic in nature.

For the ReLU function, the gradient is 0 for $x < 0$; this made the neurons die for activations in that region. Leaky ReLU is defined to address this problem. Instead of defining the ReLU function as 0 for x less than 0, we define it as a small linear component of x. The main advantage of this is to remove the zero gradient. So, in this case, the gradient of the negative values is non-zero and so we no longer encounter dead neurons in that region. Also, this speeds up the training process of the neural network.

However, in the case of a parameterized ReLU function, $y = ax$, when $x < 0$. The network learns the parameter value of 'a' for faster and more optimum convergence. The parametrized ReLU function is used when the leaky ReLU function still fails to solve the problem of dead neurons and the relevant information is not successfully passed to the next layer.

Table 11: Model Performance

Network Architectures	Sensitivity	Specificity
5-LAYER PLAIN NEURAL NETWORK	12%	96%
5-LAYER CNN	96%	18%
4-LAYER RNN	8%	97.7%

We have observed from our work that Neural Networks perform better than simpler Machine Learning Algorithms for audio classification (please see Table 11). CNN consistently performs better than RNN, if the necessary sequential information is provided. Loss functions, like LeakyRelu and Prelu, have resulted in a noticeable improvement in sensitivity and specificity of the model, which are more important than the accuracy of the model. The best model was a 5-layer CNN trained with MFCC and Mel-Spectrogram. It had a sensitivity of 96% & specificity of 18% on the test data.

CNN Mode Code

The following 5 layer CNN Code (see Table 12 and Figure 48).

Table 12: Voice Processing Python Code

```
import numpy as np
import pandas as pd
import os
from keras.models import Sequential
from keras.layers import Dense, Dropout, Activation, Flatten
from keras.layers import Convolution2D, MaxPooling2D
from keras.optimizers import SGD
from keras.regularizers import l2
from keras.utils import np_utils
from sklearn.model_selection import train_test_split
from sklearn import metrics
import librosa, glob
from sklearn.preprocessing import LabelEncoder
'''Data Wrangling for the 5-layer CNN model
Source: http://aqibsaeed.github.io/2016-09-24-urban-sound-classification-part-2/
Environmental sound classification with convolutional neural networks by Karol J. Piczak
'''
X = []
y = []
def windows(data, window_size):
    '''To generate the window range'''
    start = 0
    while start < len(data):
        yield int(start), int(start + window_size)
        start += (window_size/2)
def extract_features(parent_dir,sub_dirs,file_ext="*.wav",bands = 60, frames = 41):
    '''Wrangle audio data into 60x41x2 frames.
    Bands = 60 = number of MFCC
    frames = 41 = number of windows from the audio signal
    Number of channels (like R,G,B channels in images) = 2 = mel-spectrograms and their corresponding
delta'''
    window_size = 512 * (frames - 1)
    log_specgrams = []
    labels = []
    for l, sub_dir in enumerate(sub_dirs):
        for fn in glob.glob(os.path.join(parent_dir, sub_dir, file_ext)):
            sound_clip,s = librosa.load(fn)
            label = l
            for (start,end) in windows(sound_clip,window_size):
#                print(type(start), start, type(end), end, window_size)
                if(len(sound_clip[start:end]) == window_size):
                    signal = sound_clip[start:end] # window the original audio
                    melspec = librosa.feature.melspectrogram(signal, n_mels = bands)
                    # librosa.core.logamplitude has been removed in v0.6
                    # logspec = librosa.logamplitude(melspec)
```

Table 12 contd. ...

...Table 12 contd.

```
                logspec = librosa.amplitude_to_db(melspec)
                logspec = logspec.T.flatten()[:, np.newaxis].T
                log_specgrams.append(logspec)
                labels.append(label)
    log_specgrams = np.asarray(log_specgrams).reshape(len(log_specgrams),bands,frames,1) features = np.concatenate((log_specgrams, np.zeros(np.shape(log_specgrams))), axis = 3)
    for i in range(len(features)):
        features[i, :, :, 1] = librosa.feature.delta(features[i, :, :, 0])
    return np.array(features), np.array(labels,dtype = np.int)
train_data_path = 'C:\\Hanumayamma\\FEMH\FEMH Data\\Training Dataset Small\\'

test_data_path = 'C:\\Hanumayamma\\FEMH\\FEMH Data\\Testing Dataset Small\\'

train_csv = train_data_path + 'Demographics for Training Dataset.xlsx'

label_dict = {0 : 'Normal/', 1 : 'Pathological/Neoplasm/', 2 : 'Pathological/Phonotrauma/', 3 : 'Pathological/Vocal palsy/'}

X, y = extract_features(train_data_path,label_dict.values(), file_ext="*.wav",bands = 60, frames = 41)

'''Encode class labels - y'''

temp = np.array(y)
encoder = LabelEncoder()
Y = np_utils.to_categorical(encoder.fit_transform(temp.astype(str)))

X_train, X_test, y_train, y_test = train_test_split(X, Y, test_size=0.33)
'''
5-layer CNN implementation
Source: 5 layer CNN described in https://arxiv.org/pdf/1608.04363.pdf
'''
filter_size = 3
bands = 60
frames= 41
num_channels = 2

num_labels = y_test.shape[1]

model = Sequential()

# Layer 1 - Convolution with 24 filters + Maxpooling
model.add(Convolution2D(24, (filter_size, filter_size), border_mode='valid', input_shape=(bands, frames, num_channels)))
model.add(MaxPooling2D(pool_size=(4, 2)))
model.add(Activation('relu'))

# Layer 2 - Convolution with 48 filters + Maxpooling
model.add(Convolution2D(48, (filter_size, filter_size)))
model.add(MaxPooling2D(pool_size=(4, 2)))
model.add(Activation('relu'))

# Layer 3 - Convolution with 24 filters + Maxpooling
model.add(Convolution2D(48, (filter_size, filter_size), border_mode='valid'))
model.add(Activation('relu'))

# Flatten the tensors to feed into fully connected layers that follow
model.add(Flatten())

# Layer 4 - fully connected layer
model.add(Dense(64, W_regularizer = l2(0.001)))
model.add(Activation('relu'))
model.add(Dropout(0.5))
```

Table 12 contd. ...

...Table 12 contd.

```
# Layer 5 - output layer
model.add(Dense(num_labels, W_regularizer = l2(0.001)))
model.add(Dropout(0.5))
model.add(Activation('softmax'))

sgd = SGD(lr=0.001, momentum=0.0, decay=0.0, nesterov=False)

model.compile(loss='categorical_crossentropy', metrics=['accuracy'], optimizer=sgd)

# history = model.fit(X_pkl, Y_pkl, batch_size=32, epochs=50)
history = model.fit(X_train, y_train, batch_size=32, epochs=50)

# determine the ROC AUC score
y_prob = model.predict(X_test)
y_pred = y_prob.argmax(axis=-1)
roc = metrics.roc_auc_score(y_test, y_prob)
print ("ROC:", round(roc,3))

# determine the classification accuracy
score, accuracy = model.evaluate(X_test, y_test, batch_size=32)
print("\nAccuracy = {:.2f}".format(accuracy))
'''

Current CNN:
ROC (roc_auc_score): 0.959
Test Accuracy = 0.82
'''

# Save model
# model.save('voice_cnn_model_100k.h5')
```

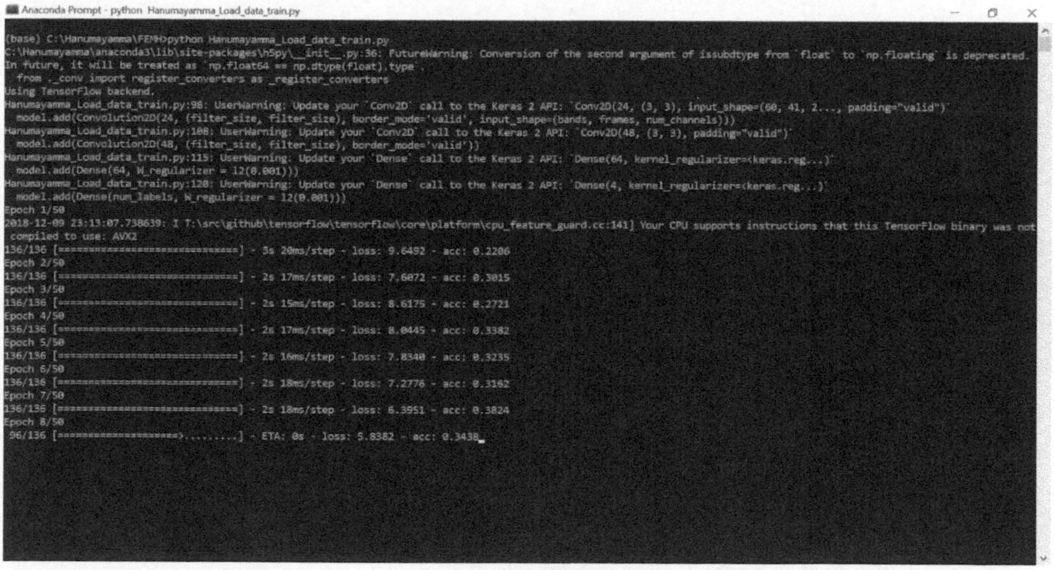

Figure 48: CNN Model output

CNN Spectrogram

Please find complete CNN Spectrogram code (see Table 13 and Figure 49).

Table 13: CNN Spectrogram Code

```
import numpy as np
import pandas as pd
from keras.models import Sequential
from keras.layers import Dense, Dropout, Activation, Flatten, LeakyReLU
from keras.layers import Convolution2D, MaxPooling2D
from keras.optimizers import SGD,Adam
from keras.regularizers import l2
from keras.utils import np_utils
from sklearn.model_selection import train_test_split
from sklearn import metrics
import librosa, glob, os
from sklearn.preprocessing import LabelEncoder
from sklearn.metrics import confusion_matrix
import pickle

'''Data Wrangling for the 5-layer CNN model
Source: http://aqibsaeed.github.io/2016-09-24-urban-sound-classification-part-2/
Environmental sound classification with convolutional neural networks by Karol J. Piczak
'''
X = []
y = []
def windows(data, window_size):
    '''To generate the window range'''
    start = 0
    while start < len(data):
        yield int(start), int(start + window_size)
        start += (window_size / 2)
def extract_features(parent_dir,sub_dirs,file_ext="*.wav",bands = 60, frames = 41):

    '''Wrangle audio data into 60x41x2 frames.
     Bands = 60 = number of MFCC
     frames = 41 = number of windows from the audio signal
     Number of channels (like R,G,B channels in images) = 2 = mel-spectrograms and their corresponding
delta'''

    window_size = 512 * (frames - 1)
    log_specgrams = []
    labels = []
    for l, sub_dir in enumerate(sub_dirs):
      for fn in glob.glob(os.path.join(parent_dir, sub_dir, file_ext)):
        sound_clip,s = librosa.load(fn)
        label = l
          for (start,end) in windows(sound_clip,window_size):
#             print(type(start), start, type(end), end, window_size)
          if(len(sound_clip[start:end]) == window_size):
              signal = sound_clip[start:end] # window the original audio
              #D = np.abs(librosa.stft(signal))**2
              #melspec = librosa.feature.melspectrogram(S=D, n_mels = bands)
              melspec = librosa.feature.melspectrogram(signal, n_mels = bands)
              #stft = np.abs(librosa.stft(signal,center=False))
              #melspec = np.array(librosa.feature.chroma_stft(S=stft, sr=s).T)
              #melspec = librosa.feature.mfcc(y=signal, sr=s, n_mfcc=bands)
              #melspec = np.array(librosa.feature.tonnetz(y=librosa.effects.harmonic(signal), sr=s).T)
              logspec = librosa.amplitude_to_db(melspec)
              logspec = melspec.T.flatten()[:, np.newaxis].T
              log_specgrams.append(logspec)
              labels.append(label)
```

Table 13 contd. ...

...Table 13 contd.

```
      log_specgrams = np.asarray(log_specgrams).reshape(len(log_specgrams),bands,frames,1)
      #features = np.concatenate((log_specgrams, np.zeros(np.shape(log_specgrams))), axis = 3)
      #for i in range(len(features)):
      # features[i, :, :, 1] = librosa.feature.delta(features[i, :, :, 0])

          return np.array(log_specgrams), np.array(labels,dtype = np.int)
```

```
train_data_path = 'C:\\Hanumayamma\\FEMH\FEMH Data\\Training Dataset Small\\'
```

```
test_data_path = 'C:\\Hanumayamma\\FEMH\\FEMH Data\\Testing Dataset Small\\'
```

```
train_csv = train_data_path + 'Demographics for Training Dataset.xlsx'
```

```
label_dict = {0 : 'Normal/', 1 : 'Pathological/Neoplasm/', 2 : 'Pathological/Phonotrauma/', 3 : 'Pathological/
Vocal palsy/'}
```

```
X, y = extract_features(train_data_path,label_dict.values(), file_ext="*.wav",bands = 128, frames = 128)
```

```
'''Encode class labels - y'''
```

```
temp = np.array(y)
encoder = LabelEncoder()
Y = np_utils.to_categorical(encoder.fit_transform(temp.astype(str)))
```

```
X_train, X_test, y_train, y_test = train_test_split(X, Y, test_size=0.33)
'''
```

```
5-layer CNN implementation
Source: 5 layer CNN described in https://arxiv.org/pdf/1608.04363.pdf
'''
```

```
# Initialize parameters for neural network
filter_size = 3
bands = 128
frames= 128
```

```
num_channels = 1
num_labels = y_test.shape[1]
regularization_rate = 0.001
```

```
# Initialize the model and its layers
```

```
model = Sequential()
```

```
# Layer 1 - Convolution with 24 filters + Maxpooling
model.add(Convolution2D(24, (filter_size, filter_size), border_mode='valid', input_shape=(bands, frames,
num_channels), strides=(1, 1)))
model.add(MaxPooling2D(pool_size=(2, 2), strides=(2, 2)))
model.add(Activation('tanh'))
```

```
# Layer 2 - Convolution with 48 filters + Maxpooling
model.add(Convolution2D(48, (filter_size, filter_size)))
model.add(MaxPooling2D(pool_size=(2, 2)))
model.add(LeakyReLU(alpha=0.01))
```

```
# Layer 3 - Convolution with 24 filters + Maxpooling
model.add(Convolution2D(48, (filter_size, filter_size), border_mode='valid'))
model.add(LeakyReLU(alpha=0.01))
```

```
# Flatten the tensors to feed into fully connected layers that follow
model.add(Flatten())
```

```
# Layer 4 - fully connected layer
model.add(Dense(64, W_regularizer = l2(regularization_rate)))
model.add(LeakyReLU(alpha=0.01))
model.add(Dropout(0.5))
```

Table 13 contd. ...

...Table 13 contd.

```
# Layer 5 - output layer
model.add(Dense(num_labels, W_regularizer = l2(regularization_rate)))
model.add(Dropout(0.5))
model.add(Activation('softmax'))

# sgd = SGD(lr=0.001, momentum=0.9, decay=0.0, nesterov=False)
adam = Adam(lr=0.001, beta_1=0.9, beta_2=0.999, epsilon=1e-08, decay=0.0)
model.compile(loss='categorical_crossentropy', metrics=['accuracy'], optimizer=adam)

# Train the neural network
history = model.fit(X_train, y_train, batch_size=32, epochs=100, validation_split= 0.1, shuffle=True)

# determine the ROC AUC score

# y_prob = model.predict(X_test)
# y_pred = y_prob.argmax(axis=-1)

y_prob = model.predict(X_test)
#y_pred = np_utils.probas_to_classes(y_prob)
#y_true = np.argmax(y_test, 1)

roc = metrics.roc_auc_score(y_test, y_prob)
print ("ROC:", round(roc,3))

# determine the classification accuracy
score, accuracy = model.evaluate(X_test, y_test, batch_size=32)
print("\nAccuracy = {:.2f}".format(accuracy))

#predictions = model.predict(X_test)

y_test = np.argmax(y_test, axis=-1)
predictions = np.argmax(y_prob, axis=-1)

c = confusion_matrix(y_test, predictions)

print('Confusion matrix:\n', c)
print('sensitivity', c[0, 0] / (c[0, 1] + c[0, 0]))
print('specificity', c[1, 1] / (c[1, 1] + c[1, 0]))

acc = history.history['acc']
output = open('cnn_accuracy.pkl', 'wb')
pickle.dump(acc, output)
output.close()
'''
Parameter values tried:
regularization_rate = 0.1, 0.01, 0.001
learning rate = 0.1, 0.01, 0.001
activation = 'tanh', 'relu'
Number_nodes = 64, 40, 100
Convolution filter size = 1,2,3
Maxpooling filter size = 2x2, 4x2
Optimizers: Adam, SGD
momentum: between 0.5 to 0.9

Original CNN:
filter_size = 3x3
ROC (roc_auc_score): 0.959
Test Accuracy = 0.82

Filter size = 2x2
ROC: 0.939
Accuracy = 0.76
```

Table 13 contd. ...

...Table 13 contd.

Filter size = 1x1
ROC: 0.741
Accuracy = 0.48

lr =0.01
ROC: 0.918
Accuracy = 0.69

Maxpooling filter size = 2x2
ROC: 0.976
Accuracy = 0.84

with conv = 3x3, pooling= 2x2, conv stride = 1x1, pooling stride = 2x2
lr = 0.001, regularization rate = 0.001, sgd_momentum=0.9:
ROC: 0.986
Accuracy = 0.90

with conv = 3x3, pooling= 2x2, conv stride = 1x1, pooling stride = 2x2
lr = 0.001, regularization rate = 0.001, adam
ROC: 0.994
Accuracy = 0.93

with conv = 3x3, pooling= 2x2, conv stride = 1x1, pooling stride = 2x2
lr = 0.001, regularization rate = 0.001, adam, all are tanh
ROC: 0.994
Accuracy = 0.94

with conv = 3x3, pooling= 2x2, conv stride = 1x1, pooling stride = 2x2
lr = 0.001, regularization rate = 0.001, adam, tanh for 1st layer, rest are relu
ROC: 0.996

Accuracy = 0.94
"""
# Save model
model.save('cnn_4.h5')

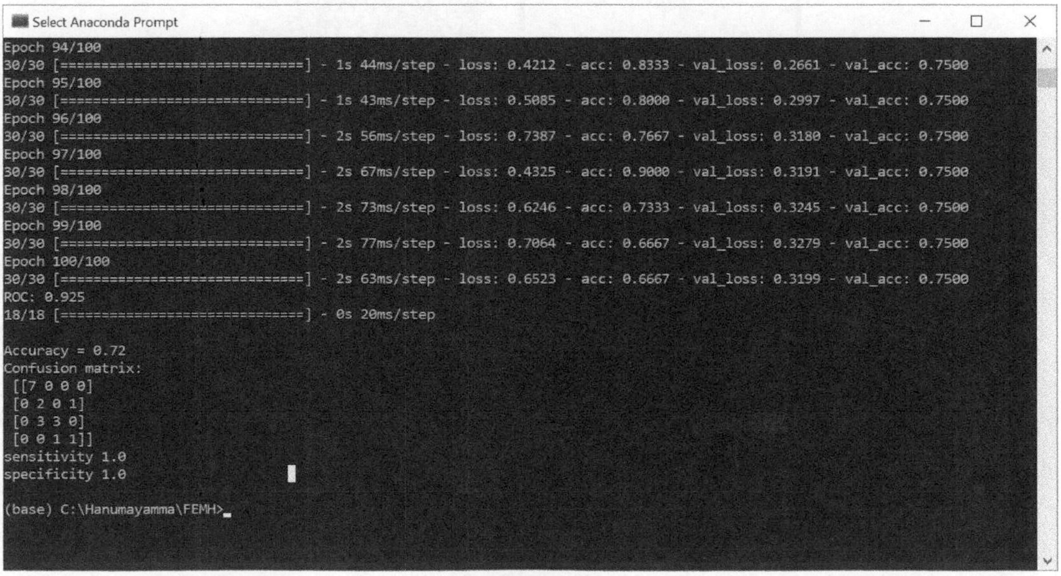

Figure 49: CNN

Tensor Flow execution on embedded Microcontroller Units (MCUs)

Tensor Flow applications can be executed using Mobile application development platforms, Android and iOS, but not ready for executing on MCUs with low constrained foot print.

In order to run the Tensor flows on MCUs, please prepare the models on desktops systems. Once, accuracy is good for the productions, shrink the model to fit into the MUC memory footprint:

- Make sure that the number of weights is small enough to fit in the RAM. Please follow the size compact process as described in https://github.com/tensorflow/tensorflow/tree/master/tensorflow/tools/graph_transforms/#inspecting-graphs

- Reduce the size of the fully-connected and convolutional layers so that the number of compute operations is small enough to run within your device's budget. The tensorflow/tools/benchmark utility with –show_flops will give you an estimate of this number. For example, I might guess that an M4 could do 2 MFLOPs/second, and so aim for a model that fits in that limit.

Edge to Cloud Amalgamation—Traffic Light Sensors as law enforcement devices

Traffic lights are classical Edge devices with advanced sensing capabilities that include speed detection, proximity detection, facial detection, and red-light violation detection.

The Edge devices capture the data and the analytics at the backend combines the sources from different data sets in order to provide a report or provide violation results to law enforcement and government agencies.

In the following violation Figure 50, the traffic camera at a stop light in Newark, California has captured the red light violation of a driver, in this case me, and reported details to the Law enforcement agencies. I was issued a traffic violation ticket [8].

Sensors and analytics are not only delivering valuable data but also helping in identifying and enforcing good governing laws.[9]

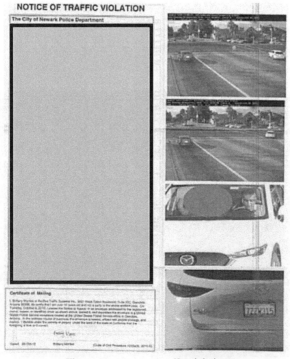

Figure 50: Notice of traffic violation

---
9   Newark Police department - http://www.newark.org/departments/police/administrative-divisions/red-light-enforcement

Hotels and Entertainment and Dynamic Pricing (Harmonious dancing for calculating the best price—both for industry and Patrons—a Win-Win)

The Hotel and Entertainment industries are very competitive. The success of these industries is dependent upon attracting new customers and maintaining the existing loyal customer base. Until recently, the pricing for the hotel industry has been very static. That is, the price for a room was same if you order in the morning, or after getting off the flight and order at check-in time, 3:00 pm, or just from country club before reaching for a stay at the hotel, late evening for example.

Edge and Cloud compute have changed the industry for good, both for the hotel management and loyal patrons. The dynamic pricing (see Figure 51) is performed edge and cloud handshake; Many hotels or entertainment places have on-site IT systems that take care of customer check-in through the Hotel Management Software (HMS) Systems. By the by, the same HMS systems are integrated with Smart Cities in order to predict the power and water consumption on a daily basis.

The on-site HMS has a complete view of available room inventory, plus upcoming check-ins and check-outs. Based on the inventory on hand and the demand (current and projected), the HMS updates the data to the central Hotel management server through Message Queues and Ni-Fi.[10] The data is consumed by Spark or backend stream processing servers in order to recommend the best pricing, based on corporate guidelines. The dynamic pricing is relayed back to the hotel on an hourly basis. It's no wonder you may be quoted two different prices on the same day.

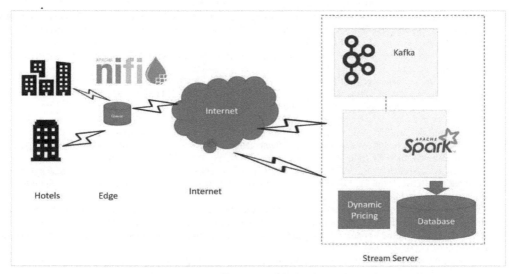

Figure 51: Dynamic Pricing

After reading this chapter, you should be able to design, implement, and deploy the following Machine Learning Algorithms into small embedded devices and mobile platforms.

- Decision Trees
- Kalman Filter
- K-means clustering
- Sliding Window
- Clustering
- Neural Network for Sound Classification
- Fuzzy Logic and Re-enforcement Learning

---

[10] Apache NiFi - https://nifi.apache.org/index.html

References

1. Gartner. Gartner Says the Internet of Things Installed Base Will Grow to 26 Billion Units By 2020. December 12, 2013, https://www.gartner.com/newsroom/id/2636073, Access Date: 14 December 2018
2. Veron Turner. Data Growth, Business Opportunities, and the IT Imperatives. April 2014, https://www.emc.com/leadership/digital-universe/2014iview/index.htm, Access Date: June 2018
3. James Manyika, Michael Chui, Peter Bisson, Jonathan Woetzel, Richard Dobbs, Jacques Bughin and Dan Aharon. Unlocking the potential of the internet of things. June 2015, https://www.mckinsey.com/business-functions/digital-mckinsey/our-insights/the-internet-of-things-the-value-of-digitizing-the-physical-world, Access Date: December 2018
4. The Apache Software Foundation, Filters. 28 Aug 2016, http://commons.apache.org/proper/commons-math/userguide/filter.html, Access Date: June 2018
5. The Silicon Laboratiroes. Si7020-A10. 2018, http://www.mouser.com/ds/2/368/Si7020-272416.pdf, Access Date: June 2018
6. Neil Tan. Why Machine Learning On the Edge? 2018, https://towardsdatascience.com/why-machine-learning-on-the-edge-92fac32105e6, Access Date: December 2018
7. Andrea Trevine. Introduction to K-meand Clustering, 6 December 2016, https://www.datascience.com/blog/k-means-clustering, Access Date: June 2018
8. Newark Police Department. Red Light Enforcement, 2018, http://www.newark.org/departments/police/administrative-divisions/red-light-enforcement, Access Date, 7 December 2018
9. The Apache Software Foundation. NiFi. 2018, https://nifi.apache.org/index.html, Access Date 12 Dec 2018
10. Nick Rose. Crowdedness at the Campus Gym. 2016, https://www.kaggle.com/nsrose7224/crowdedness-at-the-campus-gym, Access Date, 7 December 2018
11. Data Market. Daily total female births in California, 1959, 1988, https://datamarket.com/data/set/235k/daily-total-female-births-in-california-1959#!ds=235k&display=line, Access Date: 05 Dec 2018
12. Environmental sound classification with convolutional neural networks by Karol J. Piczak, 2015 IEEE INTERNATIONAL WORKSHOP ON MACHINE LEARNING FOR SIGNAL PROCESSING, SEPT. 17–20, 2015, BOSTON, USA
13. Mitrovic, D., Zeppelzauer, M. and Breiteneder, C. Discrimination and retrieval of animal sounds. 2006 12th International Multi-Media Modelling Conference, Beijing, 2006, pp. 5, doi: 10.1109 / MMMC.2006.1651344 , URL: http://ieeexplore.ieee.org/stamp/stamp.jsp?tp=&arnumber=1651344&isnumber=34625
14. Le-Qing, Z. Insect Sound Recognition Based on MFCC and PNN. 2011 International Conference on Multimedia and Signal Processing, Guilin, China, 2011, pp. 42–46. doi: 10.1109/CMSP.2011.100, URL: http://ieeexplore.ieee.org/stamp/stamp.jsp?tp=&arnumber=5957464&isnumber=5957439
15. The mel scale as a function of frequency (from Appleton and Perera, eds., The Development and Practice of Electronic Music, Prentice-Hall, 1975, p. 56; after Stevens and Davis, Hearing; used by permission
16. Deep Convolutional Neural Networks and Data Augmentation for Environmental Sound Classification Justin Salamon and Juan Pablo Bello, IEEE SIGNAL PROCESSING LETTERS, ACCEPTED NOVEMBER 2016.
17. Haytham Fayek. Speech Processing for Machine Learning: Filter banks, Mel-Frequency Cepstral Coefficients (MFCCs) and What's In-Between. Apr 21, 2016, https://haythamfayek.com/2016/04/21/speech-processing-for-machine-learning.html , accessed on: 11/11/2018
18. James Lyons. Mel Frequency Cepstral Coefficient (MFCC) tutorial, http://practicalcryptography.com/miscellaneous/machine-learning/guide-mel-frequency-cepstral-coefficients-mfccs/accessed on: 11/11/2018
19. Jordi Pons, Xavier Serra. Randomly weighted CNNs for (music) audio classification. 14 Feb 2019 https://arxiv.org/abs/1805.00237, Access Date: Feb 14, 2019
20. Annamaria Mesaros, Toni Heittola, Tuomas Virtanen. A multi-device dataset for urban acoustic scene classification. Thu, 11 Oct 2018, https://arxiv.org/abs/1807.09840, Access Date: Feb 14, 2019
21. Jongpil Lee, Taejun Kim, Jiyoung Park, Juhan Nam. 4 Dec. 2017. Sound Raw Waveform-based Audio Classification Using Sample-level CNN Architectures, https://arxiv.org/abs/1712.00866, Access Date: Feb. 14, 2019
[A] David Floyer. The Vital Role of Edge Computing in the Internet of Things. October 20 2015, https://wikibon.com/the-vital-role-of-edge-computing-in-the-internet-of-things/April 30, 2019

CHAPTER

Connectivity

This Chapter Covers

- REST Protocol
- MQTT Protocol
- LoRa Protocol
- Bluetooth Low Energy Protocol
- IoT Hardware Clocks

The focus of the chapter is to provide a foundation for all the architecture used to process the IoT Data. IoT Data exhibit 5Vs[1] (Volume, Velocity, Variability, Veracity, and Value) [1] and to process such high-density data the architectures should scale to provide actionable insights to Users.

Connectivity plays an important role in future IoT products and Services. Current carrier networks (3G & 4G) have limited capacities, bandwidth and higher latencies. The 5G network is a game changer as it brings network slices, low latency, and higher bandwidths.[2]

Applications built on underlying technologies such as augmented reality, virtual reality, telepresence and artificial intelligence will benefit from massive data pipes and ultra-low latency. Fixed-wireless services delivering fiber-like speeds will enable UHD video streaming and entirely new business models for cable companies and streaming subscription providers [2]. Cars and drones will communicate with each other and coordinate with things around them through the low-latency network. Industrial automation and robotics will finally be a reality over wide area network. And finally, 5G will get us closer to realizing the true IoT world of millions of sensor devices connected to the network.

5G enables deployment and allocation of dedicated "network slices" where, within one coverage area or functional boundary, different services and solutions could occupy their own slice of the network, with different speeds, guaranteed quality of service and reporting. For the IoT, this will create new possibilities

---

[1]   Why only one of the 5 Vs of big data really matters - https://www.ibmbigdatahub.com/blog/why-only-one-5-vs-big-data-really-matters

[2]   How The 5G Revolution Will Drive Future Innovations Of IoT - https://www.forbes.com/sites/forbestechcouncil/2018/09/07/how-the-5g-revolution-will-drive-future-innovations-of-iot/#4b83e79d637e

5G Network

Different generations of wireless technology so far are:
- The first generation (1G), cellular networks that was created in the early 1990s.
- The second generation (2G) is when text messages were invented, and everyone was actively sending them through a wireless network.
- The third generation (3G) made it possible for people to access the internet on their mobile devices.
- The fourth generation (4G) is an expanded version of the 3G's abilities, 4G is much faster than 3G in terms of internet speed.
- The fifth generation (5G) is a game changer with greater velocities.

	3G	4G	5G
Generation	Third generation of mobile broadband internet	Fourth generation of mobile broadband internet	Fifth generation of mobile broadband internet
Download Speed	Slower than 4G	10 times faster than 3G	Faster than 3G and 4G
Data transfer rate	200 kilobits/sec	100 megabits/sec for high mobile communication & 1 Gigabit/sec for low mobility communication	20 Gbits/sec
Compatibility	3G phone cannot communicate through 4G network	Backward compatible, so 4G can communicate through 3G and 2G networks	
Carrier	3G Frequencies	4G Frequencies	5G Frequencies
Latency	depends on carrier	depends on carrier	1 ms
Router	Like WIFI router but does not use ethernet cord to connect to internet. It uses 3G signal from cellular networks to get online	It is directly connected to router, switch or computer to provide instant 4G LTE network	Qualcomm is 1st company to launch 5G modem, the Snapdragon X50 modem

5G network is a game changer because it reduces latency (independent of Carrier) and has increased data transfer, comparable to edge processors. Using 5G network becomes compute platform (within the embedded device).

for connected devices across all market sectors. There could be one slice supporting the connected car services, while another is serving environmental monitoring devices. The latter might only transmit when certain conditions change, while a connected car services slice, which may be leveraging access to information from network resources near an interstate highway, might be constantly blasting traffic or route information to passing cars through new, enhanced network broadcasting modes [2,3].

Case Study: Low Power Wide Area (LPWA) and Cellular IoT

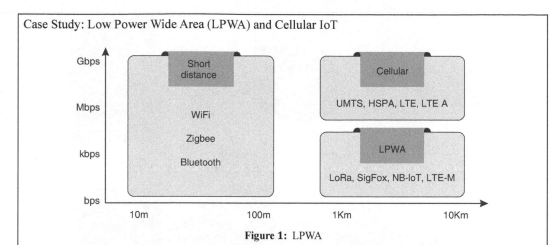

Figure 1: LPWA

Low Power Wide Area (LPWA) (see Figure 1) describes a set of technologies used for places where only small amounts of data need to be transferred but spread across a wide area. LPWA technologies can be divided between licensed and unlicensed spectrum: LoRa and SIGFOX work in unlicensed bands, while cellular IoT, such as NB-IoT and LTE-M, are typically used in the licensed spectrum[3] [3].

REST

REST stands for Representational State Transfer (REST) protocol. REST is the major application interface design used to build many enterprise applications.

Advantages of REST:

- Provides Service Oriented Architecture to call remote services
- Stateless and does not keep previous state history
- REST works on HTTP[4] and can be easily transportable

HTTP

The Hypertext Transfer Protocol (HTTP) is a stateless application level request/response protocol that uses extensible semantics and self-descriptive message payloads for flexible interaction with network-based hypertext information systems [4].

Client/Server Messaging

The Client initiates the connection to the server over a reliable transport or session-layer "Connection". An HTTP "client" is a program that establishes a connection to a server for sending one or more HTTP requests. An HTTP "server" is a program that accepts connections to service HTTP requests by sending HTTP responses (see Figure 2).

Intermediaries

HTTP enables the service of intermediaries that route the request from the Client to the Server. The intermediaries (A, B, and C) (see Figure 3) are between the user agent and origin server. A request or response message that travels the whole chain will pass through four separate connections. Some HTTP

---

[3]   Samsung IoT - https://www.samsung.com/global/business/networks/solutions/iot-solutions/

[4]   HTTP RFC 7230 - https://datatracker.ietf.org/doc/rfc7230/?include_text=1

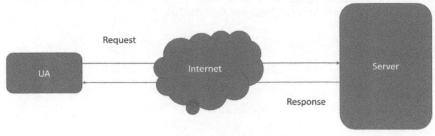

Figure 2: Internet Connectivity

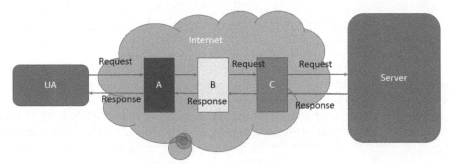

Figure 3: Internet Intermediaries

communication options might apply only to the connection with the nearest, non-tunnel neighbor, only to the endpoints of the chain, or to all connections along the chain. Although the diagram is linear, each participant might be engaged in multiple, simultaneous communications. For example, B might be receiving requests from many clients other than A, and/or forwarding requests to servers other than C, while it is handling A's request. Likewise, later requests might be sent through a different path of connections, often based on dynamic configuration for load balancing.

- Upstream & Downstream: All messages flow from upstream to downstream.
- Inbound: The requests routed to server
- Outbound: The requests routed to User Agent (UA)

Architecture

The REST Application interface sits on top of HTTP and can perform operations (PUT, POST, GET, HEAD). Six guiding principles define REST:

1. *Client-Server Architecture*: REST uses Client Server architecture with resources served by the Server, and the Client can access the server resources through remote procedure calls. The underlying principle is separation of concerns. Separating the user interface concerns from the data storage concerns improves the portability of the user interface across multiple platforms. It also improves scalability by simplifying the server components (see Figure 4).

2. *Statelessness*: In RESTful interface, there is no state representation between client and server. Like HTTP, the Restful interface maintains stateless nature. All the calls are atomic and does not carry and state representation.

3. *Cacheability*: As on the World Wide Web, clients and intermediaries can cache responses. Responses must therefore, implicitly or explicitly, define themselves as cacheable or not in order to prevent clients from getting stale or inappropriate data in response to further requests. A "cache" is a local store of previous response messages and the subsystem that controls its message storage, retrieval, and deletion. A cache stores cacheable response to reduce the response time and network bandwidth

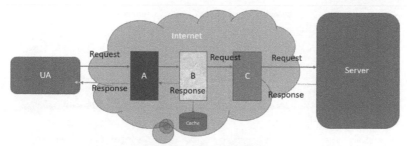

Figure 4: REST Connectivity

consumption on future, equivalent requests. Any client or server may employ a cache, though a cache cannot be used by a server while it is acting as a tunnel.[5]

In the diagram above (Figure 4), the response takes a shorter route as element B is responding from the cache. There is a constraint on what data and length of time data can be cached. In the grand scheme of request and response, a response is "cacheable" if a cache can store a copy of the response message for use in answering subsequent requests [4].

4. *Layered System*: A client cannot ordinarily tell whether it is connected directly to the end server, or to an intermediary along the way. Intermediary servers can improve system scalability by enabling load balancing and by providing shared caches.

5. *Code on demand*: Servers can temporarily extend or customize the functionality of a client by transferring executable code. For example, compiled components, such as Java applets, and client-side scripts, such as JavaScript.

6. *Uniform Interface*: The uniform interface constraint is fundamental to the design of any RESTful system. Restful interface achieves Uniform Interface though resource identifications in requests, resources manipulation through representations, Self-descriptive messages, and Hypermedia as the engine of the application state. Having accessed an initial URI for the REST application—analogous to a human Web user accessing the home page of a website—a REST client should then be able to use server-provided links dynamically in order to discover all the available actions and resources it needs.

AJAX

Ajax is a key tool in modern web application development. It allows you to send and retrieve data from a server asynchronously and process the data using JavaScript. Ajax is an acronym for *Asynchronous JavaScript and XML*. The name arose when XML was the data transfer format of choice although, given JSON, this is no longer the case [5].

Difference between REST and AJAX[6]

Using REST, we can do operations (PUT, POST, GET, HEAD) but by using AJAX we can only retrieve data from server side. AJAX is a set of (typically) client-sided web development techniques, while REST is an architecture style for sending and handling HTTP requests. So, you can use AJAX to send RESTful requests. A REST API is typically not implemented using AJAX but can be accessed by an AJAX client.

Ajax isn't a technology. It's really several technologies, each flourishing in their own right, coming together in powerful new ways (see Figure 5) [5].

---

[5] RFC 7230 HTTP - https://www.rfc-editor.org/rfc/pdfrfc/rfc7230.txt.pdf

[6] AJAX - https://courses.cs.washington.edu/courses/cse490h/07sp/readings/ajax_adaptive_path.pdf

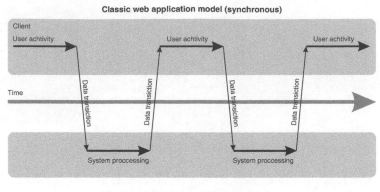

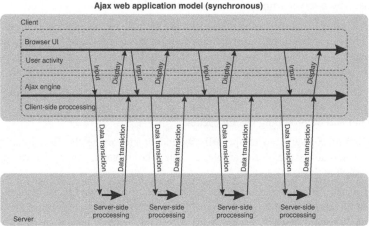

Figure 5: AJAX Request - Response

Ajax incorporates:

- standards-based presentation using XHTML and CSS
- dynamic display and interaction using the Document Object Model
- data interchange and manipulation using XML and XSLT
- asynchronous data retrieval using XMLHttpRequest
- JavaScript binding everything together

Classic Web Application Model

In a classic web application model, the user request is relayed to the web server. Until the operation is completed by the web server, the client is in wait mode. In other words, the client is locked until they have received the response from the Web Server.

In Ajax Web Application mode (see Figure 6), the blocking nature of request-response is overcome by the Ajax Compliant browsers that queues the request at the browser client. The blocking call from the User is relinquished as soon as the call is queued; once the response is received from the Web Server, the browser client routes the response to the response handler—java script code [5].

Ajax Code Demo—Web HTML

The following HTML code constructs a simple HTML Payload and calls RESTful call. The Restful call is initiated by Ajax [5]:

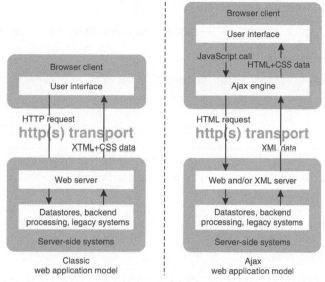

Figure 6: Ajax Architecture

```html
<!DOCTYPE HTML>
<html>
<head>
<title>Example</title>
</head>
<body>
<h3> Ajax example </h3>
<div>
<button>Apples</button>
<button>Cherries</button>
<button>Bananas</button>
</div>
<div id="target">
Press a button
</div>
<script>
var buttons = document.getElementsByTagName("button");
for (var i = 0; i < buttons.length; i++)
{
       buttons[i].onclick = handleButtonPress;
}

function handleButtonPress(e) {
       try
       {

              var httpRequest = new XMLHttpRequest();
              httpRequest.onreadystatechange = handleResponse;

              var reqInnerHTML=e.target.innerHTML + ".html";
              alert('request --' + reqInnerHTML)
              httpRequest.open("GET", reqInnerHTML);
              httpRequest.send();

       } catch (err)
       {
              alert(err);
       }
}
```

```
function handleResponse(e) {

        if (e.target.readyState == XMLHttpRequest.DONE && e.target.status ==
200) {
        alert(e.target.responseText);

        document.getElementById("target").innerHTML = e.target.responseText;
        }
}
</script>
</body>
</html>
```

The workings of Ajax to call Rest interface is in the code handle button press:

- Step 1: construct Ajax Object
- Step 2: set handler for onreadystatechange
- Step 3: Open request
- Step 4: Send HTTP request

Upon the successful completion of request by the server, the onready state change call back will be invoked. As soon as the script calls the send method, the browser makes the background request to the server. Because the request is handled in the background, Ajax relies on events to notify you about how the request progresses. In this example, these events are handled with the handleResponse function:

```
function handleResponse(e) {
        if (e.target.readyState == XMLHttpRequest.DONE &&
e.target. status == 200) { alert(e.target.                             responseText);

        document.getElementById("target").innerHTML =
e.target.responseText;
        }
}
```

When the readystatechange event is triggered, the browser passes an Event object to the specified handler function. Several different stages are signaled through the readystatechange event, and you can determine which one you are dealing with by reading the value of the XMLHttpRequest.readyState property. All the above code does is update innerHTML on target DIV upon completion of request.

Output: The output of the HTML page is Web with Ajax callable functions (see Figure 7).

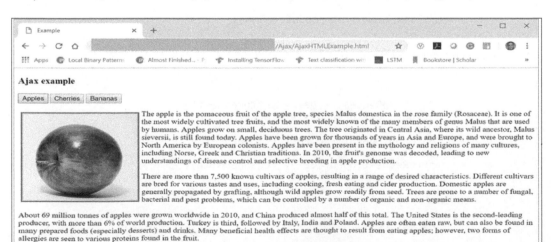

Figure 7: Ajax Example

XMLHttpRequest readyState

As soon as the script calls the send method, the browser makes the background

Value	Numeric Value	Description
UNSENT	0	The XMLHttpRequest object has been created.
OPENED	1	The open method has been called.
HEADERS_RECEIVED	2	The headers of the server response have been received.
LOADING	3	The response from the server is being received.
DONE	4	The response is complete or has failed.

Ajax Events

Most of these events are triggered at a point in the request. The exceptions are readystatechange and progress, which can be triggered several times in order to give progress updates.

Value	Description	Event Type
abort	Triggered when the requested is aborted	ProgressEvent
error	Triggered when the request fails	ProgressEvent
load	Triggered when the request completes successfully	ProgressEvent
loadend	Triggered when the request completes, either successfully or with an error	ProgressEvent
loadstart	Triggered when the request starts	ProgressEvent
progress	Triggered to indicate progress during the request	ProgressEvent
readystatechange	Triggered at different stages in the request life cycle	Event
timeout	Triggered if the request times out	ProgressEvent

Ajax Code—Web HTML Headers

The following HTML code retrieves the call headers (please see Table 1).

Table 1: AJAX HTML Code

```
<!DOCTYPE HTML>
<html>
<head>
<title>Example</title>
<style>
#allheaders, #ctheader {
border: medium solid black;
padding: 2px; margin: 2px;
}
</style>
</head>
<body>
<h3> Ajax example </h3>
<div>
<button>Apples</button>
```

Table 1 contd. ...

```
<button>Cherries</button>
<button>Bananas</button>
</div>
<div id="ctheader"></div>
<div id="allheaders"></div>
<div id="target">
Press a button
</div>
<script>
var buttons = document.getElementsByTagName("button");
for (var i = 0; i < buttons.length; i++)
{
        buttons[i].onclick = handleButtonPress;
}

var httpRequest;
function handleButtonPress(e) {
        try
        {

                httpRequest = new XMLHttpRequest();
                httpRequest.onreadystatechange = handleResponse;

                var reqInnerHTML=e.target.innerHTML + ".html";

                httpRequest.open("GET", reqInnerHTML);
                httpRequest.send();

        } catch (err)
        {
                alert(err);
        }
}
function handleResponse() {
        if (httpRequest.readyState == 2) {
                document.getElementById("allheaders").innerHTML = httpRequest.
getAllResponseHeaders();
                document.getElementById("ctheader").innerHTML =    httpRequest.
getResponseHeader("Content-Type");
                } else if (httpRequest.readyState == 4 && httpRequest.status ==
200) {
                document.getElementById("target").innerHTML = httpRequest.
responseText;
        }
}
</script>
</body>
</html>x
```

```
if (httpRequest.readyState == 2) {
        document.getElementById("allheaders").innerHTML = httpRequest.ge-
tAllResponseHeaders();
        document.getElementById("ctheader").innerHTML = httpRequest.
getResponseHeader("Content-Type");
        } else if (httpRequest.readyState == 4 && httpRequest.status == 200)
{
        document.getElementById("target").innerHTML      = httpRequest.re-
sponseText;
```

Output

Response header: includes date, Content-encoding, last-modified, Server, Accept-encoding, Content Type, and Content Length (see).

date: Fri, 23 Nov 2018 03:38:43 GMT **content-encoding**: gzip **etag**: "9d68cb63f3b7ce1:0" **last-modified:** Mon, 23 Sep 2013 00:25:21 GMT **server**: Microsoft-IIS/7.5 x-powered-by: ASP.NET vary: **Accept-Encoding content-type:** text/html accept-ranges: bytes **content-length**: 1312

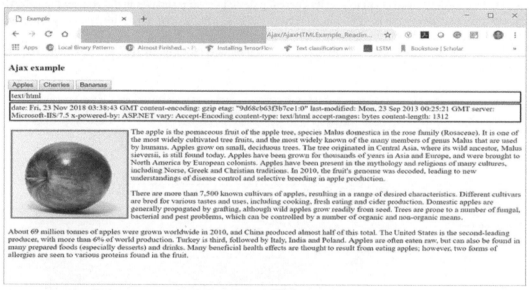

Figure 8: Ajax with Code header

CoAP (Constrained Application Protocol)

Constrained Application Protocol (CoAP)[7] is a lightweight protocol designed for Machine-to-Machine (M2M) communications within Internet of Things (IoT) applications. It allows machines to interact with each other using RESTful (HTTP-like) methods, while keeping message overhead and parsing complexity low.

DTLS (Datagram Transport Layer Security) is used to secure CoAP communication. In addition, DTLS solves the problems caused by UDP.

CoAP[8] *uses UDP*, not TCP, as the transport layer. UDP is suitable for small amounts of data, such as sensor data. However, it brings the issues of unreliable transmission and out-of-order packet delivery. DTLS implements the following mechanism to address these issues:

---

[7] CoAP - https://developer.artik.cloud/documentation/data-management/coap.html
[8] CoAP - http://coap.technology/impls.html

- packet retransmission
- assigning sequence number within the handshake
- replay detection

CoAP follows a client/server model [7]. Clients make requests to servers, servers send back responses. Clients may GET, PUT, POST and DELETE resources.

CoAP is designed to interoperate with HTTP and the RESTful web at large through simple proxies. Because CoAP is datagram based, it may be used on top of SMS and other packet-based communications protocols.[9]

Application Level QoS

Requests and response messages may be marked as "confirmable" or "nonconfirmable". Confirmable messages must be acknowledged by the receiver with an ACK packet.

Content Negotiation

Like HTTP, CoAP the uses client to specify content type and server responses in the content type. CoAP requests may use query strings in the form?a=b&c=d. These can be used to provide search, paging and other features to clients. Nonconfirmable messages are "fire and forget" [8].

As you can see, CoAP and MQTT are application protocols (see Figure 9) [7,8]. Wi-Fi and BLE are physical layer protocols.[10]

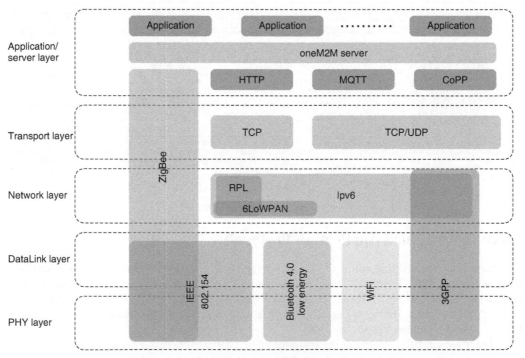

Figure 9: CoAP Architecture

---

9   CoAP-https://www.eclipse.org/community/eclipse_newsletter/2014/february/article2.php
10   https://www.slideshare.net/HamdamboyUrunov/the-constrained-application-protocol-coap-part-2?from_action=save

GET Method

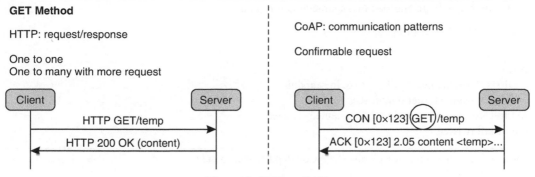

HTTP: request/response

One to one
One to many with more request

CoAP: communication patterns

Confirmable request

Figure 10: HTTP vs. CoAP

In HTTP request, one-to-one communication (see Figure 10) is established between Client and Server. In CoAP, some messages require a confirmation.

The CoAP protocol sends a one-to-one communication request. Unlike HTTP, the CoAP response is one-to-one (see Figure 11). The CoAP message contains the following header (see Figure 12):

CoAP– request response

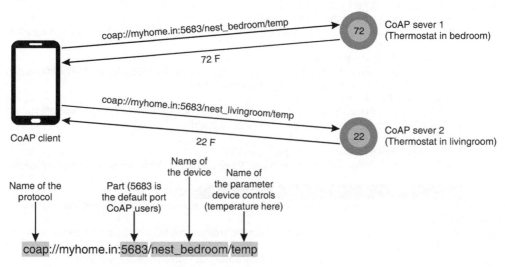

Figure 11: CoAP Request Response

```
 0                   1                   2                   3
 0 1 2 3 4 5 6 7 8 9 0 1 2 3 4 5 6 7 8 9 0 1 2 3 4 5 6 7 8 9 0 1
+-+-+-+-+-+-+-+-+-+-+-+-+-+-+-+-+-+-+-+-+-+-+-+-+-+-+-+-+-+-+-+-+
|Ver| T |  TKL  |      Code     |          Message ID           |
+-+-+-+-+-+-+-+-+-+-+-+-+-+-+-+-+-+-+-+-+-+-+-+-+-+-+-+-+-+-+-+-+
|   Token (if any, TKL bytes) ...
+-+-+-+-+-+-+-+-+-+-+-+-+-+-+-+-+-+-+-+-+-+-+-+-+-+-+-+-+-+-+-+-+
|   Options (if any) ...
+-+-+-+-+-+-+-+-+-+-+-+-+-+-+-+-+-+-+-+-+-+-+-+-+-+-+-+-+-+-+-+-+
|1 1 1 1 1 1 1 1|    Payload (if any) ...
+-+-+-+-+-+-+-+-+-+-+-+-+-+-+-+-+-+-+-+-+-+-+-+-+-+-+-+-+-+-+-+-+
```

Figure 12: CoAP Message

CoAP messages[11] are encoded in a simple binary format.

The Message Header (4 bytes).

Installation:
Error:
  File "CoAPClientGet.py", line 16, in <module>
  from twisted.internet import reactor
ModuleNotFoundError: No module named 'twisted'
pip install twisted

```
base) C:\Hanumayamma\CRCBook\Code\CoAP>pip install twisted
Collecting twisted
  Downloading https://files.pythonhosted.org/packages/5d/0e/
a72d85a55761c2c3ff1cb968143a2fd5f360220779ed90e0fadf4106d4f2/Twisted-18.9.0.tar.bz2 (3.1MB)
    100% |                                        | 3.1MB 1.7MB/s
Collecting zope.interface>=4.4.2 (from twisted)
  Downloading https://files.pythonhosted.org/packages/da/08/726e3b0e3bd9912fb530f9864bf9a3af9f9f6a1
dfd4cc7854ca14fdab441/zope.interface-4.6.0-cp36-cp36m-win_amd64.whl (133kB)
    100% |                                        | 143kB 3.3MB/s
Collecting constantly>=15.1 (from twisted)
  Downloading https://files.pythonhosted.org/packages/b9/65/48c1909d0c0aeae6c10213340ce682db01b48e
a900a7d9fce7a7910ff318/constantly-15.1.0-py2.py3-none-any.whl
Collecting incremental>=16.10.1 (from twisted)
  Downloading https://files.pythonhosted.org/packages/f5/1d/
c98a587dc06e107115cf4a58b49de20b19222c83d75335a192052af4c4b7/incremental-17.5.0-py2.py3-none-
any.whl
Collecting Automat>=0.3.0 (from twisted)
  Downloading https://files.pythonhosted.org/packages/a3/86/14c16bb98a5a3542ed8fed5d74fb064a902de3
bdd98d6584b34553353c45/Automat-0.7.0-py2.py3-none-any.whl
Collecting hyperlink>=17.1.1 (from twisted)
  Downloading https://files.pythonhosted.org/packages/a7/b6/84d0c863ff81e8e7de87cff3bd8fd8f1054c227
ce09af1b679a8b17a9274/hyperlink-18.0.0-py2.py3-none-any.whl
Collecting PyHamcrest>=1.9.0 (from twisted)
  Downloading https://files.pythonhosted.org/packages/9a/d5/
d37fd731b7d0e91afcc84577edeccf4638b4f9b82f5ffe2f8b62e2ddc609/PyHamcrest-1.9.0-py2.py3-none-
any.whl (52kB)
    100% |                                        | 61kB 1.1MB/s
Requirement already satisfied: attrs>=17.4.0 in c:\hanumayamma\anaconda3\lib\site-packages (from
twisted) (18.1.0)
Requirement already satisfied: setuptools in c:\hanumayamma\anaconda3\lib\site-packages (from zope.
interface>=4.4.2->twisted) (39.1.0)
Requirement already satisfied: six in c:\hanumayamma\anaconda3\lib\site-packages (from
Automat>=0.3.0->twisted) (1.11.0)
Requirement already satisfied: idna>=2.5 in c:\hanumayamma\anaconda3\lib\site-packages (from
hyperlink>=17.1.1->twisted) (2.6)
Building wheels for collected packages: twisted
  Running setup.py bdist_wheel for twisted ... done
  Stored in directory: C:\Users\cvuppalapati\AppData\Local\pip\Cache\wheels\57\2e\89\11ba83bc08ac30a5
e3a6005f0310c78d231b96a270def88ca0
Successfully built twisted
Installing collected packages: zope.interface, constantly, incremental, Automat, hyperlink, PyHamcrest,
twisted
Successfully installed Automat-0.7.0 PyHamcrest-1.9.0 constantly-15.1.0 hyperlink-18.0.0
incremental-17.5.0 twisted-18.9.0 zope.interface-4.6.0
You are using pip version 18.0, however version 18.1 is available.
You should consider upgrading via the 'python -m pip install --upgrade pip' command.
```

---

[11]   CoAP - https://www.slideshare.net/HamdamboyUrunov/the-constrained-application-protocol-coap-part-2?from_action=save

File "CoAPClientGet.py", line 19, in <module>

  import txthings.coap as coap

ModuleNotFoundError: No module named 'txthings'

(base) C:\Hanumayamma\CRCBook\Code\CoAP>pip install txThings

Collecting txThings

 Downloading https://files.pythonhosted.org/packages/58/32/498ddf5b5b14d2039772dc29ae7179e15c3638
4b459759c3e77eb0bf0f8f/txThings-0.3.0-py3-none-any.whl

Requirement already satisfied: six>=1.10.0 in c:\hanumayamma\anaconda3\lib\site-packages (from
txThings) (1.11.0)

Requirement already satisfied: twisted>=14.0.0 in c:\hanumayamma\anaconda3\lib\site-packages (from
txThings) (18.9.0)

Requirement already satisfied: PyHamcrest>=1.9.0 in c:\hanumayamma\anaconda3\lib\site-packages (from
twisted>=14.0.0->txThings) (1.9.0)

Requirement already satisfied: hyperlink>=17.1.1 in c:\hanumayamma\anaconda3\lib\site-packages (from
twisted>=14.0.0->txThings) (18.0.0)

Requirement already satisfied: Automat>=0.3.0 in c:\hanumayamma\anaconda3\lib\site-packages (from
twisted>=14.0.0->txThings) (0.7.0)

Requirement already satisfied: attrs>=17.4.0 in c:\hanumayamma\anaconda3\lib\site-packages (from
twisted>=14.0.0->txThings) (18.1.0)

Requirement already satisfied: constantly>=15.1 in c:\hanumayamma\anaconda3\lib\site-packages (from
twisted>=14.0.0->txThings) (15.1.0)

Requirement already satisfied: incremental>=16.10.1 in c:\hanumayamma\anaconda3\lib\site-packages
(from twisted>=14.0.0->txThings) (17.5.0)

Requirement already satisfied: zope.interface>=4.4.2 in c:\hanumayamma\anaconda3\lib\site-packages
(from twisted>=14.0.0->txThings) (4.6.0)

Requirement already satisfied: setuptools in c:\hanumayamma\anaconda3\lib\site-packages (from
PyHamcrest>=1.9.0->twisted>=14.0.0->txThings) (39.1.0)

Requirement already satisfied: idna>=2.5 in c:\hanumayamma\anaconda3\lib\site-packages (from
hyperlink>=17.1.1->twisted>=14.0.0->txThings) (2.6)

Installing collected packages: txThings

Successfully installed txThings-0.3.0

You are using pip version 18.0, however version 18.1 is available.

You should consider upgrading via the 'python-m pip install--upgrade pip' command.

Additional refernces:

https://github.com/siskin/txThings/

http://coap.technology/impls.html

https://github.com/siskin/txThings/

https://github.com/mwasilak/txThings

The following diagram (Figure 13) illustrates where CoAP and DTLS are in the protocol stack, and the DTLS handshake process between the client and server [7,8].

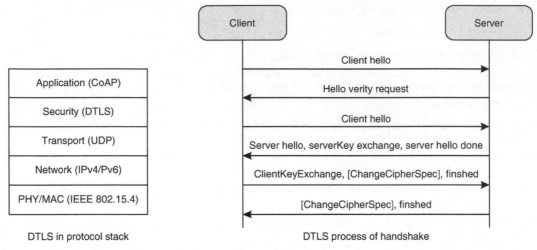

Figure 13: DTLS Protocol

CoAP Client Get

CoAP client code performs single GET request to coap.me port 5683 (official IANA assigned CoAP port), URI "test". Request is sent 1 second after initialization. Remote IP address is hardcoded—no DNS lookup is preformed (please see Table 2).

Table 2: CoAP Client Code

```
# -*- coding: utf-8 -*-
"""
Created on Sun Nov 25 01:47:16 2018
@author: cvuppalapati
"""

"""
import sys
from ipaddress import ip_address
from twisted.internet import reactor
from twisted.python import log
import txthings.coap as coap
import txthings.resource as resource
class Agent:
"""

    Example class which performs single GET request to coap.me
    port 5683 (official IANA assigned CoAP port), URI "test".
    Request is sent 1 second after initialization.
    Remote IP address is hardcoded - no DNS lookup is preformed.
    Method requestResource constructs the request message to
    remote endpoint. Then it sends the message using protocol.request().
    A deferred 'd' is returned from this operation.
    Deferred 'd' is fired internally by protocol, when complete response is received.
    Method printResponse is added as a callback to the deferred 'd'. This
    method's main purpose is to act upon received response (here it's simple print).
    """

    def __init__(self, protocol):
        self.protocol = protocol
        reactor.callLater(1, self.requestResource)
```

Table 2 contd. ...

...Table 2 contd.

```
    def requestResource(self):
        request = coap.Message(code=coap.GET)
        # Send request to "coap://coap.me:5683/test"
        request.opt.uri_path = (b'test',)
        request.opt.observe = 0
        request.remote = (ip_address("134.102.218.18"), coap.COAP_PORT)
        d = protocol.request(request, observeCallback=self.printLaterResponse)
        d.addCallback(self.printResponse)
        d.addErrback(self.noResponse)

    def printResponse(self, response):
        print('First result: ' + str(response.payload, 'utf-8'))
        # reactor.stop()

    def printLaterResponse(self, response):
        print('Observe result: ' + str(response.payload, 'utf-8'))

    def noResponse(self, failure):
        print('Failed to fetch resource:')
        print(failure)
        # reactor.stop()

log.startLogging(sys.stdout)

endpoint = resource.Endpoint(None)
protocol = coap.Coap(endpoint)

client = Agent(protocol)
reactor.listenUDP(61616, protocol) # , interface="::")
reactor.run()
```

Output

The output of the above application generates the following console output (Figure 14).

Figure 14: CoAP Client output

CoAP Server

The CoAP Server is run and the following screen (see Figure 15) provides the CoAP Server.

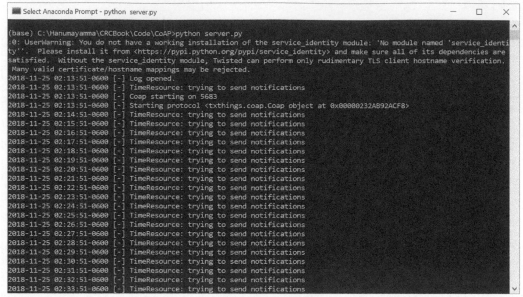

Figure 15: CoAP Server

Bluetooth Low Energy (BLE)

Bluetooth wireless technology is a short-range communications system intended to replace the cable(s) connecting portable and/or fixed electronic devices. The key features of Bluetooth wireless technology are robustness, low power consumptioh, and low cost [9]. Many features of the core specification are optional, allowing product differentiation.[12]

There are two forms of Bluetooth wireless technology systems: Basic Rate (BR) and Low Energy (LE). Both systems include device discovery, connection establishment and connection mechanisms (see Figure 16) [10].

The Basic Rate system, for instance Bluetooth available on Mac Systems, includes optional Enhanced Data Rate (EDR), Alternate Media Access Control (MAC) and Physical (PHY) layer extensions. The Basic Rate system offers synchronous and asynchronous connections with data rates of 721.2 kb/s for Basic Rate, 2.1 Mb/s for Enhanced Data Rate and high-speed operation up to 54 Mb/s with the 802.11 AMP [10].

The LE system, in iOS and Android phones, includes features designed to enable products that require lower current consumption, lower complexity and lower cost than BR/EDR. The LE system is also designed for Case Studys and applications with lower data rates and has lower duty cycles. Depending on the Case Study or application, one system including any optional parts may be more optimal than the other [10].

Bluetooth Core

The Bluetooth core system consists of a Host and one or more Controllers. A Host is a logical entity defined as all the layers below the non-core profiles and above the Host Controller Interface (HCI) (see

Figure 16: Bluetooth Connection

---

[12]   Bluetooth Core Specification - https://www.bluetooth.com/specifications/protocol-specifications

Figure 17). A Controller is a logical entity defined as all the layers below HCI. An implementation of the Host and Controller may contain the respective parts of the HCI. Two types of Controllers are defined in this version of the Core Specification: Primary Controllers and Secondary Controllers.

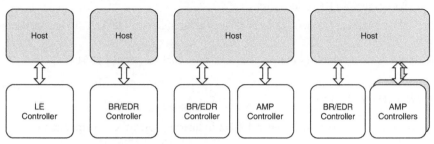

Figure 17: Bluetooth Host and Controller Combination [10]

Bluetooth Core System Architecture

The following figure (see Figure 18) provides Bluetooth core architecture:[13] The Core blocks, each with its associated communication protocol [10].

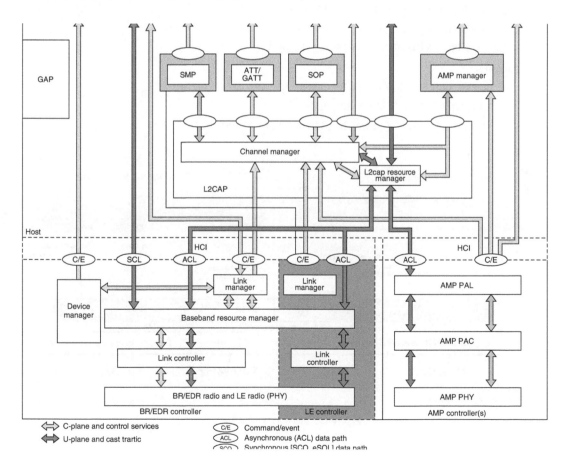

Figure 18: Bluetooth Core Architecture

---

[13] Bluetooth Specification - https://www.bluetooth.com/specifications/protocol-specifications

- Link Manager, Link Controller and BR/EDR Radio blocks comprise a BR/EDR Controller.
- An AMP PAL, AMP MAC, and AMP PHY comprise an AMP Controller.
- Link Manager, Link Controller and LE Radio blocks comprise an LE Controller. L2CAP, SDP and GAP blocks comprise a BR/EDR Host.
- L2CAP, SMP, Attribute protocol, GAP and Generic Attribute Profile (GATT) blocks comprise an LE Host.
- A BR/EDR/LE Host combines the set of blocks from each respective Host.

Bluetooth Low Energy

The LE radio operates in the unlicensed 2.4 GHz industrial, scientific and medical radio (ISM) band (see Figure 19)[14] [11].

Unlicensed ISM & Short-Range Device (SRD)

Here are the details of Unlicensed ISM & Short-Range Device (SRD) [11]:

- **USA/Canada**:
  - 260 – 470 MHz (FCC Part 15.231; 15.205)
  - 902 – 928 MHz (FCC Part 15.247; 15.249)
  - 2400 – 2483.5 MHz (FCC Part 15.247; 15.249)
- **Europe:**
  - 433.050 – 434.790 MHz (ETSI EN 300 220)
  - 863.0 – 870.0 MHz (ETSI EN 300 220)
  - 2400 – 2483.5 MHz (ETSI EN 300 440 or ETSI EN 300 328)

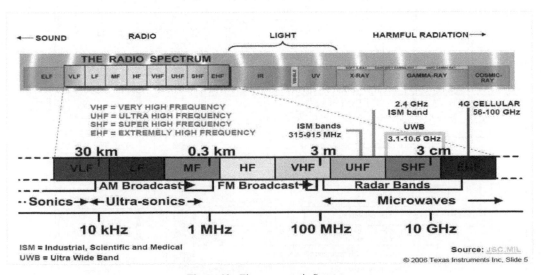

Figure 19: Electromagnetic Spectrum

---

14   RF Basics for Non-RF Engineers - http://www.ti.com/lit/ml/slap127/slap127.pdf

- **Japan:**
  - 315 MHz (Ultra low power applications)
  - 426–430, 449, 469 MHz (ARIB STD-T67)
  - 2400–2483.5 MHz (ARIB STD-T66)
  - 2471–2497 MHz (ARIB RCR STD-33)

LE employs two multiple access schemes: Frequency division multiple access (FDMA) and time division multiple access (TDMA). Forty (40) physical channels, separated by 2 MHz, are used in the FDMA scheme. Three (3) are used as primary advertising channels and 37 are used as secondary advertising channels and as data channels. A TDMA based polling scheme is used, in which one device transmits a packet at a predetermined time and a corresponding device responds with a packet after a predetermined interval [10].

Devices that transmit advertising packets on the advertising PHY channels are referred to as advertisers. Devices that receive advertising packets (see Figure 20) on the advertising channels without the intention to connect to the advertising device are referred to as scanners. Transmissions on the advertising PHY channels occur in advertising events. At the start of each advertising event, the advertiser sends an advertising packet corresponding to the advertising event type. Depending on the type of advertising packet, the scanner may make a request to the advertiser on the same advertising PHY channel, which may be followed by a response from the advertiser on the same advertising PHY channel. The advertising PHY channel changes on the next advertising packet sent by the advertiser in the same advertising event. The advertiser may end the advertising event at any time during the event. The first advertising PHY channel is used at the start of the next advertising event.

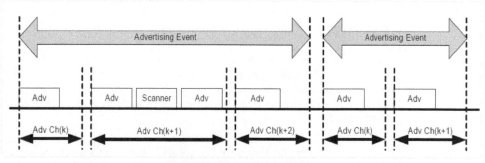

Figure 20: Bluetooth Advertising

Bluetooth Low Energy Protocol

Bluetooth Low Energy (BLE, also marketed as Bluetooth Smart) started as part of the Bluetooth 4.0 Core Specification. It's tempting to present BLE as a smaller, highly optimized version of its bigger brother, classic Bluetooth, however, BLE has an entirely different lineage and design goals.

While Bluetooth Low Energy is a good technology on its own merit, what makes BLE genuinely exciting—and what has pushed its phenomenal adoption rate so far so quickly—is that it's the right technology, with the right compromises, at the right time. For a relatively young standard (it was introduced in 2010), BLE has seen an uncommonly rapid adoption rate, and the number of product designs that already include BLE puts it well ahead of other wireless technologies at the same point of time in their release cycles.

The Bluetooth Protocol stack (see Figure 21) has GATT, ATT, and L2CAP.

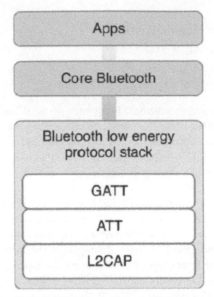

Figure 21: Bluetooth Protocol Stack

Key Terms	Description
Generic Attribute Profile (GATT)	The GATT profile is a general specification for sending and receiving short pieces of data known as "attributes" over a BLE link. All current Low Energy application profiles are based on GATT.
Attribute Protocol (ATT)	GATT is built on top of the Attribute Protocol (ATT). This is also referred to as GATT/ ATT. ATT is optimized to run on BLE devices. To this end, it uses as few bytes as possible. Each attribute is uniquely identified by a Universally Unique Identifier (UUID), which is a standardized 128-bit format for a string ID used to uniquely identify information.
L2CAP	L2CAP is used within the Bluetooth protocol stack. It passes packets to either the Host Controller Interface (HCI) or on a hostless system, directly to the Link Manager/ACL link.

Compared to other wireless standards, the rapid growth of BLE is relatively easy to explain: BLE has gone further faster because its fate is so intimately tied to the phenomenal growth in smartphones, tablets, and mobile computing. Early and active adoption of BLE by mobile industry heavyweights like Apple and Samsung broke open the doors for wider implementation of BLE. Apple has put significant effort into producing a reliable BLE stack and publishing design guidelines around BLE. This, in turn, pushed silicon vendors to commit their limited resources to the technology they felt was the most likely to succeed or flourish in the long run, and the Apple stamp of approval is clearly a compelling argument when you need to justify every research and development dollar invested. While the mobile and tablet markets become increasingly mature and costs and margins are decreasing, the need for connectivity with the outside world on these devices has a huge growth potential, and it offers peripheral vendors a unique opportunity to provide innovative solutions to problems people might not even realize that they have today.

There are two major players involved in all Bluetooth low energy communication: the central and the peripheral (see Figure 22). Based on a somewhat traditional client-server architecture, a peripheral typically has data that is needed by other devices. A central typically uses the information served up by peripherals to

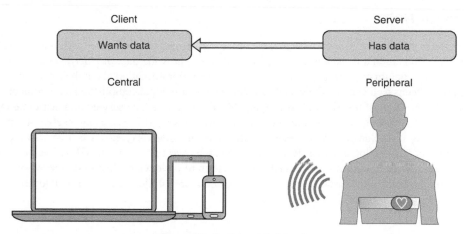

Figure 22: Central and Peripheral

accomplish some task. As Figure 22 shows, for example, a heart rate monitor may have useful information that your Mac or iOS app may need in order to display the user's heart rate in a user-friendly way.

Core Bluetooth Framework in iOS

The Core Bluetooth framework provides the classes needed for your iOS and Mac apps to communicate with devices that are equipped with BLE wireless technology. For example, your app can discover, explore, and interact with low energy peripheral devices, such as heart rate monitors and digital thermostats. As of OS X v10.9 and iOS 6, Mac and iOS devices can also function as BLE peripherals, serving data to other devices, including other Mac and iOS devices.

Centrals Discover and Connect to Peripherals that are Advertising

Peripherals broadcast some of the data they have in the form of advertising packets. An advertising packet (see Figure 23) is a relatively small bundle of data that may contain useful information about what a peripheral has to offer, such as the peripheral's name and primary functionality. For instance, a digital thermostat may advertise that it provides the current temperature of a room. In Bluetooth low energy, advertising is the primary way that peripherals make their presence known. A central, on the other hand, can scan and listen for any peripheral device that is advertising information that it's interested in, as shown in Figure 23 central can ask to connect to any peripheral that it has discovered advertising.

Figure 23: Central and Peripheral Advertisement

Peripheral Data

The purpose of connecting to a peripheral is to begin exploring and interacting with the data it has to offer. Before you can do this, however, it helps to understand how the data of a peripheral is structured. Peripherals may contain one or more services or provide useful information about their connected signal strength. A service is a collection of data and associated behaviors for accomplishing a function or feature of a device (or portions of that device) (see Figure 24). For example, one service of a heart rate monitor may be to expose heart rate data from the monitor's heart rate sensor. Services themselves are made up of characteristics or included services (that is, references to other services). A characteristic provides further details about a peripheral's service. For example, the heart rate service just described may contain one characteristic that describes the intended body location of the device's heart rate sensor and another characteristic that transmits heart rate measurement data. Figure 24 illustrates one possible structure of a heart rate monitor's service and characteristics.

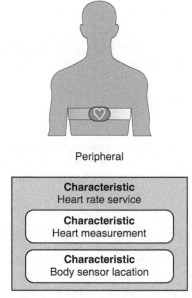

Peripheral

Characteristic
Heart rate service

Characteristic
Heart measurement

Characteristic
Body sensor lacation

Figure 24: Peripheral Characteristics

CB Peripheral

The CBPeripheral class represents remote peripheral devices that your app, by means of a central manager (an instance of CBCentralManager), has discovered advertising or is currently connected to (see Figure 25). Peripherals are identified by universally unique identifiers (UUIDs), represented by NSUUID objects. Peripherals may contain one or more services or provide useful information about their connected signal strength.

You use this class to discover, explore, and interact with the services available on a remote peripheral that supports Bluetooth low energy. A service encapsulates the way part of the device behaves. For example, one service of a heart rate monitor may be to expose heart rate data from the monitor's heart rate sensor. Services themselves are made up of characteristics or included services (references to other services).

Characteristics (see Figure 26) provide further details about a peripheral's service. For example, the heart rate service just described may contain one characteristic that describes the intended body location of the device's heart rate sensor and another characteristic that transmits heart rate measurement data. Finally, characteristics contain any number of descriptors that provide more information about the characteristic's value, such as a human-readable description and a way to format the value.

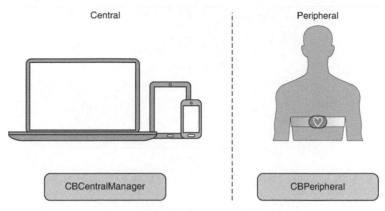

Figure 25: Data Represented

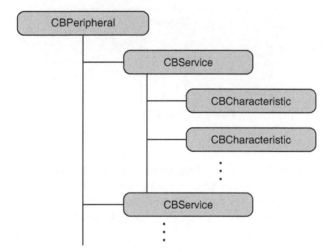

Figure 26: Characteristics

Peripheral Call Sequence

The following are the call sequence of Peripherals [12]:

- Discovering Services
- Discovering Characteristics and Characteristic Descriptors
- Reading Characteristic and Characteristic Descriptor Values
- Writing Characteristic and Characteristic Descriptor Values
- Setting Notifications for a Characteristic's Value
- Accessing a Peripheral's Received Signal Strength Indicator (RSSI) Data

iOS App—IoT made easy—Sensor Tag

Texas Instruments IoT Sensor Tag [12] provides rich combination Sensors and Connectivity features.[15]

The following code uses Sensor Tag iOS App (see Figure 27) to connect to Sensor Tag and retrieve the Sensor values [12].

---

[15]   TI Sensor Tag - http://www.ti.com/ww/en/wireless_connectivity/sensortag/

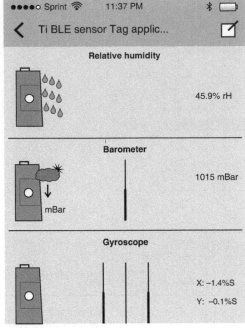

Figure 27: Sensor Tag Mobile UI

Starting Up a Central Manager

Since a CBCentralManager object is the Core Bluetooth object-oriented representation of a local central device, you must allocate and initialize a central manager instance before you can perform any Bluetooth low energy transactions (see Figure 28). You can start up your central manager by calling the initWithDelegate:queue:options: method of the CBCentralManager class, like this:

```
myCentralManager =
[[CBCentralManager alloc] initWithDelegate:self queue:niloptions:nil];
```

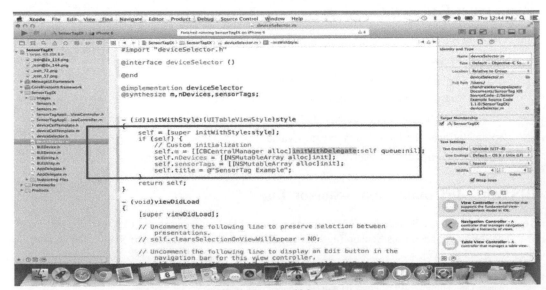

Figure 28: Central Manager

Discovering Peripheral Devices That Are Advertising

One of the first central-side tasks that you are likely to perform is to discover what peripheral devices are available for your app to connect to. As mentioned earlier in Centrals Discover and Connect to Peripherals That Are Advertising, advertising is the primary way that peripherals make their presence known. You can discover any peripheral devices that are advertising by calling the scanForPeripheralsWithServices:optio ns: method of the CBCentralManager class, like this (see Figure 29).

```
[myCentralManager scanForPeripheralsWithServices:nil options:nil];
```

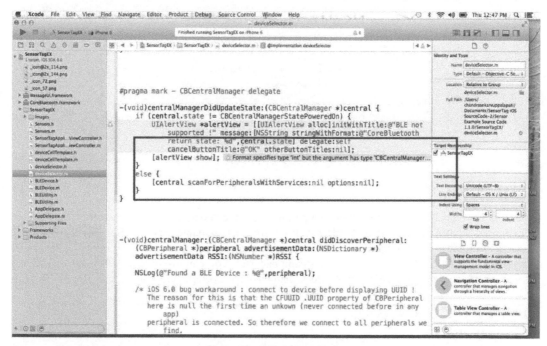

Figure 29: Scan for peripherals

Connecting to a Peripheral Device After You've Discovered it

After you have discovered a peripheral device that is advertising services you are interested in, you can request a connection to the peripheral by calling the connectPeripheral:options: method of the CBCentralManager class (see Figure 30). Simply call this method and specify the discovered peripheral that you want to connect to, like this:

```
[myCentralManager connectPeripheral:peripheral options:nil];
```

Figure 30: Connect Peripherals

Discover Services

After you have established a connection to a peripheral, you can begin to explore its data. The first step in exploring what a peripheral has to offer is discovering its available services. Because there are size restrictions on the amount of data a peripheral can advertise, you may discover that a peripheral has more services than what it advertises (in its advertising packets). You can discover all of the services that a peripheral offers by calling the discoverServices: method of the CBPeripheral class (see Figure 31), like this

```
[peripheral discoverServices:nil];
```

Figure 31: Discover Services

```
[peripheral discoverServices:nil];
```

Discover Characteristic of a Service

If you have found a service that you are interested in, the next step in exploring what a peripheral has to offer is discovering all of the service's characteristics. Discovering all of the characteristics of a service is as simple as calling the discoverCharacteristics: forService: method of the CBPeripheral class, specifying the appropriate service, like this (see Figure 32 and Figure 33):

```
NSLog(@"Discovering characteristics for service %@", interestingService);
[peripheral discoverCharacteristics:nil forService:interestingService];
```

Figure 32: Discover Characteristics

Figure 33: Characteristics

Case Study: IoT Made Easy—Texas Instruments Sensor Tag[16]

Please find technical specification of different protocols and important attributes:

	Bluetooth low energy	BLE + Sub-1 GHz	6LoWPAN	ZigBee	Wi-Fi
Battery type	Coin cell	Coin cell	Coin cell	Coin cell	AAA
Connects to Internet	Smartphone	Smartphone	BeagleBone gateway	BeagleBone gateway	Directly
BLE Beacon support	**Yes**	**Yes**			
DevPack support	**Yes**	**Yes**	**Yes**	**Yes**	
Mesh network			Yes	Yes	
Range	50 m/150 ft	100 m/300 ft (Bluetooth), 2 km (Sub-1GHz)	100 m/300 ft (extended with mesh network)	100 m/300 ft (extended with mesh network)	100 m/300 ft
Max number of devices	8	8	200	200	64, varies by Wi-Fi router
Battery lifetime*	1 year (1 second report interval)	1 year (1 second report interval)	1 year (1 second report interval)	1 year (1 second report interval)	3 months (1 minute update interval)
User Interface	App	App	App, Web	App, Web	App

---

16   TI Sensor Tag -http://www.ti.com/ww/en/wireless_connectivity/sensortag/

Android

Android 4.3 (API Level 18) introduces built-in platform support for Bluetooth Low Energy in the central role and provides APIs that apps can use to discover devices, query for services, and read/write characteristics. In contrast to Classic Bluetooth, BLE is designed to provide significantly lower power consumption. This allows Android apps to communicate with BLE devices that have low power requirements, such as proximity sensors, heart rate monitors, fitness devices, and so on.[17]

The Android platform includes support for the Bluetooth network stack, which allows a device to wirelessly exchange data with other Bluetooth devices. The application framework provides access to the Bluetooth functionality through the Android Bluetooth APIs. These APIs let applications wirelessly connect to other Bluetooth devices, enabling point-to-point and multipoint wireless features [13].

Using the Bluetooth APIs, an Android application can perform the following:

- Scan for other Bluetooth devices
- Query the local Bluetooth adapter for paired Bluetooth devices
- Establish RFCOMM channels
- Connect to other devices through service discovery
- Transfer data to and from other devices
- Manage multiple connections

This page focuses on Classic Bluetooth. Classic Bluetooth is the right choice for more battery-intensive operations, which include streaming and communicating between Android devices. For Bluetooth devices with low power requirements, Android 4.3 (API level 18) introduces API support for Bluetooth Low Energy. Android Permissions:

```
<uses-permission android:name="android.permission.BLUETOOTH"/>
<uses-permission android:name="android.permission.BLUETOOTH_ADMIN"/>
```

Scan for the device

Start LE Scan starts the peripheral scan.[18]

```
/**
 * Activity for scanning and displaying available BLE devices.
 */
public class DeviceScanActivity extends ListActivity {

    private BluetoothAdapter mBluetoothAdapter;
    private boolean mScanning;
    private Handler mHandler;

    // Stops scanning after 10 seconds.
    private static final long SCAN_PERIOD = 10000;
    ...
    private void scanLeDevice(final boolean enable) {
        if (enable) {
            // Stops scanning after a pre-defined scan period.
            mHandler.postDelayed(new Runnable() {
```

[17] Android Bluetooth Framework - https://developer.android.com/guide/topics/connectivity/bluetooth
[18] Start Bluetooth Scan - https://developer.android.com/guide/topics/connectivity/bluetooth-le#java

```
                    @Override
                    public void run() {
                        mScanning = false;
                        mBluetoothAdapter.stopLeScan(mLeScanCallback);
                    }
            }, SCAN_PERIOD);

            mScanning = true;
            mBluetoothAdapter.startLeScan(mLeScanCallback);
        } else {
            mScanning = false;
            mBluetoothAdapter.stopLeScan(mLeScanCallback);
        }
        ...
    }
    ...
}
```

Hanumayamma Dairy IoT Design

The Silicon Valley based electronic company[19] manufactures Dairy Cow necklaces for dairy farms worldwide. The Sensor works with Bluetooth Low Energy and collects Dairy cattle data across the globe and develop machine learning solutions for Dairy Farmers.

Sensor Operation

- Sensor measures skin temperature every 10 minutes.
- Each sensor has an internal data buffer, which can store up to 30 measurements.
- Each measurement data consists of data consists of
  - Timestamp
  - Skin temperature 1
  - Skin temperature 2
  - Case temperature 1
  - Case temperature 2
  - Checksum
- If the buffer is not empty, the sensor broadcast BLE packets for 5 seconds, waiting for a gateway to be connected. The BLE packet complies with iBeacon packet format.
- The sensor goes to SLEEP mode if it does not receive a connection request for the next 5 seconds after starting BLE broadcasting. The sensor wakes up every 3 minutes to broadcast if it has data in the buffer.
- If a connection is made, the gateway can request data to the sensor.

---

[19]   Silicon Valley Company – Hanumayamma Innovations and Technologies, Inc http://www.hanuinnotech.com

BLE Broadcasting

A Cow necklace sensor broadcasts BLE packets that comply with iBeacon format (except the UUID) if data is available in the buffer. Use the following UUID to scan Cow Necklace sensors.

cb30a7d0-5cf2-11e5-885d-feff819cdc9f

Each sensor has a unique pair of Major and Minor numbers. You can use these numbers as sensor IDs.

Bluetooth GATT

You can connect to a sensor using Bluetooth GATT profile. Use the following UUIDs to
- (Service) Discover Bluetooth services,
  - ○ 00002220-0000-1000-8000-00805f9b34fb
- (Characteristic) Receive data from the sensor,
  - ○ 00002221-0000-1000-8000-00805f9b34fb
- (Characteristic) Send data to the sensor,
  - ○ 00002222-0000-1000-8000-00805f9b34fb
- (Descriptor) Configure
  - ○ 00002902-0000-1000-8000-00805f9b34fb

Once a connection is made, you can receive and send bytes from/to the sensor by reading and writing the corresponding characteristics.

Reading sensor data

To read data from the sensor, send the following bytes to the sensor:

READ := { 0xC7, 0x7C, 0xC1 }

If the sensor has data, it will send 20 bytes in the following format:

Data[0] = 0xFE
Data[1] = 0xEF
Data[2] = Second
Data[3] = Minute
Data[4] = Hour
Data[5] = Month day
Data[6] = Month
Data[7] = Year
Data[8] = skin_temp_1_H
Data[9] = skin_temp_1_L
Data[10] = case_temp_1_H
Data[11] = case_temp_1_L
Data[12] = Reserved
Data[13] = Reserved
Data[14] = Battery level %
Data[15] = skin_temp_2_H
Data[16] = skin_temp_2_L
Data[17] = case_temp_2_H
Data[18] = case_temp_2_L
Data[19] = Checksum

Use 'uint8_t' or 'unsigned char' to convert data. Use Data[2]..[7] as timestamp.

Checksum: The sum of all 20 bytes should be equal to zero when data is valid. The sensor expected to receive either ACK or NACK within 200 msec after sending data.

ACK := { 0xAE, 0xEA, 0xA0 }

NACK := { 0xAE, 0xEA, 0xA1 }

Send ACK if data is valid, NACK otherwise. If sensor receives ACK, it will pop the data from the buffer. Note that a sensor can buffer up to 30 data packets. You can send a READ command to read another buffered data. If sensor has no data in its buffer, it will send the following 3 bytes on a READ command:

NO_DATA = {0xB3, 0x3B, 0xD0 }

You can disconnect if you received NO_DATA from the sensor.

Case Study: Dairy Application—Cow Necklace

Android Manifest XML

See Table 3 code for the Android Manifest file.

Table 3: Cow necklace code

```xml
<?xml version="1.0" encoding="utf-8"?>
<manifest xmlns:android="http://schemas.android.com/apk/res/android"
   package="vamshi.com.dairyapplication">

   <uses-permission android:name="android.permission.INTERNET" />
   <uses-permission android:name="android.permission.WRITE_EXTERNAL_STORAGE" />
   <uses-permission android:name="android.permission.BLUETOOTH" />
   <uses-permission android:name="android.permission.BLUETOOTH_ADMIN" />
   <uses-permission android:name="android.permission.ACCESS_COARSE_LOCATION" />
   <uses-permission android:name="android.permission.NFC" />
   <uses-permission android:name="vamshi.com.dairyapplication.LOG" />
   <uses-permission android:name="android.permission.ACCESS_FINE_LOCATION"></
       uses-permission>
 <uses-feature
     android:name="android.hardware.bluetooth_le"
     android:required="true" />

 <application
     android:allowBackup="true"
     android:icon="@mipmap/ic_launcher"
     android:label="@string/app_name"
     android:roundIcon="@mipmap/ic_launcher_round"
     android:supportsRtl="true"
     android:theme="@style/AppTheme">
     <activity
         android:name=".CattleHealthData"
         android:label="Cattle Health Data"
         android:parentActivityName=".MainActivity" />
     <activity android:name=".MainActivity">
         <intent-filter>
           <action android:name="android.intent.action.MAIN" />

           <category android:name="android.intent.category.LAUNCHER" />
         </intent-filter>
```

Table 3 contd. ...

...Table 3 contd.

```
        </activity>
        <activity android:name=".LoginActivity"></activity>
        <activity
            android:name=".RegisterActivity"
            android:label="Register"
            android:parentActivityName=".LoginActivity"></activity>
        <activity
            android:name=".activity_forgot_password"
            android:label="Reset Password"
            android:parentActivityName=".LoginActivity"></activity>
        <activity android:name=".weatheriphone"></activity>
        <activity android:name=".LineGraph" />
        <activity android:name=".barchart_activity" />
        <activity
            android:name=".MultipleGraphs"
            android:label="@string/title_activity_multiple_graphs" />
        <activity
            android:name=".CowManager"
            android:label="Cow Manager" />
        <activity android:name=".profile.BleProfileServiceReadyActivity" />
        <activity
            android:name=".dfu.DfuActivity"
            android:theme="@style/AppThemeBase" />
        <activity android:name=".profile.BleProfileActivity" />
        <activity android:name=".profile.BleProfileExpandableListActivity" />
        <activity android:name=".uart.UARTActivity"></activity>
        <activity android:name=".profile.multiconnect.BleMulticonnectProfileService
            ReadyActivity" />

        <provider
            android:name=".uart.UARTLocalLogContentProvider"
            android:authorities="vamshi.com.dairyapplication.uart.log"
            android:exported="true" />

        <service
            android:name=".uart.UARTService"
            android:label="UART SERVICE" />

        <activity android:name=".ShowLogActivity"></activity>
        </application>

</manifest>
```

Bluetooth UART

Bluetooth UART is responsible for discovery and connection of peripherals. The following Table 4 contains Bluetooth UART Code

Table 4: Bluetooth UART Code

```
public class UARTActivity extends BleProfileServiceReadyActivity<UARTService.
    UARTBinder> implements UARTInterface,UARTNewConfigurationDialogFragme
nt.NewConfigurationDialogListener, UARTConfigurationsAdapter.ActionListener,
AdapterView.OnItemSelectedListener,
    GoogleApiClient.ConnectionCallbacks {
  private final static String TAG = "UARTActivity";

  private final static String PREFS_BUTTON_ENABLED = "prefs_uart_enabled_";
  private final static String PREFS_BUTTON_COMMAND = "prefs_uart_command_";
  private final static String PREFS_BUTTON_ICON = "prefs_uart_icon_";
  /** This preference keeps the ID of the selected configuration. */
  private final static String PREFS_CONFIGURATION = "configuration_id";

  /** This preference is set to true when initial data synchronization for
wearables has been completed. */
  private final static String PREFS_WEAR_SYNCED = "prefs_uart_synced";
  private final static String SIS_EDIT_MODE = "sis_edit_mode";

  private final static int SELECT_FILE_REQ = 2678; // random
  private final static int PERMISSION_REQ = 24; // random, 8-bit

  UARTConfigurationSynchronizer mWearableSynchronizer;

  /** The current configuration. */
  private UartConfiguration mConfiguration;
  private DatabaseHelper mDatabaseHelper;
  private SharedPreferences mPreferences;
  private UARTConfigurationsAdapter mConfigurationsAdapter;
  private ClosableSpinner mConfigurationSpinner;
  private SlidingPaneLayout mSlider;
  private View mContainer;
  private UARTService.UARTBinder mServiceBinder;
  private ConfigurationListener mConfigurationListener;
  private boolean mEditMode;

  public interface ConfigurationListener {
    void onConfigurationModified();
    void onConfigurationChanged(final UartConfiguration configuration);
    void setEditMode(final boolean editMode);
  }

  public void setConfigurationListener(final ConfigurationListener listener) {
    mConfigurationListener = listener;
  }

  @Override
  protected Class<? extends BleProfileService> getServiceClass() {
    return UARTService.class;
  }
  @Override
  protected int getLoggerProfileTitle() {
  return R.string.uart_feature_title;
  }

  @Override
  protected Uri getLocalAuthorityLogger() {
    return UARTLocalLogContentProvider.AUTHORITY_URI;
  }
```

Table 4 contd. ...

...Table 4 contd.

```java
@Override
protected void setDefaultUI() {
    // empty
}

@Override
protected void onServiceBinded(final UARTService.UARTBinder binder) {
    mServiceBinder = binder;
}

@Override
protected void onServiceUnbinded() {
    mServiceBinder = null;
}

@Override
protected void onInitialize(final Bundle savedInstanceState) {
    mPreferences = PreferenceManager.getDefaultSharedPreferences(this);
    mDatabaseHelper = new DatabaseHelper(this);
    ensureFirstConfiguration(mDatabaseHelper);
    mConfigurationsAdapter = new UARTConfigurationsAdapter(this, this,
    mDatabaseHelper.getConfigurationsNames());

    // Initialize Wearable synchronizer
    mWearableSynchronizer = UARTConfigurationSynchronizer.from(this, this);
}

/**
 * Method called when Google API Client connects to Wearable.API.
 */
@Override
public void onConnected(final Bundle bundle) {
    // Ensure the Wearable API was connected
    if (!mWearableSynchronizer.hasConnectedApi())
        return;

    if (!mPreferences.getBoolean(PREFS_WEAR_SYNCED, false)) {
        new Thread(() -> {
            final Cursor cursor = mDatabaseHelper.getConfigurations();
            try {
                while (cursor.moveToNext()) {
                    final long id = cursor.getLong(0 /* _ID */);
                    try {
                        final String xml = cursor.getString(2 /* XML */);
                        final Format format = new Format(new HyphenStyle());
                        final Serializer serializer = new Persister(format);
                        final UartConfiguration configuration = serializer. read
                        (UartConfiguration.class, xml);
                        mWearableSynchronizer.onConfigurationAddedOrEdited(id,
                        configuration).await();
                    } catch (final Exception e) {
                        Log.w(TAG, "Deserializing configuration with id " + id + " failed", e);
                    }
                }
                mPreferences.edit().putBoolean(PREFS_WEAR_SYNCED, true).apply();
            } finally {
```

Table 4 contd. ...

```
        cursor.close();
      }
   }).start();
  }
}

/**
 * Method called then Google API client connection was suspended.
 * @param cause the cause of suspension
 */
@Override
public void onConnectionSuspended(final int cause) {
    // dp nothing
}

@Override
protected void onDestroy() {
    super.onDestroy();
    mWearableSynchronizer.close();
}

@Override
protected void onCreateView(final Bundle savedInstanceState) {
    setContentView(R.layout.activity_feature_uart);

    mContainer = findViewById(R.id.container);
    // Setup the sliding pane if it exists
    final SlidingPaneLayout slidingPane = mSlider = findViewById(R.id.sliding_pane);
    if (slidingPane != null) {
      slidingPane.setSliderFadeColor(Color.TRANSPARENT);
      slidingPane.setShadowResourceLeft(R.drawable.shadow_r);
      slidingPane.setPanelSlideListener(new SlidingPaneLayout.SimplePane
      lSlideListener() {
        @Override
        public void onPanelClosed(final View panel) {
          // Close the keyboard
          final UARTLogFragment logFragment = (UARTLogFragment)
getSupportFragmentManager().findFragmentById(R.id.fragment_log);
          logFragment.onFragmentHidden();
        }
      });
    }
}

@Override
protected void onViewCreated(final Bundle savedInstanceState) {
    getSupportActionBar().setDisplayShowTitleEnabled(false);
    final ClosableSpinner configurationSpinner = mConfigurationSpinner =
    findViewById(R.id.toolbar_spinner);
    configurationSpinner.setOnItemSelectedListener(this);
    configurationSpinner.setAdapter(mConfigurationsAdapter);
    configurationSpinner.setSelection(mConfigurationsAdapter.getItemPosition
(mPreferences.getLong(PREFS_CONFIGURATION, 0)));
  }
```

Table 4 contd. ...

...Table 4 contd.

```java
@Override
protected void onRestoreInstanceState(final @NonNull Bundle savedInstanceState) {
    super.onRestoreInstanceState(savedInstanceState);

    mEditMode = savedInstanceState.getBoolean(SIS_EDIT_MODE);
    setEditMode(mEditMode, false);
}

@Override
public void onSaveInstanceState(final Bundle outState) {
    super.onSaveInstanceState(outState);

    outState.putBoolean(SIS_EDIT_MODE, mEditMode);
}

@Override
public void onServicesDiscovered(final BluetoothDevice device, final boolean
optionalServicesFound) {
    // do nothing
}

@Override
public void onDeviceSelected(final BluetoothDevice device, final String name) {
    // The super method starts the service
    super.onDeviceSelected(device, name);

    // Notify the log fragment about it
    final UARTLogFragment logFragment = (UARTLogFragment)
getSupportFragmentManager().findFragmentById(R.id.fragment_log);
    logFragment.onServiceStarted();
}

@Override
protected int getDefaultDeviceName() {
return R.string.uart_default_name;
}

@Override
protected int getAboutTextId() {
    return R.string.uart_about_text;
}

@Override
protected UUID getFilterUUID() {
    return null; // not used
}

@Override
public void send(final String text) {
    if (mServiceBinder != null)
        mServiceBinder.send(text);
}

public void setEditMode(final boolean editMode) {
    setEditMode(editMode, true);
    invalidateOptionsMenu();
}
```

UART Manager

UART Manager handles the code with respect to Service and Characteristic code. See Table 5 for the service and characteristic code handling.

Table 5: UART Manager Code

```java
package vamshi.com.dairyapplication.uart;

import android.bluetooth.BluetoothGatt;
import android.bluetooth.BluetoothGattCharacteristic;
import android.bluetooth.BluetoothGattService;
import android.content.Context;
import android.text.TextUtils;

import java.io.UnsupportedEncodingException;
import java.util.Deque;
import java.util.LinkedList;
import java.util.UUID;

import no.nordicsemi.android.ble.BleManager;
import no.nordicsemi.android.log.Logger;

public class UARTManager extends BleManager<UARTManagerCallbacks> {
    /** Nordic UART Service UUID */
    private final static UUID UART_SERVICE_UUID = UUID.fromString("6E400001-
B5A3-F393-E0A9-E50E24DCCA9E");
    /** RX characteristic UUID */
    private final static UUID UART_RX_CHARACTERISTIC_UUID = UUID.
      fromString("6E400002-B5A3-F393-E0A9-E50E24DCCA9E");
    /** TX characteristic UUID */
    private final static UUID UART_TX_CHARACTERISTIC_UUID = UUID.
      fromString("6E400003-B5A3-F393-E0A9-E50E24DCCA9E");
    /** The maximum packet size is 20 bytes. */
    private static final int MAX_PACKET_SIZE = 20;

    private BluetoothGattCharacteristic mRXCharacteristic, mTXCharacteristic;
    private byte[] mOutgoingBuffer;
    private int mBufferOffset;

    public UARTManager(final Context context) {
        super(context);
    }

    @Override
    protected BleManagerGattCallback getGattCallback() {
        return mGattCallback;
    }

    /**
    * BluetoothGatt callbacks for connection/disconnection, service
      discovery, receiving indication, etc.
    */
    private final BleManagerGattCallback mGattCallback = new
      BleManagerGattCallback() {
```

Table 5 contd. ...

...Table 5 contd.

```
    @Override
    protected Deque<Request> initGatt(final BluetoothGatt gatt) {
        final LinkedList<Request> requests = new LinkedList<>();
        requests.add(Request.newEnableNotificationsRequest(mTXCharacteristic));
        return requests;
    }

    @Override
    public boolean isRequiredServiceSupported(final BluetoothGatt gatt) {
        final BluetoothGattService service = gatt.getService(UART_SERVICE_UUID);
        if (service != null) {
            mRXCharacteristic = service.getCharacteristic(UART_RX_
                CHARACTERISTIC_UUID);
            mTXCharacteristic = service.getCharacteristic(UART_TX_
                CHARACTERISTIC_UUID);
        }

        boolean writeRequest = false;
        boolean writeCommand = false;
        if (mRXCharacteristic != null) {
            final int rxProperties = mRXCharacteristic.getProperties();
            writeRequest = (rxProperties & BluetoothGattCharacteristic.
                PROPERTY_WRITE) > 0;
            writeCommand = (rxProperties & BluetoothGattCharacteristic.
                PROPERTY_WRITE_NO_RESPONSE) > 0;

            // Set the WRITE REQUEST type when the characteristic supports
            //    it. This will allow to send long write (also if the
            //    characteristic support it).
            // In case there is no WRITE REQUEST property, this manager will
            //    divide texts longer then 20 bytes into up to 20 bytes chunks.
            if (writeRequest)
                mRXCharacteristic.setWriteType(BluetoothGattCharacteristic.
                    WRITE_TYPE_DEFAULT);
        }

        return mRXCharacteristic != null && mTXCharacteristic != null &&
            (writeRequest || writeCommand);
    }
    @Override
    protected void onDeviceDisconnected() {
        mRXCharacteristic = null;
        mTXCharacteristic = null;
    }

    @Override
    public void onCharacteristicWrite(final BluetoothGatt gatt, final
        BluetoothGattCharacteristic characteristic) {
        // When the whole buffer has been sent
        final byte[] buffer = mOutgoingBuffer;
        if (mBufferOffset == buffer.length) {
            try {
            final String data = new String(buffer, "UTF-8");
            Logger.a(mLogSession, "\"" + data + "\" sent");
            mCallbacks.onDataSent(gatt.getDevice(), data);
            } catch (final UnsupportedEncodingException e) {
            // do nothing
            }
```

Table 5 contd. ...

...Table 5 contd.

```
            mOutgoingBuffer = null;
      } else { // Otherwise...
          final int length = Math.min(buffer.length - mBufferOffset, MAX_PACKET_
SIZE);
          enqueue(Request.newWriteRequest(mRXCharacteristic, buffer,
mBufferOffset, length));
          mBufferOffset += length;
      }
   }

   @Override
   public void onCharacteristicNotified(final BluetoothGatt gatt, final
BluetoothGattCharacteristic characteristic) {
      final String data = characteristic.getStringValue(0);
      Logger.a(mLogSession, "\"" + data + "\" received");
      mCallbacks.onDataReceived(gatt.getDevice(), data);
   }
};

@Override
protected boolean shouldAutoConnect() {
    // We want the connection to be kept
    return true;
}

/**
 * Sends the given text to RX characteristic.
 * @param text the text to be sent
 */
public void send(final String text) {
    // Are we connected?
    if (mRXCharacteristic == null)
      return;

    // An outgoing buffer may not be null if there is already another packet
      being sent. We do nothing in this case.
    if (!TextUtils.isEmpty(text) && mOutgoingBuffer == null) {
        final byte[] buffer = mOutgoingBuffer = text.getBytes();
        mBufferOffset = 0;

        // Depending on whether the characteristic has the WRITE REQUEST
            property or not, we will either send it as it is (hoping the long
            write is implemented),
        // or divide it into up to 20 bytes chunks and send them one by one.
        final boolean writeRequest = (mRXCharacteristic.getProperties() &
BluetoothGattCharacteristic.PROPERTY_WRITE) > 0;

        if (!writeRequest) { // no WRITE REQUEST property
            final int length = Math.min(buffer.length, MAX_PACKET_SIZE);
            mBufferOffset += length;
            enqueue(Request.newWriteRequest(mRXCharacteristic, buffer, 0,
length));
        } else { // there is WRITE REQUEST property, let's try Long Write
            mBufferOffset = buffer.length;
            enqueue(Request.newWriteRequest(mRXCharacteristic, buffer, 0,
            buffer.length));
        }
    }
  }
}
```

Database

The database code handles the Insert, Query and Edit operations (see Table 6).

Table 6: Database Code

```
package vamshi.com.dairyapplication;

import android.content.ContentValues;
import android.content.Context;
import android.database.Cursor;
import android.database.sqlite.SQLiteDatabase;
import android.database.sqlite.SQLiteOpenHelper;

import vamshi.com.dairyapplication.WeatherApiClasses.CowManagerModel;

/**
 * Created by Vamshi on 1/9/2018.
 */

public class Database extends SQLiteOpenHelper {

    private static final int DATABASE_VERSION = 2;

    // Database Name
    private static final String DATABASE_NAME = "MyDairyFarm";

    // Contacts table name
    private static final String TABLE_USERS = "USERS";
    private static final String TABLE_COWMANAGER = "COWMANAGER";

    // Contacts Table Columns names
    private static final String KEY_FIRSTNAME = "Firstname";
    private static final String KEY_LASTNAME = "Lastname";
    private static final String KEY_EMAIL = "Email";
    private static final String KEY_MOBILE = "Mobile";
    private static final String KEY_SEC_QUES = "SecurityQuestion";
    private static final String KEY_SEC_ANS = "SecurityAnswer";
    private static final String KEY_PASSWORD = "Password";


    private static final String KEY_FARMER = "Farmer";
    private static final String KEY_FARMNAME = "Farmname";
    private static final String KEY_PHONE = "Phone";
    private static final String KEY_NOOFCOW = "Noofcow";
    private static final String KEY_LOCATION = "Location";

    public Database(Context context) {
        super(context, DATABASE_NAME, null, DATABASE_VERSION);
    }
    @Override
    public void onCreate(SQLiteDatabase sqLiteDatabase) {
        String CREATE_CONTACTS_TABLE = "CREATE TABLE " + TABLE_USERS + "("
            + KEY_FIRSTNAME + " TEXT," + KEY_LASTNAME + " TEXT,"
            + KEY_EMAIL + " TEXT," + KEY_MOBILE + " TEXT PRIMARY KEY," + KEY_
                SEC_QUES + " TEXT,"
            + KEY_SEC_ANS + " TEXT," + KEY_PASSWORD + " TEXT" + ")";
```

Table 6 contd. ...

...Table 6 contd.

```java
    String CREATE_COWMANGERS_TABLE = "CREATE TABLE " + TABLE_COWMANAGER + "("
        + KEY_FARMER + " TEXT," + KEY_FARMNAME + " TEXT,"
        + KEY_PHONE + " TEXT PRIMARY KEY," + KEY_NOOFCOW + " TEXT,"
        + KEY_LOCATION + " TEXT" + ")";
    sqLiteDatabase.execSQL(CREATE_CONTACTS_TABLE);
    sqLiteDatabase.execSQL(CREATE_COWMANGERS_TABLE);
}

@Override
public void onUpgrade(SQLiteDatabase sqLiteDatabase, int i, int i1) {
// Drop older table if existed
    sqLiteDatabase.execSQL("DROP TABLE IF EXISTS " + TABLE_USERS);

    // Create tables again
    onCreate(sqLiteDatabase);
}

public long createUser(User user) {
    ContentValues values = new ContentValues();
    values.put(KEY_FIRSTNAME, user.getFirstname());
    values.put(KEY_LASTNAME, user.getLastname());
    values.put(KEY_EMAIL, user.getEmail());
    values.put(KEY_MOBILE, user.getMobile());
    values.put(KEY_SEC_QUES, user.getSec_ques());
    values.put(KEY_SEC_ANS, user.getSec_ans());
    values.put(KEY_PASSWORD, user.getPassword());

    System.out.println(user.getFirstname() + "==" + user.getLastname() + "=="
+ user.getEmail() + "==" + user.getMobile() + "==" + user.getSec_ques() + "=="
+ user.getSec_ans() + "==" + user.getPassword());

    SQLiteDatabase db = this.getWritableDatabase();
    long i = db.insert(TABLE_USERS, null, values);

    return i;

}
public long createCowmanger(CowManagerModel cow) {
    ContentValues values = new ContentValues();
    values.put(KEY_FARMER, cow.getFarmer());
    values.put(KEY_FARMNAME, cow.getFarmname());
    values.put(KEY_PHONE, cow.getPhone());
    values.put(KEY_NOOFCOW, cow.getNoofcow());
    values.put(KEY_LOCATION, cow.getLocation());

    SQLiteDatabase db = this.getWritableDatabase();
    long i = db.insert(TABLE_COWMANAGER, null, values);

    return i;
}
```

Table 6 contd. ...

...Table 6 contd.

```
public Cursor login(String email, String password) {
    SQLiteDatabase db = this.getReadableDatabase();
    System.out.print("SELECT *FROM " + TABLE_USERS + " WHERE " + KEY_EMAIL + "
= '" + email + "' AND " + KEY_PASSWORD + " = '" + password + "'");
    Cursor cursor = db.rawQuery("SELECT *FROM " + TABLE_USERS + " WHERE " + KEY_
EMAIL + " = '" + email + "' AND " + KEY_PASSWORD + " = '" + password + "'", null);

    return cursor;
}
public Cursor validateEmail(String email) {
    SQLiteDatabase db = this.getReadableDatabase();
    Cursor cursor = db.rawQuery("SELECT *FROM " + TABLE_USERS + " WHERE " +
KEY_EMAIL + " = '" + email + "'", null);
    System.out.println();
    return cursor;
}
public boolean validateAnswer(String email, String sec_ques, String secAns) {
boolean isValid = false;
    SQLiteDatabase db = this.getReadableDatabase();
    Cursor cursor = db.query(TABLE_USERS, new String[]{KEY_EMAIL},
    KEY_EMAIL + " LIKE '" + email + "' AND " + KEY_SEC_QUES + " LIKE '" + sec_
    ques + "' AND " + KEY_SEC_ANS + " LIKE '" + secAns + "'", null, null,
null, null);
    System.out.println(cursor.getCount());

    if (cursor.getCount() > 0) {
        isValid = true;
    } else {
        isValid = false;
    }
    return isValid;
}
public int updatePassword(String email, String password){
    SQLiteDatabase db=this.getWritableDatabase();
    ContentValues values=new ContentValues();
    values.put(KEY_PASSWORD,password);

    int count = db.update(TABLE_USERS,values,KEY_EMAIL+" =? ",new String[] {
email }););
return count;
}
}
```

Extended Bluetooth Device

The extended Bluetooth Device constructs Bluetooth device object (see Table 7).

Table 7: Bluetooth Code

```
package vamshi.com.dairyapplication.scanner;
import android.bluetooth.BluetoothDevice;
import no.nordicsemi.android.support.v18.scanner.ScanResult;
public class ExtendedBluetoothDevice {
    /* package */ static final int NO_RSSI = -1000;
    public final BluetoothDevice device;
    /** The name is not parsed by some Android devices, f.e. Sony Xperia Z1
with Android 4.3 (C6903). It needs to be parsed manually. */
    public String name;
    public int rssi;
    public boolean isBonded;

    public ExtendedBluetoothDevice(final ScanResult scanResult) {
        this.device = scanResult.getDevice();
        this.name = scanResult.getScanRecord() != null ? scanResult.getScanRecord().
getDeviceName() : null;
        this.rssi = scanResult.getRssi();
        this.isBonded = false;
    }
    public ExtendedBluetoothDevice(final BluetoothDevice device) {
        this.device = device;
        this.name = device.getName();
        this.rssi = NO_RSSI;
        this.isBonded = true;
    }
  public boolean matches(final ScanResult scanResult) {
        return device.getAddress().equals(scanResult.getDevice().getAddress());
    }
}
```

BLE Profile Ready Activity

The BLE Profile Ready Activity code handles the service connection and disconnect operations (see Table 8).

MQTT[20]

MQTT, a foundational Internet of Things (IoT) standard developed by the OASIS consortium, has now been approved for release by the International Organization for Standardization (ISO) and the International Electrotechnical Commission (IEC). Version 3.1.1 of MQTT was balloted through the Joint Technical Committee on Information Technology (JTC1) of ISO and IEC and given the designation 'ISO/IEC 20922'.

 MQTT is a lightweight messaging protocol. It is suitable for IoT, since it is bandwidth-efficient and uses little battery power.

MQTT is message oriented. Every message is a discrete chunk of data, opaque to the broker [14].

Every message is published to an address, known as a topic. Clients may subscribe to multiple topics. Every client subscribed to a topic receives every message published to the topic.

For example, imagine a simple network with three clients and a central broker.

All three clients open TCP connections with the broker. Clients B and C subscribe to the topic temperature.

---

[20]   MQTT - https://www.oasis-open.org/news/pr/oasis-mqtt-internet-of-things-standard-now-approved-by-iso-iec-jtc1

Table 8: Bluetooth Read

```
private ServiceConnection mServiceConnection = new ServiceConnection() {
    @SuppressWarnings("unchecked")
    @Override
    public void onServiceConnected(final ComponentName name, final IBinder
service) {
    final E bleService = mService = (E) service;
    mBluetoothDevice = bleService.getBluetoothDevice();
    mLogSession = mService.getLogSession();
    Logger.d(mLogSession, "Activity bound to the service");
    onServiceBinded(bleService);

    // Update UI
    mDeviceName = bleService.getDeviceName();
    mDeviceNameView.setText(mDeviceName);
    mConnectButton.setText(R.string.action_disconnect);

    // And notify user if device is connected
    if (bleService.isConnected()) {
       onDeviceConnected(mBluetoothDevice);
    } else {
       // If the device is not connected it means that either it is still connecting,
       // or the link was lost and service is trying to connect to it
(autoConnect=true).
    onDeviceConnecting(mBluetoothDevice);
    }
 }
 @Override
 public void onServiceDisconnected(final ComponentName name) {
    // Note: this method is called only when the service is killed by the
system,
    // not when it stops itself or is stopped by the activity.
    // It will be called only when there is critically low memory, in practice
never
    // when the activity is in foreground.
    Logger.d(mLogSession, "Activity disconnected from the service");
    mDeviceNameView.setText(getDefaultDeviceName());
    mConnectButton.setText(R.string.action_connect);

    mService = null;
    mDeviceName = null;
    mBluetoothDevice = null;
    mLogSession = null;
    onServiceUnbinded();
  }
};

@Override
protected final void onCreate(final Bundle savedInstanceState) {
    super.onCreate(savedInstanceState);

    ensureBLESupported();
    if (!isBLEEnabled()) {
       showBLEDialog();
    }
```

Table 8 contd. ...

...Table 8 contd.

```
    // Restore the old Log session
    if (savedInstanceState != null) {
        final Uri logUri = savedInstanceState.getParcelable(LOG_URI);
        mLogSession = Logger.openSession(getApplicationContext(), logUri);
    }

    // In onInitialize method a final class may register local broadcast
receivers that will Listen for events from the service
    onInitialize(savedInstanceState);
    // The onCreateView class should... create the view
    onCreateView(savedInstanceState);

    final Toolbar toolbar = findViewById(R.id.toolbar_actionbar);
    setSupportActionBar(toolbar);

    // Common nRF Toolbox view references are obtained here
    setUpView();
    // View is ready to be used
    onViewCreated(savedInstanceState);

    LocalBroadcastManager.getInstance(this).registerReceiver(mCommonBroadcastR
eceiver, makeIntentFilter());
}
```

Later, Client A publishes a value of 22.5 for topic temperature. The broker forwards the message to all subscribed clients [15].

The publisher subscriber model allows MQTT clients to communicate one-to-one, one-to-many and many-to-one (see Figure 34).

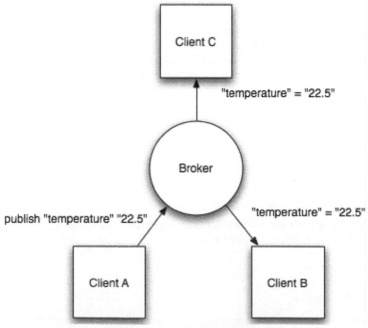

Figure 34: MQTT Architecture

MQTT defines an extremely lightweight publish/subscribe messaging transport protocol. Because it requires significantly less bandwidth and is so easy to implement, MQTT is well suited for IoT applications where resources such as battery power and bandwidth are at a premium. The range of MQTT applications continues to grow. In the healthcare sector, practitioners use the protocol to communicate with bio-medical devices such as blood pressure monitors. Oil and gas companies use MQTT to monitor thousands of miles of pipelines. MQTT is emerging as a fundamental enabler for telematics, infotainment, and other connected vehicle applications. MQTT[21] is also becoming increasingly popular for interactive mobile applications.

The original co-designer of MQTT and Distinguished Engineer for IBM Watson Internet of Things, Andy Stanford-Clark, said "I am delighted to see that, thanks to its spectacular adoption in the IoT world, MQTT has attained this significant milestone. This positions the protocol perfectly to underpin the modern IoT ecosystem, supporting businesses and organizations in their journey towards a truly connected world. The additional scrutiny required for international recognition confirms the protocol as an open and solid IoT technology."

Giles Nelson, Senior VP at Software AG, said, "MQTT being added to ISO/IEC standards is a major step forward that further solidifies its foothold as a standard messaging protocol and makes it ideal for large-scale Internet of Things applications."

MQTT protocol consists of two patterns, namely subscriber and publisher. These both require a message broker to operate. The availability of MQTT protocol generally depends on the context. A variation to MQTT protocol was termed as MQTT-SN which was aimed at embedded devices which did not work on TCP/IP network. An example of such a non-TCP/IP device was ZigBee. There are many real-world applications, such as Facebook Messenger, Amazon Web Services, McAfee Open DXL, Microsoft Azure, Open Geospatial Consortium Sensor Things API, etc. With MQTT, clients can connect as publisher or subscriber. Therefore, MQTT is basically a publish/subscribe protocol. MQTT broker is used to handle the message passing. A few challenges exist, such as the connection between software technologies and physical devices such as sensors, actuators, phones and tablets. MQTT is designed to overcome these challenges [14][15].

MQTT Benefits

Out of all the available protocols for the communication system and for the IoT network, MQ Telemetry Transport is most widely used. Mentioned below are some of the advantages of using MQTT:

- MQTT is robust, as we can encrypt the data even with unsecured TCP messaging. The partial or complete encryption is possible with the help of TLS/SSL security. Factors affecting the encryption depend on system resources and the security mandate.

- MQTT provides a central broker to minimize the packets strength in the internet. This is achieved with broker acting as server. It also helps in memory reduction required by the clients.

- MQTT provides three different types of Quality of Service(QoS) depending on importance and repetitiveness of messages, i.e.,

  o **At Most Once:** Client publishes the messages only once and is configured with level 0.

  o **At least Once:** Client publishes the messages more than once and is configured with level 1.

  o **Exactly Once:** Client publishes the messages just once and is configured with level 2.

- MQTT can predict on the availability of the client using 'Last Will'. Therefore, the extra time of waiting for a client who is actually not present is avoided. Listeners check on the availability of the publisher.

- MQTT can also help in receiving old message from the subscribers. And there is a flexible subscription pattern, i.e., depending on a particular pattern, the client can subscribe to the topics published.

---

[21] MQTT v3.1.1 - http://docs.oasis-open.org/mqtt/mqtt/v3.1.1/mqtt-v3.1.1.pdf

MQTT Highlights

Here are the highlights of MQTT:

- It is an open source and royalty-free protocol.
- It is standardized by the OASIS technical committee.
- The MQTT protocol is designed for transferring messages reliably between devices that are on low bandwidth.
- It uses TCP/IP to establish a secure connection.
- It is a client-server, publish/subscribe model that enables one-to-one and one-to-many distributions.
- It supports three qualities of service: (1) fire and forget, (2) at least once and (3) exactly one.
  - ○ "Fire and forget" does not use constant device storage—it uses only one transmission to send or receive a publication.
  - ○ "At least once" and "Exactly once" require persistent device storage to maintain the protocol state and to save a message until an acknowledgement is received.
- It has built-in support for loss of contact between client and server. The server is informed when a client connection breaks abnormally, allowing the message to be re-sent or preserved for later delivery.
- The MQTT message headers are kept as small as possible. The fixed header is just two bytes with 12 bytes of additional variable headers.
- Many applications of MQTT can be accomplished using only CONNECT, PUBLISH, SUBSCRIBE and DISCONNECT packets.
- For SUBSCRIBE and CONNECT, the protocol uses 12-byte variable headers and for PUBLISH, it uses only 2-byte variable headers.
- MQTT is adaptable for a wide variety of devices, platforms and operating systems inside an enterprise network since it connects to a message sight inside an enterprise network.
- It is very easy to use and can be implemented with a simple set of command messages.

Paho Library not found

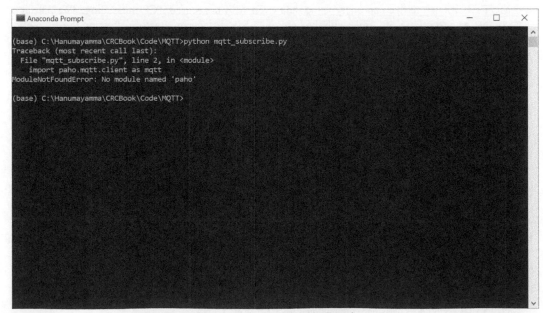

Figure 35: Paho Library Not Found

Install Paho

```
pip3 install paho-mqtt python-etcd
```

Installation of Paho library results in the following Python Installation Package UI (see Figure 36).

Figure 36: Install Paho Library

Once installed, you can see the library under your Python IDE (see Figure 37):

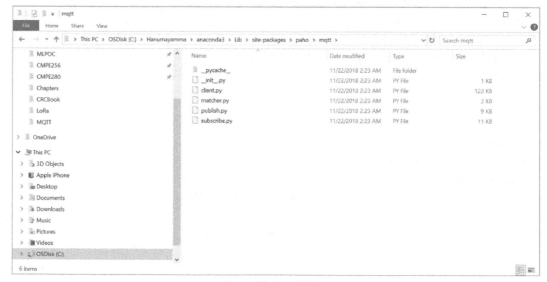

Figure 37: Install Files

Code:

Once installed, you can see the library under your Python IDE:

Code:

The following modules connect to MQTT Server and subscribe/publish messages.

mqtt_subscribe.py

```
# Import package
import paho.mqtt.client as mqtt

# Define Variables
MQTT_BROKER="broker.hivemq.com"
MQTT_HOST = "iot.eclipse.org"
MQTT_PORT = 1883
MQTT_KEEPALIVE_INTERVAL = 45
MQTT_TOPIC = "helloTopic"
MQTT_MSG = "hello MQTT"

# Define on connect event function
# We shall subscribe to our Topic in this function
def on_connect(mosq, obj, rc):
    mqttc.subscribe(MQTT_TOPIC, 0)

# Define on_message event function.
# This function will be invoked every time,
# a new message arrives for the subscribed topic
def on_message(mosq, obj, msg):
        print ("Topic: " + str(msg.topic))
        print ("QoS: " + str(msg.qos))
        print ("Payload: " + str(msg.payload))

def on_subscribe(mosq, obj, mid, granted_qos):
    print("Subscribed to Topic: " +
        MQTT_MSG + " with QoS: " + str(granted_qos))

# Initiate MQTT Client
mqttc = mqtt.Client("client-001")

# Assign event callbacks
mqttc.on_message = on_message
mqttc.on_connect = on_connect
mqttc.on_subscribe = on_subscribe

# Connect with MQTT Broker
mqttc.connect(MQTT_HOST, MQTT_PORT, MQTT_KEEPALIVE_INTERVAL)

print("Listening")
# Continue monitoring the incoming messages for subscribed topic
mqttc.loop_forever()
```

mqtt_publish.py

The following code publishes to MQTT broker and receives a message on subscribe.

```python
# Import package
import paho.mqtt.client as mqtt
# Define Variables
MQTT_BROKER="broker.hivemq.com"
MQTT_HOST = "iot.eclipse.org"
MQTT_PORT = 1883
MQTT_KEEPALIVE_INTERVAL = 45
MQTT_TOPIC = "helloTopic"
MQTT_MSG = "hello MQTT"
# Define on connect event function
# We shall subscribe to our Topic in this function
def on_connect(mosq, obj, rc):
    mqttc.subscribe(MQTT_TOPIC, 0)
# Define on_message event function.
# This function will be invoked every time,
# a new message arrives for the subscribed topic
def on_message(mosq, obj, msg):
        print ("Topic: " + str(msg.topic))
        print ("QoS: " + str(msg.qos))
        print ("Payload: " + str(msg.payload))
def on_subscribe(mosq, obj, mid, granted_qos):
    print("Subscribed to Topic: " +
        MQTT_MSG + " with QoS: " + str(granted_qos))
# Initiate MQTT Client
mqttc = mqtt.Client("client-001")
# Assign event callbacks
mqttc.on_message = on_message
mqttc.on_connect = on_connect
mqttc.on_subscribe = on_subscribe
# Connect with MQTT Broker
mqttc.connect(MQTT_HOST, MQTT_PORT, MQTT_KEEPALIVE_INTERVAL)
print("Listening")
# Continue monitoring the incoming messages for subscribed topic
mqttc.loop_forever()
```

The combined code (mqtt_publish_and_subscribe.py)

```
# -*- coding: utf-8 -*-
"""
Created on Thu Nov 22 02:39:46 2018
@author: cvuppalapati
"""
import time
import paho.mqtt.client as paho
broker="broker.hivemq.com"
broker="iot.eclipse.org"
#define callback
def on_message(client, userdata, message):
    time.sleep(1)
    print("received message =",str(message.payload.decode("utf-8")))

client= paho.Client("client-001") #create client object client1.on_publish = on_publish #assign function
to callback client1.connect(broker,port) #establish connection client1.publish("house/bulb1","on")
######Bind function to callback
client.on_message=on_message
#####
print("connecting to broker",broker)
client.connect(broker)#connect
client.loop_start() #start loop to process received messages
print("subscribing")
client.subscribe("house/bulb1")#subscribe
time.sleep(2)
print("publishing")
client.publish("house/bulb1","on")#publish
time.sleep(4)
client.disconnect() #disconnect
client.loop_stop() #stop loop
```

Library: MQTT Paho

This is an MQTT v3.1 client module. MQTT is a lightweight pub/sub messaging protocol that is easy to implement and suitable for low powered devices.

Client.py

The Client.py code is the central to establishing MQTT on Python platform. It uses systems resources and OS platform capabilities to establish socket, subscribe socket, and publish socket.

In order to perform packet exchange, every message payload has to be time stamped. In the case of Paho Python implementation, the time function used is time monotomic.

Under the Windows, the time tick for the time.monotomic is derived from an on board processor timer and it's the most accurate.

```
"""
This is an MQTT v3.1 client module. MQTT is a lightweight pub/sub messaging
protocol that is easy to implement and suitable for low powered devices.
"""
import collections
import errno
import os
import platform
import select
import socket

try:
    import ssl
except ImportError:
    ssl = None

import struct
import sys
import threading

import time
import uuid
import base64
import string
import hashlib
import logging

try:
    # Use monotonic clock if available
    time_func = time.monotonic
except AttributeError:
    time_func = time.time

try:
    import dns.resolver
except ImportError:
    HAVE_DNS = False
else:
    HAVE_DNS = True

from .matcher import MQTTMatcher

if platform.system() == 'Windows':
    EAGAIN = errno.WSAEWOULDBLOCK
else:
    EAGAIN = errno.EAGAIN

MQTTv31 = 3
MQTTv311 = 4

if sys.version_info[0] >= 3:
    # define some alias for python2 compatibility
    unicode = str
    basestring = str
```

Timer functions in Python

Python offers several timer functions[22] (PEP 418). The following five timer functions[23] are supported in Python 3.3:

1. time.clock() deprecated in 3.3

---

[22] PEP 418 -- Add monotonic time, performance counter, and process time functions - https://www.python.org/dev/peps/pep-0418/#rationale

[23] Python Clocks Explained - https://www.webucator.com/blog/2015/08/python-clocks-explained/

2. time.monotonic()
3. time.perf_counter()
4. time.process_time()
5. time.time()

Time Functions in Python:

Get Clock Information API[24] (time.get_clock_info()) provides you the information on the specified clock as a namespace object. Supported clock names and the corresponding functions to read their value are:

- 'clock': time.clock()
- 'monotonic': time.monotonic()
- 'perf_counter': time.perf_counter()
- 'process_time': time.process_time()
- 'thread_time': time.thread_time()
- 'time': time.time()

Please time_thread_time is not supported.

```
# -*- coding: utf-8 -*-
"""
Created on Thu Nov 22 11:56:01 2018
@author: cvuppalapati
"""
import collections
import errno
import os
import platform
import select
import socket

try:
    import ssl
except ImportError:
    ssl = None

import struct
import sys
import threading

import time
import uuid

print('\n clock \n')
print(time.get_clock_info('clock'))

print('\n Monotonic \n')
print(time.get_clock_info('monotonic'))

print('\n Monotonic \n')
print(time.get_clock_info('monotonic'))

print('\n Perf_Counter \n')
print(time.get_clock_info('perf_counter'))

print('\n process_time \n')
print(time.get_clock_info('process_time'))

print('\n time \n')
print(time.get_clock_info('time'))
```

The output has following key response attributes [16]:

---

24   Get Clock Information - https://docs.python.org/3/library/time.html#time.get_clock_info

Attribute	Description
Adjustable	True if the clock can be changed automatically (e.g., by a NTP daemon) or manually by the system administrator, False otherwise
Implementation	The name of the underlying C function used to get the clock value. Refer to Clock ID Constants for possible values.
Monotonic	True if the clock cannot go backward, False otherwise
Resolution	The resolution of the clock in seconds (float)

The timer outputs:

Timer	Adjustable	Implementation	Monotonic	Resolution	Tick Rate
Clock	False	QueryPerformance Counter()	True	3.64672032207 8339e-07	2,143,482
Monotonic	False	GetTickCount64()	True	0.015625	64
Perf_Counter	False	QueryPerformance Counter()	True	3.64672032207 8339e-07	2,143,482
process_time	False	GetProcessTimes()	True	1e-07	10,000,000
Time	True	GetSystemTimeAsFile Time()	False	0.015625	64

Please note: Tick Rate is the number of ticks per second. This is *the inverse of Resolution*. A high-resolution clock has a high tick rate.

IoT and Hardware Clocks

The success of IoT messaging is very tightly coupled with the timestamps from the client (be it, MQTT, LoRa, HTTP, REST, AMP or other protocols). It's essential, when looking at the hardware capabilities and performance needs of the clocks, how IoT connectivity and middleware could be designed to process the data at IoT Clock rate [16].

List of hardware Clocks:[25]

- HPET: A High Precision Event Timer (HPET) chip consists of a 64-bit up-counter (main counter) counting at least at 10 MHz and a set of up to 256 comparators (at least 3). Each HPET can have up to 32 timers. HPET can cause around 3 seconds of drift per day.

https://arstechnica.com/gadgets/2017/03/intel-still-beats-ryzen-at-games-but-how-much-does-it-matter/

- TSC (Time Stamp Counter): Historically, the TSC has increased with every internal processor clock cycle, but now the rate is usually constant (even if the processor changes frequency) and usually equals the maximum processor frequency. Multiple cores have different TSC values. Hibernation of system will reset the TSC value. The RDTSC instruction can be used to read this counter [17].

Source: https://www.febo.com/pages/TICC/

- ACPI Power Management Timer: ACPI 24-bit timer with a frequency of 3.5 MHz (3,579,545 Hz).

https://en.wikipedia.org/wiki/Power_management_integrated_circuit

---

[25]   List of hardware clocks - https://www.python.org/dev/peps/pep-0418/#rationale

- Cyclone: The Cyclone timer uses a 32-bit counter on IBM Extended X-Architecture (EXA) chipsets which include computers that use the IBM "Summit" series chipsets (ex: x440). This is available in IA32 and IA64 architectures.

Source: https://www.dhgate.com/product/altera-cyclone-iv-ep4ce6-fpga-development/392819262.html

- PIT (programmable interrupt timer): Intel 8253/8254 chipsets with a configurable frequency in range 18.2 Hz–1.2 MHz. This uses a 16-bit counter.

- RTC (Real-time clock). Most RTCs use a crystal oscillator with a frequency of 32,768 Hz.

Source: https://www.maximintegrated.com/en/products/digital/real-time-clocks.html

Linux Clock Sources

Android Used Linux Kernel. Understanding Linux Clock (see Figure 38) sources gives the Android intenral platform event generation resolution. Additionally, the open source protocol libraries, MQTT/LoRA, build on top of Linux branch libraries.

There were 4 implementations of the time in the Linux kernel: UTIME (1996), timer wheel (1997), HRT (2001) and hrtimers (2007). The latter is the result of the "high-res-timers" project started by George Anzinger in 2001, with contributions by Thomas Gleixner and Douglas Niehaus. The hrtimers implementation was merged into Linux 2.6.21, released in 2007.[26]

Linux supports following Clock Sources [18]:

- tsc
- hpet
- pit
- pmtmr: ACPI Power Management Timer
- cyclone

High-resolution timers are not supported on all hardware architectures. They are at least provided on x86/x86_64, ARM and PowerPC.

Figure 38: Linux Clock

---

[26]   Linux Clock Sources - https://www.python.org/dev/peps/pep-0418/#rationale

Free BSD Timer Counters

Figure 39: Free BSD Timers

kern.timecounter.choice lists available hardware clocks with their priority (see Figure 39). The sysctl program can be used to change the timecounter. Example [19]:

# dmesg	grep Timecounter
Timecounter	"i8254" frequency 1193182 Hz quality 0
Timecounter "ACPI-safe"	frequency 3579545 Hz quality 850
Timecounter "HPET"	frequency 100000000 Hz quality 900
Timecounter "TSC"	frequency 3411154800 Hz quality 800
Timecounters tick	every 10.000 msec
# sysctl kern.timecounter.choice kern.timecounter.choice: TSC(800) HPET(900) ACPI-safe(850) i8254(0) dummy(-1000000) # sysctl kern.timecounter.hardware="ACPI-fast" kern.timecounter.hardware: HPET -> ACPI-fast	

Available clocks:

- "TSC": Time Stamp Counter of the processor
- "HPET": High Precision Event Timer
- "ACPI-fast": ACPI Power Management timer (fast mode)
- "ACPI-safe": ACPI Power Management timer (safe mode)
- "i8254": PIT with Intel 8254 chipset

Finally, Monotonic Clock resolutions:

Name	Operating system	OS Resolution	Python Resolution
QueryPerformanceCounter	Windows Seven	10 ns	10 ns
CLOCK_HIGHRES	SunOS 5.11	2 ns	265 ns
CLOCK_MONOTONIC	Linux 3.0	1 ns	322 ns
CLOCK_MONOTONIC_RAW	Linux 3.3	1 ns	628 ns
CLOCK_BOOTTIME	Linux 3.3	1 ns	628 ns
mach_absolute_time() Mac OS 10.6		1 ns	3 µs
CLOCK_MONOTONIC	FreeBSD 8.2	11 ns	5 µs
CLOCK_MONOTONIC	OpenBSD 5.0	10 ms	5 µs
CLOCK_UPTIME	FreeBSD 8.2	11 ns	6 µs
CLOCK_MONOTONIC_COARSE	Linux 3.3	1 ms	1 ms
CLOCK_MONOTONIC_COARSE	Linux 3.0	4 ms	4 ms
GetTickCount64()	Windows Seven	16 ms	15 ms

Please note: The event sources rages from nano seconds to milli seconds on the edge level (see Figure 40).

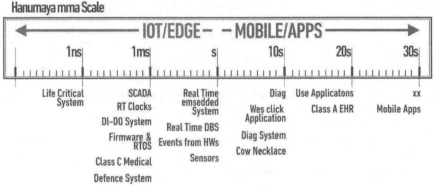

Figure 40: Hanumayamma IoT Scale

Clock Resolution

Please check Appendix A: bench_time.C for C program version of Clock resolutions computed.
    The following code has Python version [20] (see Table 9):

Table 9: Clock Resolution Code

```
# -*- coding: utf-8 -*-
"""
Created on Thu Nov 22 14:00:54 2018
@author: cvuppalapati
 clock_resolution.py

 Reference: PEP0418
    https://www.python.org/dev/peps/pep-0418/#rationale

    Code: https://hg.python.org/peps/file/tip/pep-0418/clock_resolution.py

"""
import time
try:
    from time import timeout_time
except ImportError:
    from time import time as timeout_time
def compute_resolution(func):
    resolution = None
    points = 0
    timeout = timeout_time() + 1.0
    previous = func()
    while timeout_time() < timeout or points < 3:
        for loop in range(10):
            t1 = func()
            t2 = func()
            dt = t2 - t1
            if 0 < dt:
                break
        else:
            dt = t2 - previous
            if dt <= 0.0:
                continue
        if resolution is not None:
            resolution = min(resolution, dt)
        else:
            resolution = dt
        points += 1
        previous = func()
    return resolution

def format_duration(dt):
    if dt >= 1e-3:
        return "%.0f ms" % (dt * 1e3)
    if dt >= 1e-6:
        return "%.0f us" % (dt * 1e6)
    else:
        return "%.0f ns" % (dt * 1e9)
```

Table 9 contd. ...

...Table 9 contd.

```
def test_clock(name, func):
    print("%s:" % name)
    resolution = compute_resolution(func)
    print("- resolution in Python: %s" % format_duration(resolution))

clocks = ['clock', 'perf_counter', 'process_time']
if hasattr(time, 'monotonic'):
    clocks.append('monotonic')
clocks.append('time')
for name in clocks:
    func = getattr(time, name)
    test_clock("%s()" % name, func)
    info = time.get_clock_info(name)
    print("- implementation: %s" % info.implementation)
    print("- resolution: %s" % format_duration(info.resolution))

clock_ids = [name for name in dir(time) if name.startswith("CLOCK_")]
clock_ids.sort()
for clock_id_text in clock_ids:
    clock_id = getattr(time, clock_id_text)
    name = 'clock_gettime(%s)' % clock_id_text
    def gettime():
        return time.clock_gettime(clock_id)
    try:
        gettime()
    except OSError as err:
        print("%s failed: %s" % (name, err))
        continue
    test_clock(name, gettime)
    resolution = time.clock_getres(clock_id)
    print("- announced resolution: %s" % format_duration(resolution))
```

The application call time function for each time features and take snap shot at time 1 and time 2 and difference the performance.

```
clock():
- resolution in Python: 365 ns
- implementation: QueryPerformanceCounter()
- resolution: 365 ns
perf_counter():
- resolution in Python: 365 ns
- implementation: QueryPerformanceCounter()
- resolution: 365 ns
process_time():
- resolution in Python: 16 ms
- implementation: GetProcessTimes()
- resolution: 100 ns
monotonic():
- resolution in Python: 15 ms
- implementation: GetTickCount64()
- resolution: 16 ms
time():
- resolution in Python: 441 us
- implementation: GetSystemTimeAsFileTime()
- resolution: 16 ms
```

Please see Figure 41:

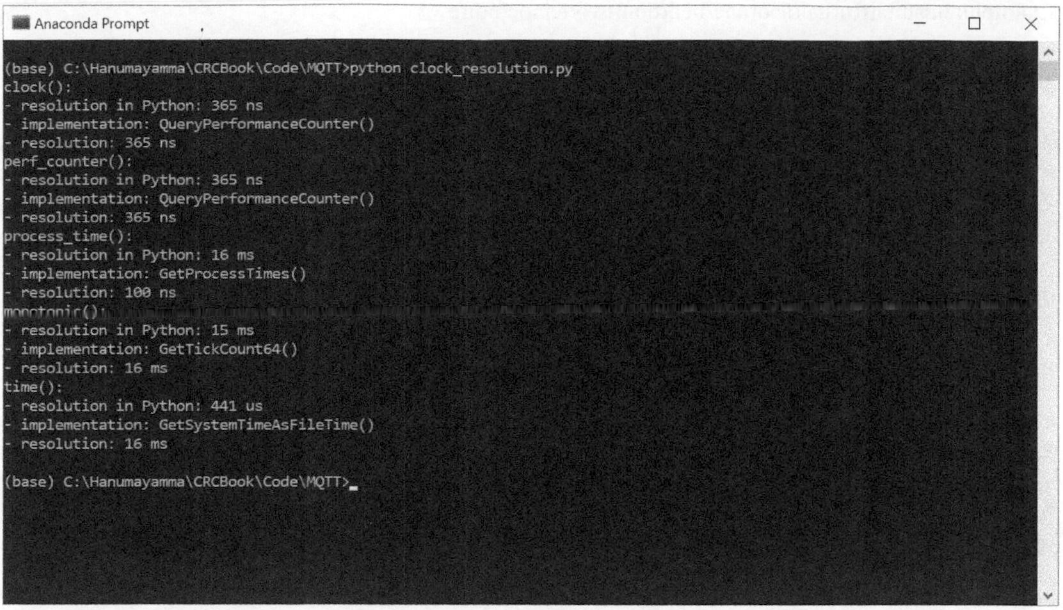

Figure 41: Clock resolution output

Message Types

Connection Acknowledgement Codes

Sockets

Base 62 Conversion (please see Table 10)

Table 10: Base 62 Code

```
def base62(num, base=string.digits + string.ascii_letters, padding=1):
    """"Convert a number to base-62 representation."""
    assert num >= 0
    digits = []
    while num:
        num, rest = divmod(num, 62)
        digits.append(base[rest])
    digits.extend(base[0] for _ in range(len(digits), padding))
    return ''.join(reversed(digits))
```

Payload
if len(local_payload) > 268435455:
        raise ValueError('Payload too large.')

MQTT Signals

The publish and subscribe messages distinguishes a publisher sending a message from the subscriber that receives the message.

MQTT uses topics to identify which message reaches which client. A broker uses topic (UTF-8 string) to filter messages for each client which is connected. A topic is of very light weight.

Example: hospitalRurual/floorone/bedroomtwo/temperature

MQTT Client

A publisher and a subscriber make an MQTT client. An MQTT can be any device that has its library running and should be able to connect to an MQTT broker on any network. MQTT libraries support a huge variety of programming languages, for example android, C, C++, iOS, java, JavaScript, etc.

During the client construction, the following constructor parameters are set:

def __init__(self, client_id="", clean_session=True, userdata=None,
    protocol=MQTTv311, transport="tcp"):

Parameters:

client_id: is the unique client id string used when connecting to the broker. If client_id is zero length or None, then the behaviour is defined by which protocol version is in use. If using MQTT v3.1.1, then a zero-length client id will be sent to the broker and the broker will generate a random for the client. If using MQTT v3.1 then an id will be randomly generated. In both cases, clean_session must be True. If this is not the case a ValueError will be raised.

clean_session is a boolean that determines the client type. If True the broker will remove all information about this client when it disconnects. If False, the client is a persistent client and subscription information and queued messages will be retained when the client disconnects. Note that a client will never discard its own outgoing messages on disconnect. Calling connect() or reconnect() will cause the messages to be resent. Use reinitialise() to reset a client to its original state.

userdata is user-defined data of any type that is passed as the "userdata" parameter to callbacks. It may be updated at a later point with the user_data_set() function.

The protocol argument allows explicit setting of the MQTT version to use for this client. Can be paho. mqtt.client.MQTTv311 (v3.1.1) or paho.mqtt.client.MQTTv31 (v3.1), with the default being v3.1.1 If the broker reports that the client connected with an invalid protocol version, the client will automatically attempt to reconnect using v3.1 instead.

Set transport to "websockets" to use WebSockets as the transport mechanism. Set to "tcp" to use raw TCP, which is the default.

General usage flow

- Use connect()/connect_async() to connect to a broker
- Call loop() frequently to maintain network traffic flow with the broker
- Or use loop_start() to set a thread running to call loop() for you.
- Or use loop_forever() to handle calling loop() for you in a blocking function.
- Use subscribe() to subscribe to a topic and receive messages
- Use publish() to send message
- Use disconnect() to disconnect from the broker

Data returned from the broker is made available with the use of callback functions as described below (see Figure 42) [21].

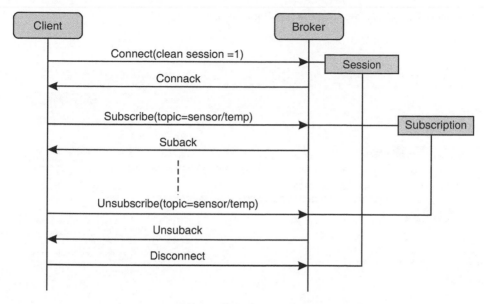

Figure 42: MQTT Sequence flow

Callbacks

- A number of callback functions are available to receive data back from the broker. To use a callback, define a function and then assign it to the client.

def on_connect(client, userdata, flags, rc):
    print("Connection returned " + str(rc))

client.on_connect = on_connect

All of the callbacks as described below have a "client" and an "userdata" argument. "client" is the Client instance that is calling the callback.

"userdata" is user data of any type and can be set when creating a new client

instance or with user_data_set(userdata).

The callbacks:

On_connect(client, userdata, flags, rc): Called when the broker responds to our connection request.
    flags is a dict that contains response flags from the broker:
        flags['session present'] - this flag is useful for clients that are
            using clean session set to 0 only. If a client with clean
            session=0, that reconnects to a broker that it has previously
            connected to, this flag indicates whether the broker still has the
            session information for the client. If 1, the session still exists.
    The value of rc determines success or not:
    0: Connection successful
    1: Connection refused - incorrect protocol version
    2: Connection refused - invalid client identifier
    3: Connection refused - server unavailable
    4: Connection refused - bad username or password
    5: Connection refused - not authorised
    6-255: Currently unused.

Please find MQTT publish/subscribe code (see Figure 43).

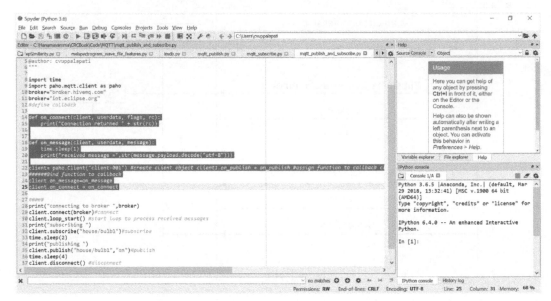

Figure 43: MQTT Pub/Sub method

on_disconnect(client, userdata, rc): Called when the client disconnects from the broker. The rc parameter indicates the disconnection state. If MQTT_ERR_SUCCESS (0), the callback was called in response to a disconnect() call. If any other value the disconnection was unexpected, possibly caused by a network error.

on_message(client, userdata, message): Called when a message has been received on a topic that the client subscribes to. The message variable is a MQTTMessage that describes all of the message parameters.

on_publish(client, userdata, mid): Called when a message that was to be sent using the publish() call has completed transmission to the broker. For messages with QoS levels 1 and 2, this means that the appropriate handshakes have completed. For QoS 0, this simply means that the message has left the client. The mid variable matches the mid variable returned from the corresponding publish() call, to allow outgoing messages to be tracked. This callback is important because even if the publish() call returns success, it does not always mean that the message has been sent.

on_subscribe(client, userdata, mid, granted_qos): Called when the broker responds to a subscribe request. The mid variable matches the mid variable returned from the corresponding subscribe() call. The granted_qos variable is a list of integers that give the QoS that level the broker has granted for each of the different subscription requests.

on_unsubscribe(client, userdata, mid): Called when the broker responds to an unsubscribe request. The mid variable matches the mid variable returned from the corresponding unsubscribe() call.

on_log(client, userdata, level, buf): Called when the client has log information. Define to allow debugging. The level variable gives the severity of the message and will be one of MQTT_LOG_INFO, MQTT_LOG_NOTICE, MQTT_LOG_WARNING, MQTT_LOG_ERR, or MQTT_LOG_DEBUG. The message itself is in buf.

on_socket_open(client, userdata, sock): Called when the socket has been opened. Use this to register the socket with an external event loop for reading.

on_socket_close(client, userdata, sock): Called when the socket is about to be closed. Use this to unregister a socket from an external event loop for reading.

on_socket_register_write(client, userdata, sock): Called when a write operation to the socket failed because it would have blocked, e.g. output buffer full. Use this to register the socket with an external event loop for writing.

on_socket_unregister_write(client, userdata, sock): Called when a write operation to the socket succeeded after it had previously failed. Use this to unregister the socket from an external event loop for writing.
"""""

MQTT Control Packets

Below are the parts of MQTT protocol [22]:

1. *Fixed Header*: The length of this packet in MQTT is fixed (see Figure 44)

Bit	7	6	5	4	3	2	1	0
byte 1	MQTT Control Packet type				Flags specific to each MQTT Control Packet type			
byte 2…	Remaining Length							

Figure 44: MQTT Header

2. *Variable header*: The header contains some variable length components and depends on the packet type. Most variables have packet identifiers in them. Both client and broker can have a separate packet identifier that helps them to maintain concurrent message transfer (see Figure 45).

Bit	7	6	5	4	3	2	1	0
byte 1	Packet Identifier MSB							
byte 2	Packet Identifier LSB							

Figure 45: MQTT Variable Header

3. *Payloads*: Generally, it is the last part of a control packet. For Publish packets, payloads act as an application (see Figure 46).

CONTROL PACKET	Payload
CONNECT	Required
CONNACK	None
PUBLISH	Optional
PUBACK	None
PUBREC	None
PUBREL	None
PUBCOMP	None
SUBSCRIBE	Required
SUBACK	Required
UNSUBSCRIBE	Required
UNSUBACK	None
PINGREQ	None
PINGRESP	None
DISCONNECT	None

Figure 46: MQTT Packet and Payload

MQTT Wireshark Captures

Install Wireshark to capture protocol level messages. Installation of Wireshark (see Figure 47).

Figure 47: Wireshark Installation

Successful install enables capture of MQTT (see Figure 48)

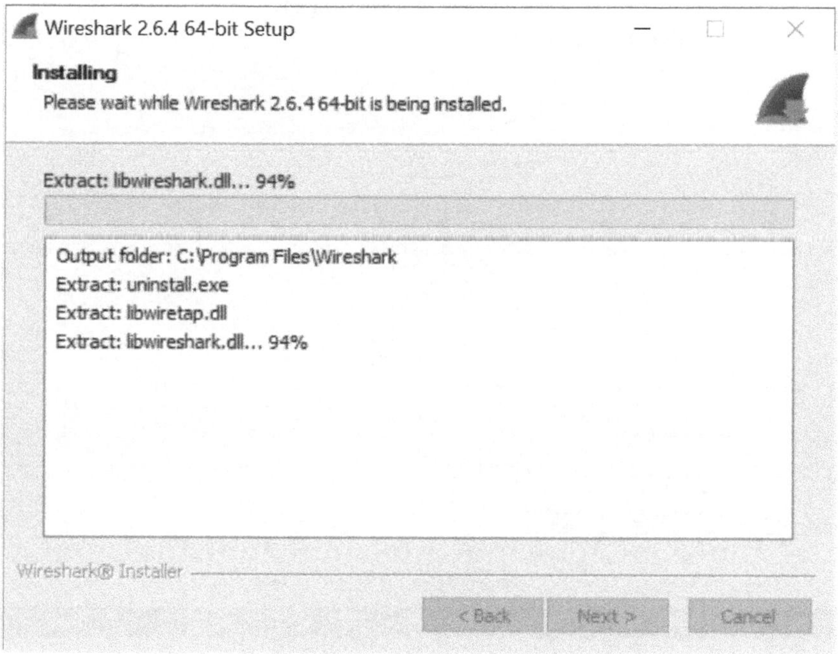

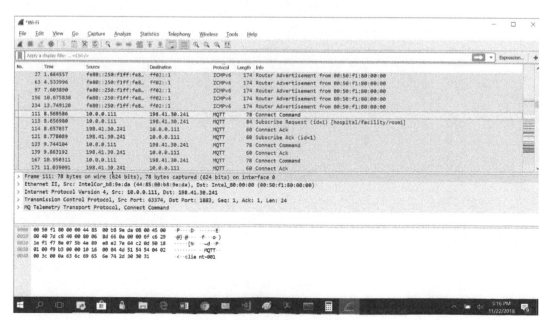

Figure 48: Installation in progress

WS - MQ Telemetry Transport Protocol, Connect Command

Capture MQTT Transport Protocol by running Wireshark (see Figure 49).

From Source: 10.0.0111   Destination: 198.41.30.241

Figure 49: Wireshark Telemetry Command

WS - MQ Telemetry Transport Protocol, Subscribe Request

From Source: 10.0.0111   Destination: 198.41.30.241
Telemetry Subscript details (see Figure 50)

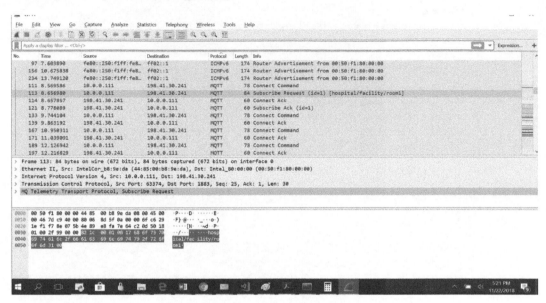

Figure 50: Subscribe request

WS - MQ Telemetry Transport Protocol, Subscribe Ack

From Source: 10.0.0111   Destination: 198.41.30.241
Telemetry Subscribe acknowledgement (see Figure 51)

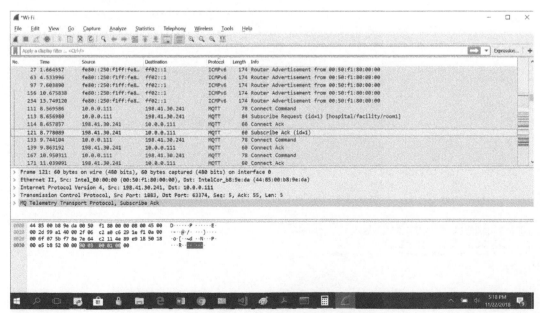

Figure 51: Wireshark Subscribe Ack

WS - MQ Telemetry Transport Protocol, Connect Command

From Source: 198.41.30.241   Destination: 10.0.0111 See Figure 52

WS - MQ Telemetry Transport Protocol, Disconnect Req

From Source: 10.0.0111   Destination: 198.41.30.241 see Figure 52 and Figure 53.

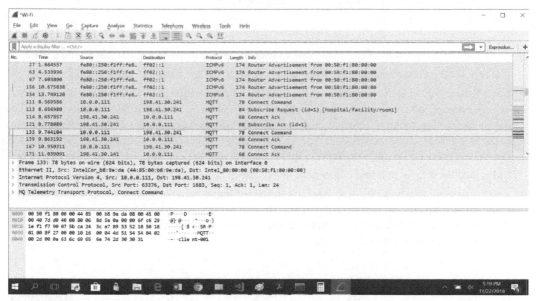

Figure 52: Wireshark Connect Command

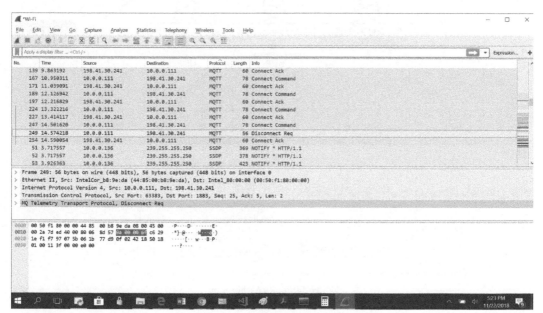

Figure 53: Wireshark Disconnect Command

LoRa

LoRaWAN is a long range, low power, wireless protocol that is intended for use in building IoT networks. IoT devices ("nodes") send small data packets to any number of "gateways" that may be in the several-kilometer range of a node via the LoRaWAN wireless protocol. The gateways then use more traditional communications, such as wired Internet connections ,to forward the messages to a network-server which validates the packets and forwards the application payload to an application-server.[27]

- LoRa Gateway Bridge: The LoRa gateway bridge takes the Packet forward message (UDP) and translates it into MQTT. The advantage of using MQTT is:

  ○ It makes debugging easier

  ○ Sending downlink data only requires knowledge about the corresponding MQTT topic of the gateway, the MQTT broker will route it to the LoRa Gateway Bridge instance responsible for the gateway

  ○ It enables a secure connection between your gateways and the network (using MQTT over TLS)

- LoRa Server: The LoRa Server manages the WAN and provides the network. It also maintains registry of active devices on the network. The LoRa Server performs one more additional functionality: When data is received by multiple gateways, the LoRa Server will de-duplicate this data and forward it once to the LoRaWAN application-server. When an application-server needs to send data back to a device, the LoRa Server will keep these items in queue, until it is able to send to them one of the gateways.

- LoRa Geo Server: An optional component that provides geo resolution of Gateways.

- LoRa App Server: The LoRa App Server component implements a LoRaWAN application-server compatible with the LoRa Server component. To receive the application payload sent by one of your devices, you can use one of the integrations that the LoRa App Server provides (e.g., MQTT, HTTP or directly write to an InfluxDB database).

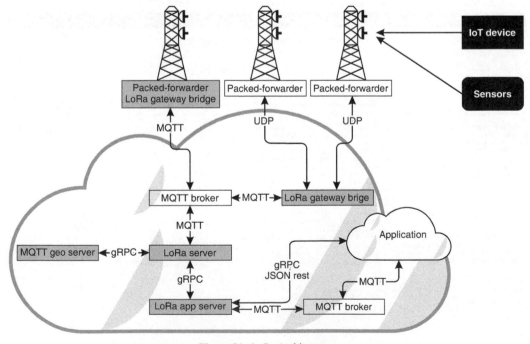

Figure 54: LoRa Architecture

---

27   LoRa - https://www.loraserver.io/overview/architecture/

Data Flow

On the left side of LoRa implementation, the sensor connects to the LoRa Gateway via UDP protocol. The LoRa Gateway connects to the LoRa Gateway bridge and relays the connection to using UDP. The LoRa gateway bridge translates the UDP into either MQTT or HTTP and sends the data on MQTT or HTTP respectively. Finally, the LoRa Server processes the pay load and executes the business application.

Code: https://github.com/Lora-net/packet_forwarder

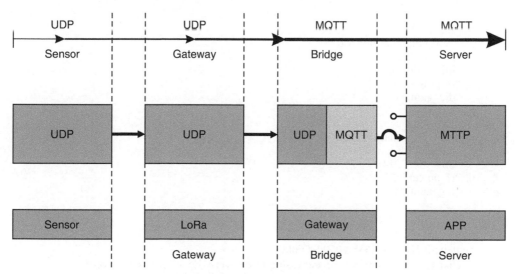

Figure 55: Protocol Data Flow

The protocol exchange is in the following sequences (see Figure 55):
    UDP to UDP [21]
                UDP to MQTT
                        MQTT and Application Process

Configuration (see Table 11) for important socket and LoRa.

Table 11: Configuration

```
{
    "SX1301_conf": {
        "lorawan_public": true,
        "clksrc": 1, /* radio_1 provides clock to concentrator */
        "lbt_cfg": {
            "enable": false,
            "rssi_target": -80, /* dBm */
            "chan_cfg":[ /* 8 channels maximum */
                { "freq_hz": 867100000, "scan_time_us": 128 },
                { "freq_hz": 867300000, "scan_time_us": 5000 },
                { "freq_hz": 867500000, "scan_time_us": 128 },
                { "freq_hz": 869525000, "scan_time_us": 128 }
            ],
            "sx127x_rssi_offset": -4 /* dB */
        },
```

Table 11 contd. ...

```
            "antenna_gain": 0, /* antenna gain, in dBi */
         "radio_0": {
            "enable": true,
            "type": "SX1257",
            "freq": 867500000,
            "rssi_offset": -166.0,
            "tx_enable": true,
            "tx_notch_freq": 129000, /* [126..250] KHz */
            "tx_freq_min": 863000000,
            "tx_freq_max": 870000000
         },
         "tx_lut_0": {
            /* TX gain table, index 0 */
            "pa_gain": 0,
            "mix_gain": 8,
            "rf_power": -6,
            "dig_gain": 0
         },
         "tx_lut_15": {
            /* TX gain table, index 15 */
            "pa_gain": 3,
            "mix_gain": 14,
            "rf_power": 27,
            "dig_gain": 0
         }
      },
      "gateway_conf": {
         "gateway_ID": "AA555A0000000000",
         /* change with default server address/ports, or overwrite in local_conf.json */
         "server_address": "localhost",
         "serv_port_up": 1680,
         "serv_port_down": 1680,
         /* adjust the following parameters for your network */
         "keepalive_interval": 10,
         "stat_interval": 30,
         "push_timeout_ms": 100,
         /* forward only valid packets */
         "forward_crc_valid": true,
         "forward_crc_error": false,
         "forward_crc_disabled": false
      }
}
```

LoRa Gateway Bridge Architecture

The Gateway Bridge's role is twofold: (a) receiving packets from Sensor/Gateway (downstream) and (b) Sending data to the LoRa Server, the upstream (see Figure 56). To perform these essential operations, Gateway startup code, in addition to setup, initiates the following threads (in the order):

- Socket initiation (downstream & upstream)
- void thread_up(void);
- void thread_down(void);
- void thread_gps(void);
- void thread_valid(void);

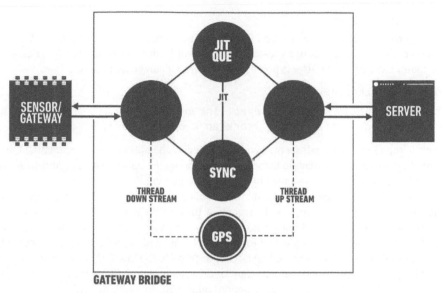

Figure 56: LoRa composite

- void thread_jit(void);
- void thread_timersync(void);

Socket Operations

```
/* connect so we can send/receive packet with the server only */
   i = connect(sock_up, q->ai_addr, q->ai_addrlen);
   if (i != 0) {
       MSG("ERROR: [up] connect returned %s\n", strerror(errno));
       exit(EXIT_FAILURE);
   }
   freeaddrinfo(result);
```

Starting GPS Thread

Start GPS Thread to send data packets to Upstream.

```
/* Start GPS a.s.a.p., to allow it to lock */
    if (gps_tty_path[0] != '\0') { /* do not try to open GPS device if no
path set */
    i = lgw_gps_enable(gps_tty_path, "ubx7", 0, &gps_tty_fd); /* HAL only
supports u-blox 7 for now */
    if (i != LGW_GPS_SUCCESS) {
        printf("WARNING: [main] impossible to open %s for GPS sync (check
permissions)\n", gps_tty_path);
        gps_enabled = false;
        gps_ref_valid = false;
    } else {
        printf("INFO: [main] TTY port %s open for GPS synchronization\n",
gps_tty_path);
        gps_enabled = true;
        gps_ref_valid = false;
    }
 }
```

Thread Up

The Thread Up routine receives data packets from downstream and sends the data to the Upstream.

At the core, the thread_up routine uses set socket and open the socket option till the data is read. As part of the routine, the payload buffer is populated with headers and data details.

```
#include <sys/socket.h>
int setsockopt(int socket, int level, int option_name,
      const void *option_value, socklen_t option_len);
```

The setsockopt() function shall set the option specified by the option_name argument, at the protocol level specified by the level argument, to the value pointed to by the option_value argument for the socket associated with the file descriptor specified by the socket argument.[28]

SO_RCVTIMEO[29]— Sets the timeout value that specifies the maximum amount of time an input function waits until it completes. It accepts a timeval structure with the number of seconds and microseconds specifying the limit on how long to wait for an input operation to complete (see Table 12). If a receive operation has blocked for this much time without receiving additional data, it shall return with a partial count or errno set to [EAGAIN] or [EWOULDBLOCK] if no data is received. The default for this option is zero, which indicates that a receive operation shall not time out. This option takes a timeval structure. Note that not all implementations allow this option to be set.

Table 12: LoRa Socket Operations (Threads)

```
/* --- --------------------- */
void thread_up(void) {
    int i, j; /* loop variables */
    unsigned pkt_in_dgram; /* nb on Lora packet in the current datagram */

    /* allocate memory for packet fetching and processing */
    struct lgw_pkt_rx_s rxpkt[NB_PKT_MAX]; /* array containing inbound packets
+ metadata */
    struct lgw_pkt_rx_s *p; /* pointer on a RX packet */
    int nb_pkt;

    /* local copy of GPS time reference */
    bool ref_ok = false; /* determine if GPS time reference must be used or
not */
    struct tref local_ref; /* time reference used for UTC <-> timestamp
conversion */
    /* data buffers */
    uint8_t buff_up[TX_BUFF_SIZE]; /* buffer to compose the upstream packet */
    int buff_index;

    uint8_t buff_ack[32]; /* buffer to receive acknowledges */

    /* protocol variables */
    uint8_t token_h; /* random token for acknowledgement matching */
    uint8_t token_l; /* random token for acknowledgement matching */

    /* ping measurement variables */
    struct timespec send_time;
    struct timespec recv_time;

    /* GPS synchronization variables */
    struct timespec pkt_utc_time;
    struct tm * x; /* broken-up UTC time */
    struct timespec pkt_gps_time;
    uint64_t pkt_gps_time_ms;
    /* report management variable */
```

Table 12 contd. ...

---

[28] Set_socket http://pubs.opengroup.org/onlinepubs/009695399/functions/setsockopt.html

[29] So_rcvtimeo - http://pubs.opengroup.org/onlinepubs/009695399/functions/setsockopt.html

```
  bool send_report = false;
/* mote info variables */
uint32_t mote_addr = 0;
uint16_t mote_fcnt = 0;
/* set upstream socket RX timeout */
i = setsockopt(sock_up, SOL_SOCKET, SO_RCVTIMEO, (void *)&push_timeout_
half, sizeof push_timeout_half);
if (i != 0) {
  MSG("ERROR: [up] setsockopt returned %s\n", strerror(errno));
  exit(EXIT_FAILURE);
}
/* pre-fill the data buffer with fixed fields */
buff_up[0] = PROTOCOL_VERSION;
buff_up[3] = PKT_PUSH_DATA;
*(uint32_t *)(buff_up + 4) = net_mac_h;
*(uint32_t *)(buff_up + 8) = net_mac_l;
while (!exit_sig && !quit_sig) {
  /* fetch packets */
  pthread_mutex_lock(&mx_concent);
  nb_pkt = lgw_receive(NB_PKT_MAX, rxpkt);
  pthread_mutex_unlock(&mx_concent);
  if (nb_pkt == LGW_HAL_ERROR) {
    MSG("ERROR: [up] failed packet fetch, exiting\n");
    exit(EXIT_FAILURE);
  }
  /* check if there are status report to send */
  send_report = report_ready; /* copy the variable so it doesn't change mid-
function */
  /* no mutex, we're only reading */
  /* wait a short time if no packets, nor status report */
  if ((nb_pkt == 0) && (send_report == false)) {
    wait_ms(FETCH_SLEEP_MS);
    continue;
  }
  /* get a copy of GPS time reference (avoid 1 mutex per packet) */
  if ((nb_pkt > 0) && (gps_enabled == true)) {
    pthread_mutex_lock(&mx_timeref);
    ref_ok = gps_ref_valid;
    local_ref = time_reference_gps;
    pthread_mutex_unlock(&mx_timeref);
  } else {
    ref_ok = false;
  }
  /* start composing datagram with the header */
  token_h = (uint8_t)rand(); /* random token */
  token_l = (uint8_t)rand(); /* random token */
  buff_up[1] = token_h;
  buff_up[2] = token_l;
  buff_index = 12; /* 12-byte header */
  /* start of JSON structure */
  memcpy((void *)(buff_up + buff_index), (void *)"{\"rxpk\":[", 9);
  buff_index += 9;
  /* serialize Lora packets metadata and payload */
  pkt_in_dgram = 0;
  for (i=0; i < nb_pkt; ++i) {
```

Table 12 contd. ...

...Table 12 contd.

```
        p = &rxpkt[i];
        /* Get mote information from current packet (addr, fcnt) */
        /* FHDR - DevAddr */
        mote_addr = p->payload[1];
        mote_addr |= p->payload[2] << 8;
        mote_addr |= p->payload[3] << 16;
        mote_addr |= p->payload[4] << 24;
        /* FHDR - FCnt */
        mote_fcnt = p->payload[6];
        mote_fcnt |= p->payload[7] << 8;

        /* basic packet filtering */
        pthread_mutex_lock(&mx_meas_up);
        meas_nb_rx_rcv += 1;
        switch(p->status) {
            case STAT_CRC_OK:
                meas_nb_rx_ok += 1;
                printf( "\nINFO: Received pkt from mote: %08X (fcnt=%u)\n", mote_
addr, mote_fcnt );
                if (!fwd_valid_pkt) {
                    pthread_mutex_unlock(&mx_meas_up);
                    continue; /* skip that packet */
                }
                break;
            case STAT_CRC_BAD:
                meas_nb_rx_bad += 1;
                if (!fwd_error_pkt) {
                    pthread_mutex_unlock(&mx_meas_up);
                    continue; /* skip that packet */
                }
                break;
            case STAT_NO_CRC:
                meas_nb_rx_nocrc += 1;
                if (!fwd_nocrc_pkt) {
                    pthread_mutex_unlock(&mx_meas_up);
                    continue; /* skip that packet */
                }
                break;
        default:
                MSG("WARNING: [up] received packet with unknown status %u
(size %u, modulation %u, BW %u, DR %u, RSSI %.1f)\n", p->status, p->size,
p->modulation, p->bandwidth, p->datarate, p->rssi);
                pthread_mutex_unlock(&mx_meas_up);
                continue; /* skip that packet */
                // exit(EXIT_FAILURE);
    }
    meas_up_pkt_fwd += 1;
    meas_up_payload_byte += p->size;
    pthread_mutex_unlock(&mx_meas_up);

    /* Start of packet, add inter-packet separator if necessary */
    if (pkt_in_dgram == 0) {
        buff_up[buff_index] = '{';
        ++buff_index;
    } else {
        buff_up[buff_index] = ',';
        buff_up[buff_index+1] = '{';
        buff_index += 2;
    }
```

Table 12 contd. ...

...Table 12 contd.

```
      /* RAW timestamp, 8-17 useful chars */
      j = snprintf((char *)(buff_up + buff_index), TX_BUFF_SIZE-buff_index,
   "\"tmst\":%u", p->count_us);
      if (j > 0) {
         buff_index += j;
      } else {
         MSG("ERROR: [up] snprintf failed line %u\n", (__LINE__ - 4));
         exit(EXIT_FAILURE);
      }
      /* Packet RX time (GPS based), 37 useful chars */
      if (ref_ok == true) {
         /* convert packet timestamp to UTC absolute time */
         j = lgw_cnt2utc(local_ref, p->count_us, &pkt_utc_time);
         if (j == LGW_GPS_SUCCESS) {
            /* split the UNIX timestamp to its calendar components */
            x = gmtime(&(pkt_utc_time.tv_sec));
            j = snprintf((char *)(buff_up + buff_index), TX_BUFF_SIZE-buff_index,
   ",\"time\":\"%04i-%02i-%02iT%02i:%02i:%02i.%06liZ\"", (x->tm_year)+1900,
   (x->tm_mon)+1, x->tm_mday, x->tm_hour, x->tm_min, x->tm_sec, (pkt_utc_time.
   tv_nsec)/1000); /* ISO 8601 format */
            if (j > 0) {
               buff_index += j;
            } else {
               MSG("ERROR: [up] snprintf failed line %u\n", (__LINE__ - 4));
               exit(EXIT_FAILURE);
            }
         }
         /* convert packet timestamp to GPS absolute time */
         j = lgw_cnt2gps(local_ref, p->count_us, &pkt_gps_time);
         if (j == LGW_GPS_SUCCESS) {
            pkt_gps_time_ms = pkt_gps_time.tv_sec * 1E3 + pkt_gps_time.tv_nsec /
   1E6;
            j = snprintf((char *)(buff_up + buff_index), TX_BUFF_SIZE-buff_index,
   ",\"tmms\":%llu",
                  pkt_gps_time_ms); /* GPS time in milliseconds since 06.Jan.1980 */
            if (j > 0) {
               buff_index += j;
            } else {
               MSG("ERROR: [up] snprintf failed line %u\n", (__LINE__ - 4));
               exit(EXIT_FAILURE);
            }
         }
      }
      /* Packet concentrator channel, RF chain & RX frequency, 34-36 useful chars
   */
      j = snprintf((char *)(buff_up + buff_index), TX_BUFF_SIZE-buff_index,
   ",\"chan\":%1u,\"rfch\":%1u,\"freq\":%.6lf", p->if_chain, p->rf_chain,
   ((double)p->freq_hz / 1e6));
      if (j > 0) {
         buff_index += j;
      } else {
         MSG("ERROR: [up] snprintf failed line %u\n", (__LINE__ - 4));
         exit(EXIT_FAILURE);
      }
      /* Packet status, 9-10 useful chars */
      switch (p->status) {
         case STAT_CRC_OK:
```

Table 12 contd. ...

...Table 12 contd.

```
        memcpy((void *)(buff_up + buff_index), (void *)",\"stat\":1", 9);
        buff_index += 9;
        break;
   case STAT_CRC_BAD:
     memcpy((void *)(buff_up + buff_index), (void *)",\"stat\":-1", 10);
        buff_index += 10;
        break;
   case STAT_NO_CRC:
     memcpy((void *)(buff_up + buff_index), (void *)",\"stat\":0", 9);
        buff_index += 9;
        break;
   default:
     MSG("ERROR: [up] received packet with unknown status\n");
     memcpy((void *)(buff_up + buff_index), (void *)",\"stat\":?", 9);
   buff_index += 9;
   exit(EXIT_FAILURE);
 }
```

Thread Down

The Thread down (see Table 13) functionalities are (a) polling the server and enqueuing packets in the JIT (Just in Time) Queue to send them to the server. Thread down uses a Radio Frequency channel and sends one PPS pulse. A pulse per second (PPS or 1PPS) is an electrical signal that has a width of less than one second and a sharply rising or abruptly falling edge that accurately repeats once per second. PPS signals are output by radio beacons, frequency standards, other types of precision oscillators and some GPS receivers. The goal of the PPS is to time the signals.

 The 1 PPS output has a much lower jitter than anything a Microcontroller can do. In some more demanding applications this pulse can be used to time things very accurately. With some scientific grade GPS's this 1 PPS output might be accurate to better than 1 nS.

Thread JIT

Checks Packets from the Thread Queue and sends them (see Table 14 and Table 15). In order to send the packets, first, it acquires the lock to then collects the data and sends it (see table).

```
/* hardware access control and correction */
pthread_mutex_t mx_concent = PTHREAD_MUTEX_INITIALIZER; /* control access
to the concentrator */
```

Table 13: Thread down Code

```
/* --- ---------- */
void thread_down(void) {
    int i; /* loop variables */
    /* configuration and metadata for an outbound packet */
    struct lgw_pkt_tx_s txpkt;
    bool sent_immediate = false; /* option to sent the packet immediately */
    /* local timekeeping variables */
    struct timespec send_time; /* time of the pull request */
    struct timespec recv_time; /* time of return from recv socket call */
    /* data buffers */
    uint8_t buff_down[1000]; /* buffer to receive downstream packets */
    uint8_t buff_req[12]; /* buffer to compose pull requests */
    int msg_len;
    /* protocol variables */
    uint8_t token_h; /* random token for acknowledgement matching */
    uint8_t token_l; /* random token for acknowledgement matching */
    bool req_ack = false; /* keep track of whether PULL_DATA was acknowledged
or not */
    /* JSON parsing variables */
    JSON_Value *root_val = NULL;
    JSON_Object *txpk_obj = NULL;
    JSON_Value *val = NULL; /* needed to detect the absence of some fields */
    const char *str; /* pointer to sub-strings in the JSON data */
    short x0, x1;
    uint64_t x2;
    double x3, x4;
    /* variables to send on GPS timestamp */
    struct tref local_ref; /* time reference used for GPS <-> timestamp
conversion */
    struct timespec gps_tx; /* GPS time that needs to be converted to
timestamp */
    /* beacon variables */
    struct lgw_pkt_tx_s beacon_pkt;
    uint8_t beacon_chan;
    uint8_t beacon_loop;
    size_t beacon_RFU1_size = 0;
    size_t beacon_RFU2_size = 0;
    uint8_t beacon_pyld_idx = 0;
    time_t diff_beacon_time;
    struct timespec next_beacon_gps_time; /* gps time of next beacon packet */
    struct timespec last_beacon_gps_time; /* gps time of last enqueued beacon
packet */
int retry;
    /* beacon data fields, byte 0 is Least Significant Byte */
    int32_t field_latitude; /* 3 bytes, derived from reference latitude */
    int32_t field_longitude; /* 3 bytes, derived from reference longitude */
    uint16_t field_crc1, field_crc2;
    /* auto-quit variable */
    uint32_t autoquit_cnt = 0; /* count the number of PULL_DATA sent since the
latest PULL_ACK */
    /* Just In Time downlink */
    struct timeval current_unix_time;
    struct timeval current_concentrator_time;
    enum jit_error_e jit_result = JIT_ERROR_OK;
    enum jit_pkt_type_e downlink_type;
    /* set downstream socket RX timeout */
```

Table 13 contd. ...

...Table 13 contd.

```
   i = setsockopt(sock_down, SOL_SOCKET, SO_RCVTIMEO, (void *)&pull_timeout,
sizeof pull_timeout);
   if (i != 0) {
   MSG("ERROR: [down] setsockopt returned %s\n", strerror(errno));
   exit(EXIT_FAILURE);
 }
   /* pre-fill the pull request buffer with fixed fields */
   buff_req[0] = PROTOCOL_VERSION;
   buff_req[3] = PKT_PULL_DATA;
   *(uint32_t *)(buff_req + 4) = net_mac_h;
   *(uint32_t *)(buff_req + 8) = net_mac_l;

   /* beacon variables initialization */
   last_beacon_gps_time.tv_sec = 0;
   last_beacon_gps_time.tv_nsec = 0;

   /* beacon packet parameters */
   beacon_pkt.tx_mode = ON_GPS; /* send on PPS pulse */
   beacon_pkt.rf_chain = 0; /* antenna A */
   beacon_pkt.rf_power = beacon_power;
   beacon_pkt.modulation = MOD_LORA;
   switch (beacon_bw_hz) {
```

Table 14: Thread JIT Code

```
void thread_jit(void) {
    int result = LGW_HAL_SUCCESS;
    struct lgw_pkt_tx_s pkt;
    int pkt_index = -1;
    struct timeval current_unix_time;
    struct timeval current_concentrator_time;
    enum jit_error_e jit_result;
    enum jit_pkt_type_e pkt_type;
    uint8_t tx_status;

    while (!exit_sig && !quit_sig) {
        wait_ms(10);
    /* transfer data and metadata to the concentrator, and schedule TX */
    gettimeofday(&current_unix_time, NULL);
    get_concentrator_time(&current_concentrator_time, current_unix_time);
    jit_result = jit_peek(&jit_queue, &current_concentrator_time, &pkt_index);
    if (jit_result == JIT_ERROR_OK) {
      if (pkt_index > -1) {
          jit_result = jit_dequeue(&jit_queue, pkt_index, &pkt, &pkt_type);
          if (jit_result == JIT_ERROR_OK) {
              /* update beacon stats */
              if (pkt_type == JIT_PKT_TYPE_BEACON) {
                  /* Compensate breacon frequency with xtal error */
                  pthread_mutex_lock(&mx_xcorr);
                  pkt.freq_hz = (uint32_t)(xtal_correct * (double)pkt.freq_hz);
                  MSG_DEBUG(DEBUG_BEACON, "beacon_pkt.freq_hz=%u (xtal_
correct=%.15lf)\n", pkt.freq_hz, xtal_correct);
                  pthread_mutex_unlock(&mx_xcorr);

                  /* Update statistics */
                  pthread_mutex_lock(&mx_meas_dw);
                  meas_nb_beacon_sent += 1;
                  pthread_mutex_unlock(&mx_meas_dw);
```

Table 14 contd. ...

...Table 14 contd.

```
                 MSG("INFO: Beacon dequeued (count_us=%u)\n", pkt.count_us);
             }
         /* check if concentrator is free for sending new packet */
         pthread_mutex_lock(&mx_concent); /* may have to wait for a fetch to
finish */
         result = lgw_status(TX_STATUS, &tx_status);
         pthread_mutex_unlock(&mx_concent); /* free concentrator ASAP */
         if (result == LGW_HAL_ERROR) {
             MSG("WARNING: [jit] lgw_status failed\n");
         } else {
             if (tx_status == TX_EMITTING) {
                 MSG("ERROR: concentrator is currently emitting\n");
                 print_tx_status(tx_status);
                 continue;
             } else if (tx_status == TX_SCHEDULED) {
                 MSG("WARNING: a downlink was already scheduled, overwritting
it...\n");
                 print_tx_status(tx_status);
             } else {
                 /* Nothing to do */
             }
         }
     /* send packet to concentrator */
     pthread_mutex_lock(&mx_concent); /* may have to wait for a fetch to
finish */
     result = lgw_send(pkt);
     pthread_mutex_unlock(&mx_concent); /* free concentrator ASAP */
     if (result == LGW_HAL_ERROR) {
         pthread_mutex_lock(&mx_meas_dw);
         meas_nb_tx_fail += 1;
         pthread_mutex_unlock(&mx_meas_dw);
         MSG("WARNING: [jit] lgw_send failed\n");
         continue;
     } else {
         pthread_mutex_lock(&mx_meas_dw);
         meas_nb_tx_ok += 1;
         pthread_mutex_unlock(&mx_meas_dw);
         MSG_DEBUG(DEBUG_PKT_FWD, "lgw_send done: count_us=%u\n", pkt.
count_us);
         }
     } else {
         MSG("ERROR: jit_dequeue failed with %d\n", jit_result);
     }
   }
 } else if (jit_result == JIT_ERROR_EMPTY) {
   /* Do nothing, it can happen */
 } else {
   MSG("ERROR: jit_peek failed with %d\n", jit_result);
 }
 }
}
```

THREAD 4: PARSE GPS MESSAGE AND KEEP GATEWAY IN SYNC

Table 15: Thread 4 - GPS Sync

```
/* -------------------------------------------------------------------------
*/
/* --- ---------------- */
static void gps_process_sync(void) {
   struct timespec gps_time;
   struct timespec utc;
   uint32_t trig_tstamp; /* concentrator timestamp associated with PPM pulse */
   int i = lgw_gps_get(&utc, &gps_time, NULL, NULL);

   /* get GPS time for synchronization */
   if (i != LGW_GPS_SUCCESS) {
      MSG("WARNING: [gps] could not get GPS time from GPS\n");
      return;
   }
   /* get timestamp captured on PPM pulse */
   pthread_mutex_lock(&mx_concent);
   i = lgw_get_trigcnt(&trig_tstamp);
   pthread_mutex_unlock(&mx_concent);
   if (i != LGW_HAL_SUCCESS) {
     MSG("WARNING: [gps] failed to read concentrator timestamp\n");
     return;
   }
   /* try to update time reference with the new GPS time & timestamp */
   pthread_mutex_lock(&mx_timeref);
   i = lgw_gps_sync(&time_reference_gps, trig_tstamp, utc, gps_time);
   pthread_mutex_unlock(&mx_timeref);
   if (i != LGW_GPS_SUCCESS) {
     MSG("WARNING: [gps] GPS out of sync, keeping previous time reference\n");
   }
}
```

Thread Valid

Checks the time reference and calculate XTAL correction. In simple form, it performs validation.

```
/* -------------------------------------------------------------------------
*/
/* --- --------- */
void thread_valid(void) {
   /* GPS reference validation variables */
   long gps_ref_age = 0;
   bool ref_valid_local = false;
   double xtal_err_cpy;

   /* variables for XTAL correction averaging */
   unsigned init_cpt = 0;
   double init_acc = 0.0;
   double x;

   /* main loop task */
   while (!exit_sig && !quit_sig) {
     wait_ms(1000);
     /* calculate when the time reference was last updated */
     pthread_mutex_lock(&mx_timeref);
     gps_ref_age = (long)difftime(time(NULL), time_reference_gps.systime);
```

```
        if ((gps_ref_age >= 0) && (gps_ref_age <= GPS_REF_MAX_AGE)) {
            /* time ref is ok, validate and */
            gps_ref_valid = true;
            ref_valid_local = true;
            xtal_err_cpy = time_reference_gps.xtal_err;
            //printf("XTAL err: %.15lf (1/XTAL_err:%.15lf)\n", xtal_err_cpy, 1/
xtal_err_cpy); // DEBUG
        } else {
            /* time ref is too old, invalidate */
            gps_ref_valid = false;
            ref_valid_local = false;
        }
        pthread_mutex_unlock(&mx_timeref);

        /* manage XTAL correction */
        if (ref_valid_local == false) {
            /* couldn't sync, or sync too old -> invalidate XTAL correction */
            pthread_mutex_lock(&mx_xcorr);
            xtal_correct_ok = false;
            xtal_correct = 1.0;
            pthread_mutex_unlock(&mx_xcorr);
            init_cpt = 0;
            init_acc = 0.0;
        } else {
            if (init_cpt < XERR_INIT_AVG) {
                /* initial accumulation */
                init_acc += xtal_err_cpy;
                ++init_cpt;
            } else if (init_cpt == XERR_INIT_AVG) {
                /* initial average calculation */
                pthread_mutex_lock(&mx_xcorr);
                xtal_correct = (double)(XERR_INIT_AVG) / init_acc;

                xtal_correct_ok = true;
                pthread_mutex_unlock(&mx_xcorr);
                ++init_cpt;

            } else {
                /* tracking with low-pass filter */
                x = 1 / xtal_err_cpy;
                pthread_mutex_lock(&mx_xcorr);
                xtal_correct = xtal_correct - xtal_correct/XERR_FILT_COEF + x/XERR_
FILT_COEF;
                pthread_mutex_unlock(&mx_xcorr);

            }
        }

    }

}
```

Build LoRa on Windows Systems

(a) You need UNIX libraries to compile the code (see Figure 57) [23]

(b) Install Cygwin or Install UNIIX Support on Windows

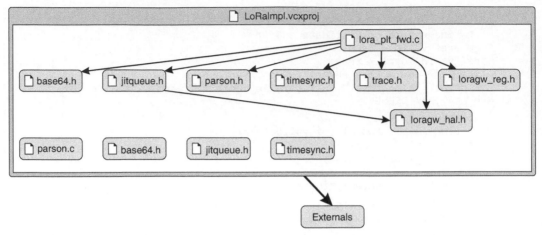

Figure 57: LoRa Project Structure

Cygwin Install

Download the latest version of Cygwin (see figure Cygwin) from https://www.cygwin.com/

Installing and Updating Cygwin for 64-bit versions of Windows

https://cygwin.com/install.html

Installation Notes[30]

- When installing packages for the first time, setup*.exe does not install every package. Only the minimal base packages from the Cygwin distribution are installed by default, taking up about 100 MB.

- Clicking on categories and packages in the setup*.exe package installation screen allows you to select what is installed or updated.

- Individual packages, like bash, gcc, less, etc., are released independently of the Cygwin DLL, so the Cygwin DLL version is not useful as a general Cygwin release number. The setup*.exe utility tracks the versions of all installed components and provides the mechanism for installing or updating everything available from this site for Cygwin.

Figure 58: Cygwin Distribution

---

[30]   Cygwin Installation Notes: https://cygwin.com/install.html

Installation Steps

Once you've installed your desired subset of the Cygwin distribution, setup*.exe will remember what you selected, so rerunning the program will update your system with any new package releases.

Cygwin Setup

Continue to accept default Cygwin inputs: (see Figure 59)

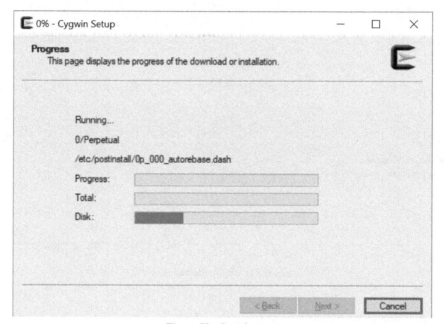

Figure 59: Cygwin setup

Once the installation is complete, you should see: Cygwon install folder (see Figure 60 and Figure 61)

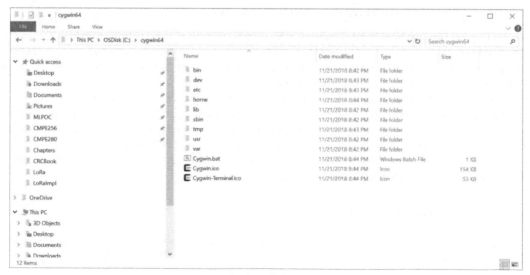

Figure 60: Cygwin install folder

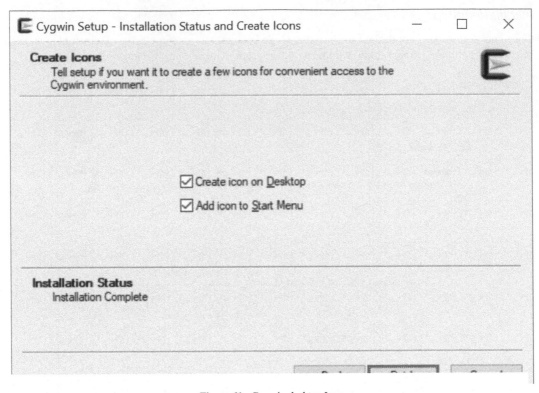

Figure 61: Cygwin desktop Icon

Libraries available on Github[31]

Cygwin distributed libraries are available on GitHub and we can build by coping libraries from Git (this is in lieu of Cygwin install).

Remote Electrocardiogram (ECG or EKG) Monitoring

IoT will transform the world. The data generated by the sensors in our house, work, car and all other places we pass through on a daily basis will provide insights into how we live and can improve our environment. IoT will also enable the gathering of a specific type of data that is just on another level of importance and that will literally transform our lives: Health data.

Our body is the most complex machine ever built and generates an incredible amount of data every second. Being able to capture and analyze that data will provide a path to what is called Precision Medicine, which is the application of medicine tailored to your own unique characteristics. Examples of health data include our body temperature, weight, blood glucose concentration, EMG and ECG.

In the next sections, we will explore a practical example: A low-cost IoT ECG system, which enables doctors to have live remote access to a patient's ECG.

ECG or Electrocardiogram Sesnors

Our heart has one main objective: Make the blood flow inside our body. It achieves that by contracting and expanding its chambers with the passing of an electrical current. The measurement of that current

---

[31] Github Cygwin references - https://github.com/openunix/cygwin

is what is called Electrocardiogram (also known as ECG or EKG). Figure 62 below illustrates what an ECG looks like. We can capture a person's ECG by placing electrodes in different positions on the body.

As you may have noted from Figure 62, the ECG is a wave that has a pattern that repeats itself. The shape of that pattern will vary depending on the position of the electrodes on the body, but a very common pattern is the one known as PQRST complex, shown below Figure 63.

By measuring some characteristics of that shape (and also between consecutives shapes) we can have access to data that provides information about our heart, such as heart rate and heart health condition. Therefore, monitoring a person's ECG can provide extremely useful information, such as when a person is having a heart attack or is showing some abnormal condition that can lead to severe outcomes if not treated quickly.

ECG data, due to its person uniqueness, can also be used as a biometric identification, the same way that fingerprints and the iris are used today.

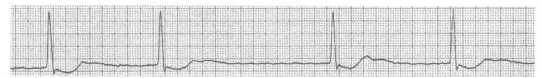

Figure 62: ECG

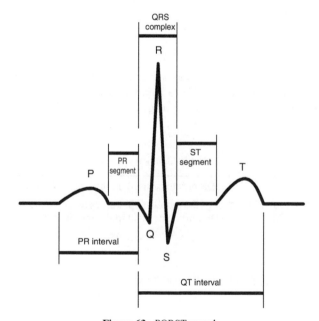

Figure 63: PQRST complex

Creating a Remote ECG monitoring system

Architecture

Let's assume that we want to create a cloud monitoring system where doctors could monitor the ECG of patients in real time. We can have the system broken down in 3 components: Sensor (Data Acquisition), Cloud (Data Storage/Transformation), Front-end (Data Visualization).

The Sensor component is mainly responsible for the ECG data acquisition. In this example we will be using an AD8232 Heart monitor sensor with 3 electrodes. The data must be read by a microcontroller to be then transmitted to the cloud. We will be using a Raspberry Pi 3B+ to read the AD8232 data. Different from Arduino and other microcontrollers, Raspberry Pi does not have analog pins, so we need to have

Analog-to-Digital converter connected between the AD8232 and the Raspberry Pi. We will be using an ADS1115 for that purpose. Figure—Sensor component wiring illustrates the sensor component wiring Table 16 below describes all items used in the sensor component.

Table 16: Sensor Components

Quantity	Name	Estimated Cost ($)
1	Raspberry Pi 3 B+	40
1	AD8232 Heart Monitor	20
1	ADS1115	7
8	Jumper wires	1
Total		68

The cloud component will receive the data sent by the Raspberry Pi, store it (for posterior access) and forward it so the front-end component for visualization. The data can also be processed (transformed) in order to remove noise and identify desired characteristics of the signal (such as normal/abnormal heart beat). In this example, to make it simple for academic purposes, the server will receive the data and forward it to the front-end.

The front-end is where the ECG can be visualized by the user. Our front-end will be a mobile app that shows the data in a graphic.

Code

The code used in the Raspberry Pi and the Server (see Table 17 and Table 18) was created using Python 3.6 and requires RPi.GPIO and Adafruit-ADS1x15 libraries (which can be easily installed using pip).

The mobile app was developed using Lua language and Corona SDK framework.

Server Code

Table 17: ECG Server Code

```
SERVER_PORT = 3421                    # UDP port to be used on this server to receive the data

MOBILE_APP_IP = '192.168.1.59'   # IP of the mobile app that will receive the data
MOBILE_APP_PORT = 4532           # UDP port being used in the mobile app

# loading libraries
import socket

# initialization
s = socket.socket(socket.AF_INET,socket.SOCK_DGRAM)
s.bind(('', SERVER_PORT))

while True:
                    # receiving data
                    data, address = s.recvfrom(1024)
                    value = data.decode()
                    print("Received=", value)

                    # saving data in database (not implemented in this example)
                    # add code to persist data here

                    # forwarding the data to the front end
                    byt=str(value).encode()
                    try:

                        s.sendto(byt, (MOBILE_APP_IP, MOBILE_APP_PORT))
                    except:
                        print("Failed to send data to mobile app. Please check address/port and if it is
reachable.")
```

RaspberryPi Code

```
# Server information
SERVER_IP = '192.168.1.51'  # IP of the server that will receive the data
SERVER_PORT = 3421                   # UDP port being used in the server

# loading libraries
import time
import Adafruit_ADS1x15
import RPi.GPIO as GPIO
import socket

# initialization
GPIO.setmode(GPIO.BCM)
GPIO.setup(4, GPIO.IN,GPIO.PUD_DOWN)
adc = Adafruit_ADS1x15.ADS1115()
gain = 2/3
bitToMilivoltRatio = 0.1875 # this value depends on the gain used.
adcChannel = 0
samplingRate = 60 # number of samples taken per second

s = socket.socket(socket.AF_INET,socket.SOCK_DGRAM)

while True:

        # Reading ECG data
        value = adc.read_adc(adcChannel, gain=gain, data_rate=860) * bitToMilivoltRatio
        print(value)

        try:
                # Sending data to server using UDP
                byt=str(value).encode()
                s.sendto(byt, (SERVER_IP, SERVER_PORT))
        except:
                print("Failed to send. Please check server address/port and if it is reachable.")
        # Pause for before continuing the loop
        time.sleep(1/samplingRate)
```

The communication between the components were done using UDP socket.

UI

The system presented shows a simple and low-cost option to remotely measure a person's ECG. A person connected to the sensors would have the ECG data acquired and submitted to a cloud server. The server then redirects the data to the front-end (a mobile app). See Figure 64 below illustrates the mobile app showing a data acquired from a real person using this system.

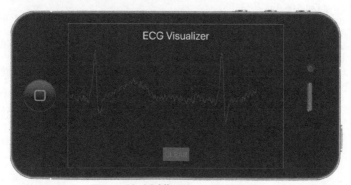

Figure 64: Mobile App - capture ECG

Due to academic purposes, the code and system were designed to be minimal and as simple as possible in order to facilitate its understanding. Several improvements can and should be made for actual deployment.

After Reading this Chapter, you should be able to

- Create and use REST Protocol for connecting web applications and web servers
- Customize MQTT Protocol for sending data between edge devices
- Create Android and IOS applications and connecting hardware sensors
- Use LoRa Protocol for constrained IoT devices
- Create mixed Protocol based hardware and software IoT products

---

References

1   Bernard Marr. Why only one of the 5 Vs of big data really matters, MARCH 19, 2015, https://www.ibmbigdatahub.com/blog/why-only-one-5-vs-big-data-really-matters, Access Date: July 2018

2   Yosi Fait. How The 5G Revolution Will Drive Future Innovations Of IoT, Sep 7, 2018, https://www.forbes.com/sites/forbestechcouncil/2018/09/07/how-the-5g-revolution-will-drive-future-innovations-of-iot/#4b83e79d637e, Access Date: December 2018

3   Samsung. IoT Solutions, 2018, https://www.samsung.com/global/business/networks/solutions/iot-solutions/Access Date: December 2018

4   Mark Nottingham, RFC 7230, Hypertext Transfer Protocol (HTTP/1.1): Message Syntax and Routing, June 2014, https://datatracker.ietf.org/doc/rfc7230/?include_text=1, Access Date: August 2018

5   Jesse James Garrett, Ajax: A new approach to Web Applications, 18 Feb 2015, https://courses.cs.washington.edu/courses/cse490h/07sp/readings/ajax_adaptive_path.pdf, Access date: 08 Aug 2018

6   Eclipse Foundation, MQTT, CoAP and IoT Protocols, 2014, https://www.eclipse.org/community/eclipse_newsletter/2014/february/article2.php, Access Date 18 Sep 2018

7   Carsten Bormann, CoAP Technologies, Implementations, 2016, http://coap.technology/impls.html, Access Date: 08 Nov 2018

8   H. Urunov, The constrained application protocol (coap) part 2, 01 June 2016, https://www.slideshare.net/HamdamboyUrunov/the-constrained-application-protocol-coap-part-2?from_action=save, Access Date: 18 Sep 2018

9   Robert Davidson, Akiba, Carles Cufi, Kevin Townsend, Getting Started with Bluetooth Low Energy, May 2014, ISBN: 9781491900550, https://www.oreilly.com/library/view/getting-started-with/9781491900550/

10 Bluetooth, Protocol Specification, https://www.bluetooth.com/specifications/protocol-specifications, 2018

11 Dag Grini, RF Basics, RF For Non-Engineers, 2006, http://www.ti.com/lit/ml/slap127/slap127.pdf, Access Date: 12/02/2018

12 Texas Instruments, IoT Made Easy, http://www.ti.com/ww/en/wireless_connectivity/sensortag/

13 Android, Bluetooth Overview, https://developer.android.com/guide/topics/connectivity/bluetooth

14 Carol Geyer, OASIS, OASIS MQTT Internet of Things Standard Now Approved by ISO/IEC JTC1, 19 July 2016, https://www.oasis-open.org/news/pr/oasis-mqtt-internet-of-things-standard-now-approved-by-iso-iec-jtc1, Access Date: 08 Nov 2018

15 Carol Geyer,OASIS, MQTT Version 3.1.1 Plus Errata 01, 10 December 2015, http://docs.oasis-open.org/mqtt/mqtt/v3.1.1/mqtt-v3.1.1.pdf, Access date: 30 Apr 2018

16 Python Software Foundation, PEP 418, Add monotonic time, performance counter, and process time functions, 26-March-2012, https://www.python.org/dev/peps/pep-0418/#rationale, Access Date: 06 Aug 2018

17 Nat Dunn, Python Clocks Explained, Aug 14, 2015, https://www.webucator.com/blog/2015/08/python-clocks-explained/, Access Date: 06 Aug 2018

18 Peter Bright, Intel still beats Ryzen at games, but how much does it matter?, 17 Mar 2017, https://arstechnica.com/gadgets/2017/03/intel-still-beats-ryzen-at-games-but-how-much-does-it-matter/, Access Date: 08 Nov 2018

19 John Ackermann, The TICC Timestamping/Time Interval Counter, 06-Feb-2017, https://www.febo.com/pages/TICC/, Access Date: 08 Nov 2018

20 Maximum Integrated, Real-Time Clocks (ICs), 2018, https://www.maximintegrated.com/en/products/digital/real-time-clocks.html, Access Date: 08 Nov 2018

21 LoRa Server, System Architecture, https://www.loraserver.io/overview/architecture/, Access Date: 08 Nov 2018

22 GITHub, LoRa Packet Forward, https://github.com/Lora-net/packet_forwarder, Access Date: 06 Aug 2018

23 Mark Funkenhauser, Pthread Support in Microsoft Windows Services for UNIX Version 3.5, 12/04/2007, https://docs.microsoft.com/en-us/previous-versions/tn-archive/bb463209(v=technet.10), 06 Aug 2018

Middleware

Real time processing deals with streams of data that are captured in real-time and processed with minimal latency. The processing of the streams data is split on a distance basis: Closer to the source, Edge level, and traditional cloud or high compute levels. Nevertheless, both processing architectures need a message ingestion store to act as a buffer for messages, and to support scale-out processing, reliable delivery, and other message queuing semantics.

For IoT events, IoT Data sources are ingested to Cloud systems for processing at Cloud level. The Data sources are ingested into either topic based middle ware, such as Kafka, or inserted into database for processing (see Figure 1).

In this chapter, we will go through the creation of simple Kafka based message system that ingests data into and is processed by Spark or Scala based systems.

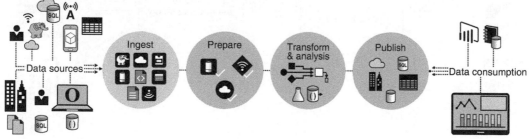

Figure 1: Data Ingestion

Message Architectures

The Message Event architectures use central messaging stream Hubs, such as the IoT Hub, to ingest data in real-time. The events ingested into the IoT Hub are relayed to Stream servers such as Kafka (see Figure 2) to be queued in the event architecture.[1]

---

[1]   Azure Reference Architecture - https://azure.microsoft.com/en-us/services/hdinsight/apache-kafka/

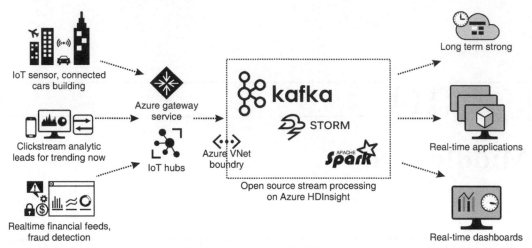

Figure 2: Message Event Architectures

The data from the Kafka topics are consumed by backend processors such as Spark to route analytics engine and thereafter to database or insights delivery to the client.

Streaming Patterns

Streaming patterns relay on high throughput ingestion and Complex event processors. Events are ingested into Stream processors at a high frequency and the events are processed through Complex Event processors with very low latency. The subsequent operations results in event processing and analytics (see Figure 3).

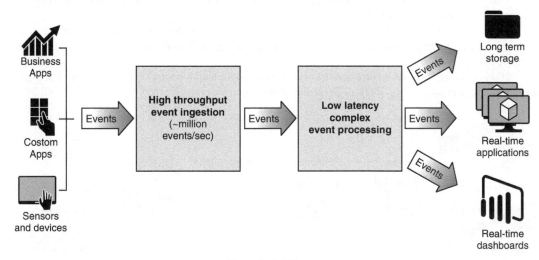

Figure 3: IoT Events

Breaking down streaming space (see Table 1):

Table 1: Streaming Technologies

Purpose	Technology
High throughput Event Ingestion	Apache Kafka
	Azure Event Hubs
	Amazon Kinesis Firehose
Complex Event processing	Apache Storm
	Apache Heron
	Apache Spark Streaming
	Azure Stream Analytics
	Microsoft Orleans
	Apache Samza
	Apache Flink
	Apache Kafka Streams
	Google Millwheel
	Google Cloud Dataflow
	Amazon Kinesis Analytics

Apache Kafka

Apache Kafka is a ***distributed streaming platform***. A Streaming platform has the following characteristics (see Figure 4) [1]:

- Publish and Subscribe to stream of records. For instance, a central message topic publishes restful services to be consumed by clients. In our case, Hanumayamma Dairy IoT Sensor uses Kafka to publish real-time events.

- Persistent of Stream records in a fault-tolerant durable way.

- Process stream of records as they occur.

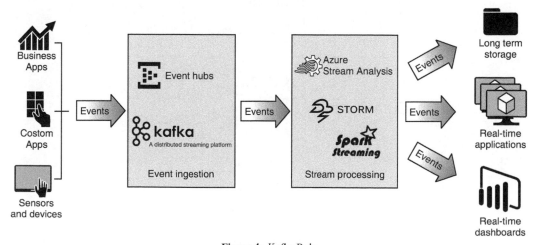

Figure 4: Kafka Role

Streaming vs. Message Systems

In traditional message processing systems, generally referred to as message-oriented middleware (MOM), you apply simple computations on the messages, in most cases individually per message. Remote Procedure Calls (RPC) and Object Request Broker (ORB) are good examples of MOM. For instance, the computation takes place on RPC call—between RPC client and RPC Server.

Remote Procedure Call (RPC) Specifications[2]
Bruce Jay Nelson is generally credited with coining the term "remote procedure call" (1981), and the first practical implementation was by Andrew Birrel and Bruce Nelson, called Lupine, in the Cedar environment at Xerox PARC. Lupine automatically generated stubs, providing type-safe bindings, and used an efficient protocol for communication. One of the first business uses of RPC was by Xerox under the name "Courier" in 1981. The first popular implementation of RPC on Unix was Sun's RPC (now called ONC RPC), used as the basis for Network File System (NFS).

In stream processing, you apply complex operations on multiple input streams and multiple records (i.e., messages) at the same time (like aggregations and joins).

	Traditional messaging systems cannot go "back in time", i.e., they automatically delete messages after they have been delivered to all subscribed consumers. In contrast, distributed streaming keeps the messages, as it uses a pull-based model (i.e., consumer pull data out of distributed streaming) for a configurable amount of time.

Furthermore, the distributed streaming systems allows consumers to "***rewind***" and consume messages multiple times. Importantly, if you add a new consumer, it can read the complete history. This makes stream processing possible, because it allows for more complex applications.

	Stream Processing is more about processing infinite input streams and less about real-time processing.

Broad class of applications:

Generally, two kinds of applications are built on distributed streaming platforms (see Figure 5).

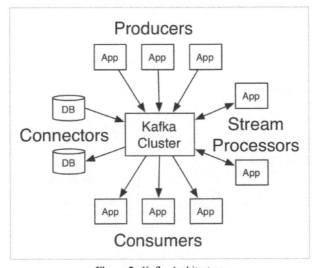

Figure 5: Kafka Architecture

---

2   RPC - https://en.wikipedia.org/wiki/Remote_procedure_call

- Real-time Data Pipelines: Building real-time data pipelines that reliably get data between two data platforms or two applications.

- Real-time Streaming Applications: Building real-time streaming applications that can respond or provide closed loop behavior to streams of data.

Topics

Topics store streams of records in categories called topics (see Figure 6). Each record contains a key, a value and a timestamp. Topics in Kafka are always multi-subscriber, with 0, 1, 2 or many consumers that subscribe to the data written to it.

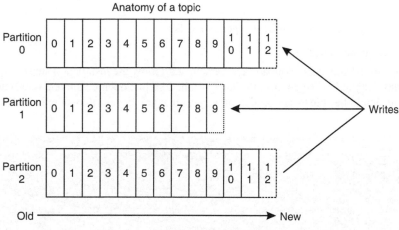

Figure 6: Kafka Topic

Partition and offset

Partition has two unique attributes: (1) it is ordered and (2) it has an immutable sequence of records that are continually added to. The records are identified by a unique ID, known as an offset (see Figure 7).

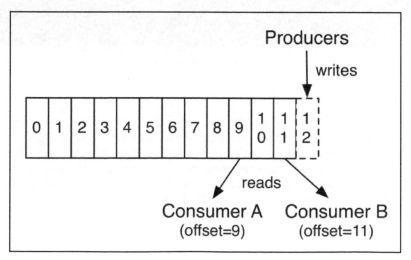

Figure 7: Kafka Partition

Installation of Kafka

Download Required Files

- Download Server JRE according to your OS and CPU architecture from http://www.oracle.com/technetwork/java/javase/downloads/jre8-downloads-2133155.html
- Download and install 7-zip from http://www.7-zip.org/download.html
- Download and extract Zookeeper using 7-zip from http://zookeeper.apache.org/releases.html
- Download and extract Kafka using 7-zip from http://kafka.apache.org/downloads.html

Java Install

Java installation required for setting Zookeeper and Kafka.

Step 1: go to Java Install location

Go to Java Download site and download Java Version and follow the installation steps.

Successful Installation: To confirm the Java installation, just open Command (cmd) and type "java –version", you should be able to see version of the java that you have just installed (see Figure 8).

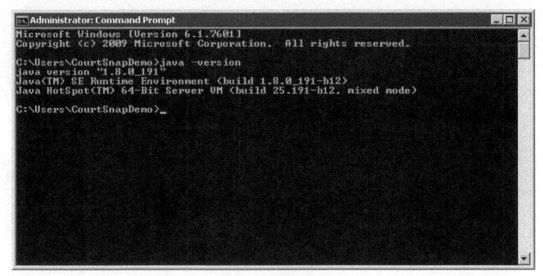

Figure 8: Java Version

Zookeeper Installation[3]

Apache ZooKeeper is an effort to develop and maintain an open-source server which enables highly reliable distributed coordination.[4] ZooKeeper is a centralized service for maintaining configuration information, naming, providing distributed synchronization, and providing group services. Distributed applications, for instance Kafka, use some or all forms of Zookeeper. Each time they are implemented, there is a lot of work that goes into fixing the bugs and race conditions that are inevitable. Because of the difficulty of implementing these kinds of services, applications initially usually skimp on them, which make them brittle

---

[3]   Zookeeper installation: https://dzone.com/articles/running-apache-kafka-on-windows-os
[4]   Zookeeper Web reference: https://zookeeper.apache.org/

in the presence of change and difficult to manage. Even when done correctly, different implementations of these services lead to management complexity when the applications are deployed.

Please follow the steps to install Zookeeper:

1. Download Zookeeper from http://zookeeper.apache.org/releases.html
   After successful download, please extract zookeeper-3.4.13.tar.gz file.
2. Go to your Zookeeper config directory. For instance, C:\zookeeper\zookeeper-3.4.13\conf
3. Rename file "zoo_sample.cfg" to "zoo.cfg"
4. Open zoo.cfg in any text editor (see Table 2), like notepad, but preferbly notepad++.
5. Find & edit dataDir=/tmp/zookeeper to ;\zookeeper-3.4.7\data

Table 2: Zookeeper Configuration

Zookeeper Configuration file

# The number of milliseconds of each tick 	
tickTime=2000
# The number of ticks that the initial 	
# synchronization phase can take 	
initLimit=10
# The number of ticks that can pass between 	
# sending a request and getting an acknowledgement 	
syncLimit=5
# the directory where the snapshot is stored. 	
# do not use /tmp for storage, /tmp here is just 	
# example sakes. 	
dataDir=\zookeeper\zookeeper-3.4.13\data
# the port at which the clients will connect 	
clientPort=2181
# the maximum number of client connections. 	
# increase this if you need to handle more clients 	
#maxClientCnxns=60
#
# Be sure to read the maintenance section of the 	
# administrator guide before turning on autopurge. 	
#
# http://zookeeper.apache.org/doc/current/zookeeperAdmin.html#sc_ 	
  maintenance
#
# The number of snapshots to retain in dataDir 	
#autopurge.snapRetainCount=3
# Purge task interval in hours 	
# Set to "0" to disable auto purge feature 	
#autopurge.purgeInterval=1

6. Add entry in System Environment Variables, as we did for Java
7. Add in System Variables ZOOKEEPER_HOME = C:\zookeeper-3.4.7

8. Edit System Variable named "Path" add; %ZOOKEEPER_HOME%\bin;

 Deleting or incorrectly overriding values of environmental variable "Path" value would have negative effect on the workings of the system.

9. You can change the default Zookeeper port in zoo.cfg file (Default port 2181).
10. Run Zookeeper by opening a new cmd and type zkserver.

Congratulations, your Zookeeper is up and running on port 2181 (see Figure 9)!

```
Administrator: Command Prompt - zkserver
2018-10-21 07:44:57,315 [myid:] - INFO  [main:Environment@100] - Server environm
ent:java.compiler=<NA>
2018-10-21 07:44:57,315 [myid:] - INFO  [main:Environment@100] - Server environm
ent:os.name=Windows Server 2008 R2
2018-10-21 07:44:57,315 [myid:] - INFO  [main:Environment@100] - Server environm
ent:os.arch=amd64
2018-10-21 07:44:57,315 [myid:] - INFO  [main:Environment@100] - Server environm
ent:os.version=6.1
2018-10-21 07:44:57,331 [myid:] - INFO  [main:Environment@100] - Server environm
ent:user.name=CourtSnapDemo
2018-10-21 07:44:57,331 [myid:] - INFO  [main:Environment@100] - Server environm
ent:user.home=C:\Users\CourtSnapDemo
2018-10-21 07:44:57,331 [myid:] - INFO  [main:Environment@100] - Server environm
ent:user.dir=C:\Users\CourtSnapDemo
2018-10-21 07:44:57,331 [myid:] - INFO  [main:ZooKeeperServer@836] - tickTime se
t to 2000
2018-10-21 07:44:57,331 [myid:] - INFO  [main:ZooKeeperServer@845] - minSessionT
imeout set to -1
2018-10-21 07:44:57,331 [myid:] - INFO  [main:ZooKeeperServer@854] - maxSessionT
imeout set to -1
2018-10-21 07:44:57,725 [myid:] - INFO  [main:ServerCnxnFactory@117] - Using org
.apache.zookeeper.server.NIOServerCnxnFactory as server connection factory
2018-10-21 07:44:57,726 [myid:] - INFO  [main:NIOServerCnxnFactory@89] - binding
 to port 0.0.0.0/0.0.0.0:2181
```

Figure 9: ZooKeeper Install

Setting Kafka

1. Download Kafka Binaries from http://kafka.apache.org/downloads.html
   a. Download and Unzip Kafka tarzip file.
   b. Go to your Kafka config directory. For me its C:\kafka\kafka_2.12-2.0.0\config
2. Edit file "server.properties" (see Table 3)

Table 3: Kafka Server Properties

Kafka Server Properties
Licensed to the Apache Software Foundation (ASF) under one or more # contributor license agreements. See the NOTICE file distributed with # this work for additional information regarding copyright ownership. # The ASF licenses this file to You under the Apache License, Version 2.0 # (the "License"); you may not use this file except in compliance with # the License. You may obtain a copy of the License at # # http://www.apache.org/licenses/LICENSE-2.0 # # Unless required by applicable law or agreed to in writing, software # distributed under the License is distributed on an "AS IS" BASIS, # WITHOUT WARRANTIES OR CONDITIONS OF ANY KIND, either express   or implied. # See the License for the specific language governing permissions and # limitations under the License. # see kafka.server.KafkaConfig for additional details and defaults
########################### Server Basics ###########################

# The id of the broker. This must be set to a unique integer for each broker. 	
  broker.id=0

########################### Socket Server Settings ###########################
The address the socket server listens on. It will get the value returned from # java.net.InetAddress.getCanonicalHostName() if not configured. # FORMAT: # listeners = listener_name://host_name:port # EXAMPLE: # listeners = PLAINTEXT://your.host.name:9092 #listeners=PLAINTEXT://:9092
Hostname and port the broker will advertise to producers and consumers. If   not set, # it uses the value for "listeners" if configured. Otherwise, it will use the value # returned from java.net.InetAddress.getCanonicalHostName(). # advertised.listeners=PLAINTEXT://your.host.name:9092
Maps listener names to security protocols, the default is for them to be the same. See the config documentation for more details # listener.security.protocol.map=PLAINTEXT:PLAINTEXT,SSL:SSL,SASL_PLAINTEXT:SASL_PLAINTEXT,SASL_SSL:SASL_SSL

# The number of threads that the server uses for receiving requests from the network and sending responses to the network 	
num.network.threads=3

The number of threads that the server uses for processing requests, which may include disk I/O

Table 3 contd. ...

...Table 3 contd.

Kafka Server Properties
num.io.threads=8

# The send buffer (SO_SNDBUF) used by the socket server 	
socket.send.buffer.bytes=102400

# The receive buffer (SO_RCVBUF) used by the socket server 	
socket.receive.buffer.bytes=102400

# The maximum size of a request that the socket server will accept (protection against OOM) 	
socket.request.max.bytes=104857600

########################### Log Basics ###########################

# A comma separated list of directories under which to store log files 	
log.dirs=C:\kafka\kafka_2.12-2.0.0\kafka-logs

# The default number of log partitions per topic. More partitions allow greater 	
# parallelism for consumption, but this will also result in more files across 	
# the brokers. 	
num.partitions=1

# The number of threads per data directory to be used for log recovery at startup and flushing at shutdown. 	
# This value is recommended to be increased for installations with data dirs located in RAID array. 	
num.recovery.threads.per.data.dir=1

########################### Internal Topic Settings ###########################

# The replication factor for the group metadata internal topics "__consumer_offsets" and "__transaction_state" 	
# For anything other than development testing, a value greater than 1 is recommended for to ensure availability such as 3. 	
offsets.topic.replication.factor=1
transaction.state.log.replication.factor=1
transaction.state.log.min.isr=1

########################### Log Flush Policy ###########################
Messages are immediately written to the filesystem but by default we only fsync() to sync # the OS cache lazily. The following configurations control the flush of data to disk. # There are a few important trade-offs here: # 1. Durability: Unflushed data may be lost if you are not using replication. # 2. Latency: Very large flush intervals may lead to latency spikes when the flush does occur as there will be a lot of data to flush. # 3. Throughput: The flush is generally the most expensive operation, and a small flush interval may lead to excessive seeks. # The settings below allow one to configure the flush policy to flush data after a period of time or # every N messages (or both). This can be done globally and overridden on a per-topic basis. # The number of messages to accept before forcing a flush of data to disk #log.flush.interval.messages=10000
The maximum amount of time a message can sit in a log before we force a flush #log.flush.interval.ms=1000
########################### Log Retention Policy ###########################

# The following configurations control the disposal of log segments. The policy can 	
# be set to delete segments after a period of time, or after a given size has accumulated. 	
# A segment will be deleted whenever *either* of these criteria are met. Deletion always happens 	
# from the end of the log. 	
# The minimum age of a log file to be eligible for deletion due to age 	
log.retention.hours=168

Table 3 contd. ...

...Table 3 contd.

Kafka Server Properties
A size-based retention policy for logs. Segments are pruned from the log unless the remaining # segments drop below log.retention.bytes. Functions independently of log.retention.hours. #log.retention.bytes=1073741824

# The maximum size of a log segment file. When this size is reached a new log segment will be created. 	
log.segment.bytes=1073741824

# The interval at which log segments are checked to see if they can be deleted according 	
# to the retention policies 	
log.retention.check.interval.ms=300000

############################# Zookeeper #############################

# Zookeeper connection string (see zookeeper docs for details). 	
# This is a comma separated host:port pairs, each corresponding to a zk 	
# server. e.g. "127.0.0.1:3000,127.0.0.1:3001,127.0.0.1:3002". 	
# You can also append an optional chroot string to the urls to specify the 	
# root directory for all kafka znodes. 	
zookeeper.connect=127.0.0.1:2181,0.0.0.0:2181,localhost:2181

# Timeout in ms for connecting to zookeeper 	
zookeeper.connection.timeout.ms=6000

############################# Group Coordinator Settings #############################
The following configuration specifies the time, in milliseconds, that the GroupCoordinator will delay the initial consumer rebalance. # The rebalance will be further delayed by the value of group.initial.rebalance.delay.ms as new members join the group, up to a maximum of max.poll.interval.ms. # The default value for this is 3 seconds. # We override this to 0 here as it makes for a better out-of-the-box experience for development and testing. # However, in production environments the default value of 3 seconds is more suitable as this will help to avoid unnecessary, and potentially expensive, rebalances during application startup. group.initial.rebalance.delay.ms=0

3. Find & edit line "log.dirs=/tmp/kafka-logs" to "log.dirs=C:\kafka\kafka_2.12-2.0.0\kafka-logs".

4. If your Zookeeper is running on some other machine or cluster you can edit "zookeeper. connect:2181" to your custom IP and port. If we are using same machine there no need to change. Also, Kafka port and broker.id are configurable in this file. Leave other settings as they are.

5. Your Kafka will run on default port 9092 and connect to zookeeper's default port, which is 2181.

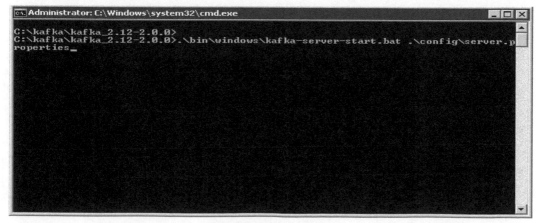

Figure 10: Start Kafka Server

```
Administrator: C:\Windows\system32\cmd.exe - .\bin\windows\kafka-server-start.bat .\config\serv...  _ □ ×
[2018-10-22 00:08:53,139] INFO [GroupMetadataManager brokerId=0] Finished loadin
g offsets and group metadata from __consumer_offsets-45 in 0 milliseconds. (kafk
a.coordinator.group.GroupMetadataManager)
[2018-10-22 00:08:53,139] INFO [GroupMetadataManager brokerId=0] Finished loadin
g offsets and group metadata from __consumer_offsets-48 in 0 milliseconds. (kafk
a.coordinator.group.GroupMetadataManager)
[2018-10-22 00:09:04,553] INFO [GroupCoordinator 0]: Member consumer-1-13aa3474-
2015-4362-a5ab-09e14fb9a613 in group console-consumer-73829 has failed, removing
 it from the group (kafka.coordinator.group.GroupCoordinator)
[2018-10-22 00:09:04,569] INFO [GroupCoordinator 0]: Preparing to rebalance grou
p console-consumer-73829 with old generation 1 (__consumer_offsets-20) (kafka.co
ordinator.group.GroupCoordinator)
[2018-10-22 00:09:04,580] INFO [GroupCoordinator 0]: Group console-consumer-7382
9 with generation 2 is now empty (__consumer_offsets-20) (kafka.coordinator.grou
p.GroupCoordinator)
[2018-10-22 00:09:04,669] INFO [GroupCoordinator 0]: Member consumer-1-f69e5d9f-
49ee-4255-b014-6f9278cdc9ca in group console-consumer-99993 has failed, removing
 it from the group (kafka.coordinator.group.GroupCoordinator)
[2018-10-22 00:09:04,669] INFO [GroupCoordinator 0]: Preparing to rebalance grou
p console-consumer-99993 with old generation 1 (__consumer_offsets-30) (kafka.co
ordinator.group.GroupCoordinator)
[2018-10-22 00:09:04,669] INFO [GroupCoordinator 0]: Group console-consumer-9999
3 with generation 2 is now empty (__consumer_offsets-30) (kafka.coordinator.grou
p.GroupCoordinator)
```

Figure 11: Kafka Server (running)

Running Kafka Server

> Important: Please ensure that your Zookeeper instance is up and running before starting a Kafka server.

1. Go to your Kafka installation directory C:\kafka\kafka_2.12-2.0.0\

2. Open a command prompt here by pressing Shift + right click and choose "Open command window here" option, Now, enter comman:

3. Now type .\bin\windows\kafka-server-start.bat.\config\server.properties and press Enter (see Figure 10).

.\bin\windows\kafka-server-start.bat .\config\server.properties

4. If everything went fine, your command prompt will look like Figure 11. It will start Kafka Server.

5. Now your Kafka is up and running, you can create topics to store messages. Also, we can produce or consume data from Java or Scala code or directly from the command prompt.

Creating Topic

1. Now create a topic with name "ent_iot_middleware" and replication factor 1, as we have only one Kafka server running. If you have a cluster with more than 1 Kafka server running, you can increase the replication-factor accordingly , which will increase the data availability and act like a fault-tolerant system.

2. Open a new command prompt in the location C:\kafka_2.11-0.9.0.0\bin\windows

3. Type the following command and hit Enter (see Figure 12):

kafka-topics.bat --create --zookeeper localhost:2181 --replication-factor 1 --partitions 1 --topic ent_iot_middleware

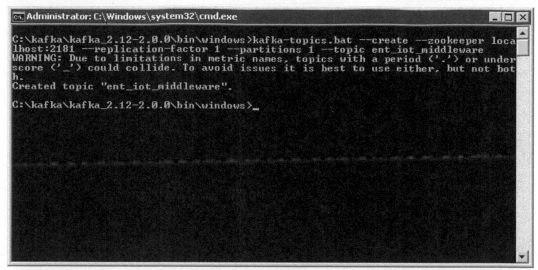

Figure 12: Kafka Topic

Creating a Console Producer and Consumer to Test Server

1. Open a new command prompt in the location C:\kafka\kafka_2.12-2.0.0\bin\windows
2. To start a producer, type the following command:

```
kafka-console-producer.bat --broker-list localhost:9092 --topic ent_iot_middleware
```

The producer is ready! (see Figure 13)

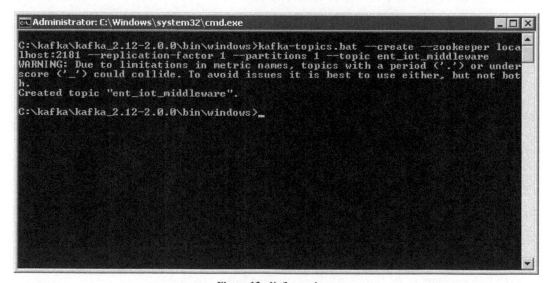

Figure 13: Kafka producer

3. Again, open a new command prompt in the same location as C:\kafka\kafka_2.12-2.0.0\bin\windows

4. Now start a consumer by typing the following command (see Figure 14):

kafka-console-consumer.bat --bootstrap-server localhost:9092 --topic ent_iot_middleware --from-beginning

Figure 14: Kafka Consumer

5. Now type anything in the producer command prompt and press Enter, and you should be able to see the message in the other consumer command prompt.

There is as old Kafka consumer (0.8.2 and earlier) and a new Kafka consumer (0.9 and above). For a great description of how the new consumer works see the original announcement blog here
https://www.confluent.io/blog/tutorial-getting-started-with-the-new-apache-kafka-0-9-consumer-client/
The old consumer would connect to zookeeper but still would fetch all messages from Kafka. The new consumer has no zookeeper dependency and does not connect to zookeeper at all. The console-producer and console-consumer supports both the old an new apis depending on which options you give it. The example console-consumer you have provided is the old consumer because it specifies --zookeeper instead of --bootstrap-server

Creating a Python Producer and Consumer to Test Server

Python Code—Producer

Error:
base) C:\Hanumayamma\CRCBook\Code\Kafka>python kafka_producer.py
Traceback (most recent call last):
   File "kafka_producer.py", line 12, in <module>
     from kafka import KafkaProducer
ModuleNotFoundError: No module named 'kafka'

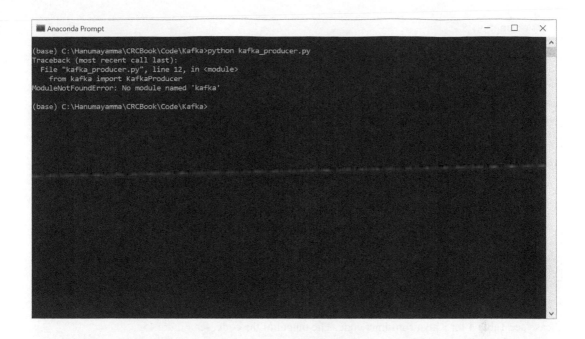

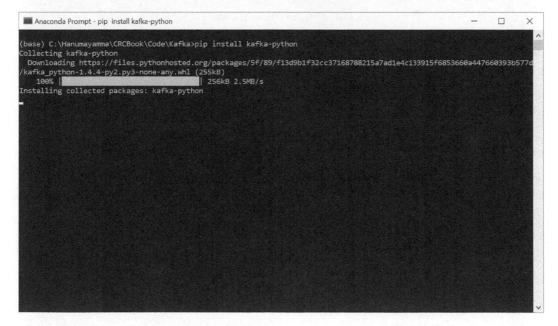

Installing collected packages: kafka-python
Successfully installed kafka-python-1.4.4

```
# -*- coding: utf-8 -*-
"""
Created on Thu Nov 22 22:47:36 2018
@author: cvuppalapati
"""
import threading, logging, time
import multiprocessing
from kafka import KafkaConsumer, KafkaProducer

class Producer():
    producer = KafkaProducer(bootstrap_servers='courtsnapdemo.
    cloudapp.net:9092')
    producer.send('ent_iot_middleware', b"Hello World")
    time.sleep(1)
    producer.close()
def main():
    Producer()
```

Python Code—Consumer

Please see Table 4 for Kafka consumer code. The output of the code, see Figure 15.

Table 4: Kafka Consumer Code

```
# -*- coding: utf-8 -*-
"""
Created on Thu Nov 22 22:48:12 2018
@author: cvuppalapati
"""
from kafka import KafkaConsumer

print("To consume latest messages and auto-commit offsets")
consumer = KafkaConsumer(bootstrap_servers='courtsnapdemo.cloudapp.net:9092',
            auto_offset_reset='earliest',
            consumer_timeout_ms=1000)
consumer.subscribe(['ent_iot_middleware'])

print("Connection estalished")
for message in consumer:
    print("inside message")
    # message value and key are raw bytes -- decode if necessary!
    # e.g., for unicode: `message.value.decode('utf-8')`
    print ("%s:%d:%d: key=%s value=%s" % (message.topic, message.partition,
                message.offset, message.key,
                message.value))
```

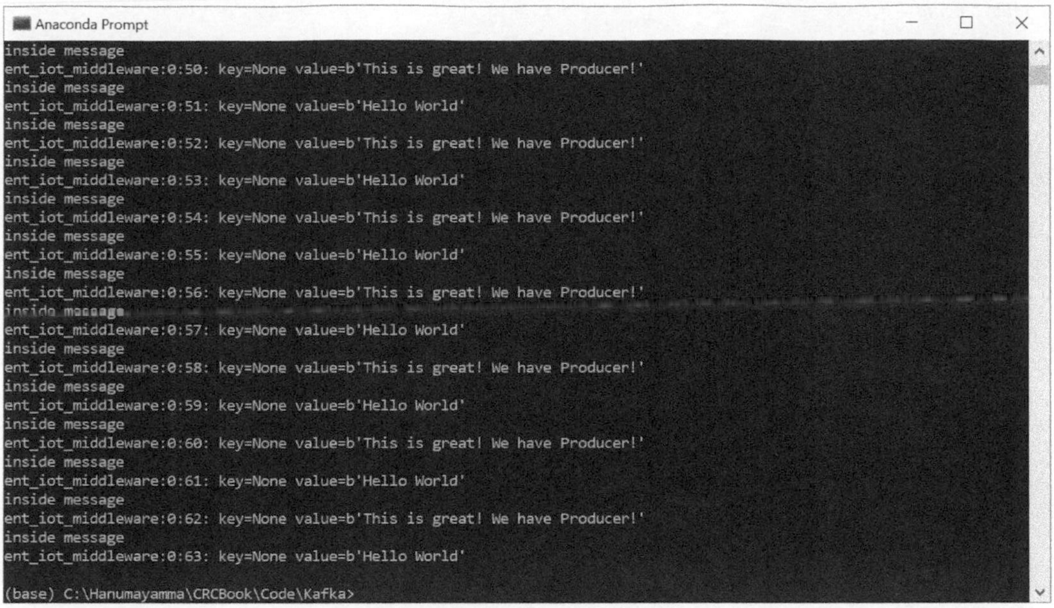

Figure 15: Kafka Consumer

Producer and Consumer

The following code[5] has producer and consumer (see Table 5)

Table 5: Producer and consumer full code

```
# -*- coding: utf-8 -*-
"""
Created on Thu Nov 22 22:59:18 2018

@author: cvuppalapati
"""
import threading, logging, time
import multiprocessing

from kafka import KafkaConsumer, KafkaProducer

class Producer(threading.Thread):
  def __init__(self):
      threading.Thread.__init__(self)
      self.stop_event = threading.Event()

  def stop(self):
      self.stop_event.set()
  def run(self):
      producer = KafkaProducer(bootstrap_servers='courtsnapdemo. cloudapp.net:9092')

      while not self.stop_event.is_set():
        producer.send('ent_iot_middleware', b"Hello World")
        producer.send('ent_iot_middleware', b"This is great! We have
        Producer!")
```

Table 5 contd. ...

---

[5]   https://kafka.apache.org/quickstart

...Table 5 contd.

```
      time.sleep(1)

    producer.close()
class Consumer(multiprocessing.Process):
 def __init__(self):
    multiprocessing.Process.__init__(self)
    self.stop_event = multiprocessing.Event()

  def stop(self):
    self.stop_event.set()

  def run(self):
      consumer = KafkaConsumer(bootstrap_servers='courtsnapdemo. cloudapp.net:9092',
                 auto_offset_reset='earliest',
                 consumer_timeout_ms=1000)
      consumer.subscribe(['ent_iot_middleware'])

      while not self.stop_event.is_set():
          for message in consumer:
              print(message)
              if self.stop_event.is_set():
                  break

      consumer.close()
def main():
  tasks = [
      Producer(),
      Consumer()
  ]
  for t in tasks:
      t.start()
  time.sleep(10)

  for task in tasks:
      task.stop()
  for task in tasks:
      task.join()

if __name__ == "__main__":
  logging.basicConfig(
format='%(asctime)s.%(msecs)s:%(name)s:%(thread)d:%(levelname)s:%(process)d:%(message)s',
      level=logging.INFO
      )
  main()
```

Apache Spark

Apache Spark is a unified analytics engine for big data processing. Apache Spark extends the MapReduce model in order to provide efficient support for computing, interactive queries and streaming process. The computing is platform designed to provide high speed computation and general purpose.

One of the main features that Spark offers for speed is the ability to run competition in memory and avoid unnecessary I/O operations with disk.

On the generality side, Spark offers is the ability to chain the tasks even at an application programming level without writing onto the disks at all or minimizing the number of writes to the disks. In addition, it reduces the management burden of maintaining separate tools.

Spark at Core

Spark comes with a very advanced Directed acyclic graph(DAG) data processing engine. The DAG in mathematical idiom consists of a set of vertices and directed edges connecting them. The tasks are executed as per the DAG layout. The in-memory data processing, combined with its DAG-based data processing engine, makes Spark very efficient

Every spark application consisting of a driver program launches various parallel operations on a cluster. Driver programs access Spark through a Spark Context Object (which establishes a connection between application and computing cluster. In the shell, a Spark Context is automatically created for you as the variable called sc).

The driver program creates the DAG (directed acyclic graph) or execution plan (job) for your program. Once the DAG is created, the driver divides this DAG into several tasks. The driver program is responsible for converting a user program into units of physical execution called tasks. These tasks are given to executors for execution. At a high level, all Spark programs follow the same structure.

The following are the steps to install spark and run spark through python, using jupyter notebook and PySpark (see Figure 16)

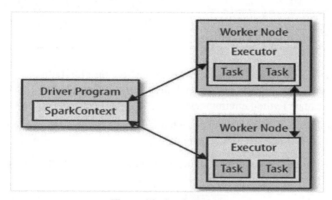

Figure 16: Spark Context

Steps to set up Spark on Windows System

Download and install Anaconda

Follow instructions to install Anaconda on the Windows system - https://docs.anaconda.com/anaconda/install/windows/

Download Apache Spark

Go to the Apache Spark website (http://spark.apache.org/downloads.html) follow the instructions to download Apache Spark.

Move Files and Unzip

Move the file to where you want to unzip it

a. mkdir C:\Spark

b. Move **spark-2.1.0-bin-hadoop2.7.tgz** to Spark folder (**mv C:\Users\ \Downloads\spark-2.1.0-bin-hadoop2.7.tgz C:\ \spark\spark-2.1.0-bin-hadoop2.7.tgz**)

WinUtils

Download winutils.exe from https://github.com/steveloughran/winutils/blob/master/hadoop 2.6.0/bin/winutils.exe?raw=true

To setup HDFS

Environment Variables Setup

Create environment variables (see Table 6) for SPARK_HOME and HADOOP_HOMEand related PATH variables. You can do this in the Command Prompt
setx SPARK_HOME C:\opt\spark\spark-2.1.0-bin-hadoop2.7

- setx HADOOP_HOME C:\opt\spark\spark-2.1.0-bin-hadoop2.7
- setx PYSPARK_DRIVER_PYTHON ipython
- setx PYSPARK_DRIVER_PYTHON_OPTS notebook

Table 6: Environment Variables

Name	Value
SPARK_HOME	D:\spark\spark-2.2.1-bin-hadoop2.7
HADOOP_HOME	D:\spark\spark-2.2.1-bin-hadoop2.7
PYSPARK_DRIVER_PYTHON	jupyter
PYSPARK_DRIVER_PYTHON_OPTS	notebook

6) Activate your Anaconda virtual environment with the version of Python that you'd like to use. You could make a new environment (recommended!) to try things out, or YOLO just > conda install pyspark into your main Python environment.

7) You need to enable access to the default scratch directory for Hive. First, make sure the directory C:\tmp\hive is created; if it doesn't exist, create it.

Second, you need to give it permission to access winutils.exe. Navigate back to where you put this .exe file, then run the permission command

> cd c:\hadoop\bin
> winutils.exe chmod -R 777 C:\tmp\hive

8) (Optional, if see Java related error in step C) Find the installed Java JDK folder from step A5, for example, D:\Program Files\Java\jdk1.8.0_121, and add the following environment variable

Name   Value

JAVA_HOME   D:\Progra~1\Java\jdk1.8.0_121

9) Running PySpark in Jupyter Notebook
    a. To run Jupyter notebook, open Windows command prompt or Git Bash and run jupyter notebook. If you use Anaconda Navigator to open Jupyter Notebook instead, you might see a Java gateway process exited before sending the driver its port number error from PySpark in step C. Fall back to Windows cmd if it happens.
    b. Once inside Jupyter notebook, open a Python 3 notebook (see Figure 17)

Standalone Application

Spark Standalone cluster (aka Spark deploy cluster or standalone cluster) is the default distribution of Apache Spark and it is the easiest way to run your Spark applications in a clustered environment in many cases. Spark can be linked into standalone application is either using java, Scala, or Python. The main

```
In [1]:  import pyspark # only run after findspark.init()
         from pyspark.sql import SparkSession
         spark = SparkSession.builder.getOrCreate()

         df = spark.sql('''select 'spark' as hello ''')
         df.show()

         +-----+
         |hello|
         +-----+
         |spark|
         +-----+
```

Figure 17: Successful Spark

difference from using it in the shell is that you need to initialize your own Spark Context. After that, the API is the same.

Initializing a SparkContext:

The first step to use spark is configuring the spark by importing spark packages in your program and create a SparkContext. It allows your Spark Application to access Spark Cluster with the help of Resource Manager. The resource manager can be one of these three—Spark Standalone, YARN, Apache Mesos.

If you want to create SparkContext, first SparkConf should be made. The SparkConf has a configuration parameter that our Spark driver application will pass to SparkContext. Some of these parameters defines properties of Spark driver application, while some are used by Spark to allocate resources on the cluster, like the number, memory size, and cores used by the executor running on the worker nodes.

Example: Initializing Spark in Python

from pyspark import SparkConf, SparkContext

conf = SparkConf().setMaster("local").setAppName("My App")

sc = SparkContext(conf = conf)

1) setMaster API allows us to connect to the cluster. Local runs Spark on the local machine, without connecting to a cluster.

2) An application name will identify your application on the cluster manager's UI.

Apache Spark APIs

Spark RDD APIs: RDD stands for Resilient Distributed Datasets. It is a Read-only partition collection of records. RDD is the fundamental data structure of Spark. It allows a programmer to perform in-memory computations on large clusters in a fault tolerant manner, thus, speed up the task. Each RDD is split into multiple partitions, which may be computed on different nodes of the cluster.

Features of RDD:

1. In Memory Computation
2. Lazy Evaluations
3. Fault Tolerance
4. Immutability
5. Partitioning
6. Persistence

Ways to Create RDD:

1) Parallelized Collection:

Spark Context provides the parallelize () method. It is a quick way to create a your own RDDs in spark shell and perform operations on them. This method is rarely used outside testing and prototyping because it requires having the entire dataset on one machine.

```
import pyspark
from pyspark.sql import SparkSession
spark = SparkSession.builder.getOrCreate()
df = spark.sparkContext.parallelize([("user1",52),("user2",75),("us
er3",82), ("user4",65),("user5",85)])
df.collect()
    Output:[('user1', 52), ('user2', 75), ('user3', 82), ('user4', 65),
('user5', 85)]
```

2) External Datasets (Referencing a dataset)

Spark supports different data sources, such as local system, HDFS, Cassandra, HBase, etc. External Datasets can be a

*   csv
*   JSON
*   Text file

```
csvRDD = spark.read.csv("path/of/csv/file").rdd
textRDD = spark.read.textFile("path/of/text/file").rdd
jsonRDD = spark.read.json("path/of/json/file").rdd
```

Once created, RDDs offer two types of operations:

1) Transformations: Spark RDD Transformations are functions that take an RDD as the input and produce one or many RDDs as the output. They do not change the input RDD (since RDDs are immutable and, hence, cannot be changed), but always produce one or more new RDDs by applying the computations they represent, e.g., Map(), filter(), reduceByKey(), etc.

2) Actions: Compute a result based on an RDD, and either return it to the driver program or save it to an external storage system (e.g., HDFS). Actions are RDD operations that produce non-RDD values. They materialize a value in a Spark program. An Action is one of the ways to send results from executors to the driver. First(), take(), reduce(), collect(), the count() is some of the Actions in spark.

DataFrame

DataFrames generally refer to a data structure, which is tabular in nature. DataFrames are designed to process a large collection of structured and semi-structured data. Observations in Spark DataFrame are organized under named columns, which helps Apache Spark understand the schema of a Dataframe. This helps Spark optimize the execution plan on these queries. It can also handle petabytes of data.

DataFrame provides different operations similar to a table in a relational database. DataFrame supports operations like insert, create, update, delete, join, group by, filter, Sort and different mathematical operations.

```
import pyspark
from pyspark.sql import SparkSession
spark = SparkSession.builder.getOrCreate()
l = [('Alice', 1)]
df = spark.createDataFrame(l, ['name', 'age'])
df.show()
```

How Spark Streaming Works?

Spark Streaming is Spark's model for acting on data as soon as it arrives. For example, if you want to track real time behavior of a user based on sensor data and provide recommendations about his/her health in real time.

Like spark, spark streaming uses an abstraction called DStreams or discretized streams. A DStream is a sequence of data arriving over time. Internally, each DStream is represented as a sequence of RDDs arriving at each time step. When we apply a transformation on Dstreams, this yields a new DStream, and output operations, which write data to an external system. DStreams provide many of the same operations available on RDDs, plus new operations related to time, such as sliding windows.

Architecture and Abstraction

Spark Streaming implements a "batch" architecture, where the streaming computation is treated as a continuous series of batch computations on small batches of data. Spark Streaming receives data from various input sources and groups it into small batches. New batches are created at regular time intervals. The size of batches is determined by a parameter called batch interval. The batch interval is typically 500 milliseconds and several seconds. DStreams support output operations. Output operations are like RDD actions and output is periodically written to the external system.

For each input source, Spark Streaming launches task running components called receivers, which run within the application's executors that collect data from the input source and save it as RDDs. These receive the input data and replicate it (by default) to another executor for fault tolerance. This data is stored in the memory of the executors and the driver program then Spark periodically runs jobs to process this data and combine it with RDDs from previous time steps (see Figure 18).

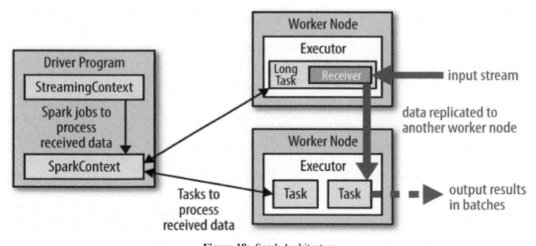

Figure 18: Spark Architecture

After completing the chapter, you should have learned

- Setting up of Apache Kafka
- Creation of Kafka Topic
- Creation of Kafka Producer and Consumer
- Apache Spark

References

1. The Apache Software Foundation, Apache Kafka Documentation, 2016, https://kafka.apache.org/documentation/ Access Date12/25/2018
2. Jay Kreps. Benchmarking Apache Kafka: 2 Million Writes Per Second (On Three Cheap Machines). April 27 2014, https://engineering.linkedin.com/kafka/benchmarking-apache-kafka-2-million-writes-second-three-cheap-machines, Access Date12/25/2018
3. Raghav Mohan. Streaming Big Data on Azure with HDInsight Kafka, Storm and Spark. September 2018, https://www.google.com/url?sa=t&rct=j&q=&esrc=s&source=web&cd=4&ved=2ahUKEwi_2P-W75jeAhWoITQIHcimAcoQFjADegQIBxAC&url=https%3A%2F%2F8gportalvhdsf9v440s15hrt.blob.core.windows.net%2Fignite2017%2Fsession-presentations%2FBRK3320.PPTX&usg=AOvVaw2i9Of6LtC9g7qpPmfnSENG , Access Date12/25/2018
4. Michael Galarnyk. Install Spark on Windows (PySpark). Apr 2, 2017, https://medium.com/@GalarnykMichael/install-spark-on-windows-pyspark-4498a5d8d66c, Access date: August 06, 2018
5. Lauren Oldja. Installing Apache Spark (PySpark): The missing "quick start" guide for Windows, Jan 28, 2018, https://medium.com/@loldja/installing-apache-spark-pyspark-the-missing-quick-start-guide-for-windows-ad81702ba62d, Access Date12/25/2018
6. Chang Hsin Lee. How to Install and Run PySpark in Jupyter Notebook on Windows. December 30, 2017, https://changhsinlee.com/install-pyspark-windows-jupyter/, Access date: December 27, 2018
7. Dataflair Team. Spark Streaming Tutorial for Beginners. MARCH 11, 2017, https://data-flair.training/blogs/apache-spark-dstream-discretized-streams/, January 08, 2018

## CHAPTER 11

Cloud and IoT

This Chapter Covers

- Basics of Cloud Computing
- NIST Standards for Cloud & Fog Compute
- IoT Analytics Platform
- Data Processing architectures

Cloud computing is a model for enabling ubiquitous, convenient, on-demand network access to a shared pool of configurable computing resources (e.g., networks, servers, storage, applications, and services) that can be rapidly provisioned and released with minimal management effort or service provider interaction [1] (please Figure 1). This cloud model is composed of five essential characteristics, three service models, and four deployment models.[1]

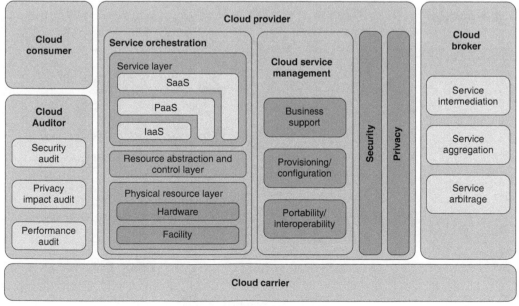

Figure 1: NIST Reference Model

---

[1] NIST Cloud Reference models - https://ws680.nist.gov/publication/get_pdf.cfm?pub_id=909505

Five essential characteristics

- On-demand self-service
- Broad network access
- Resource pooing
- Rapid elasticity
- Measured service

Three Cloud Service Models for Cloud Consumers

- Cloud Software as a Service (SaaS)
- Cloud Platform as a Service (PaaS)
- Cloud Infrastructure as a Service (IaaS)

Cloud Deployment Models

- Public Cloud
- Private Cloud
- Hybrid Cloud
- Community Cloud

Cloud Consumption Model

Based on the IoT devices, need, the IoT processing architecture appropriates the consumption model[2] (please see Figure 2) [2].

The Cloud enabled Data Analytics platform is the backbone for the IoT Integration with enterprise compute. IoT Data could take any shape and form with varying speeds and velocities. The successful adoption to IoT device platform is directly dependent on Cloud platform [2]. Additionally, the success of IoT Cloud platform also depends upon the following Edge/Fog attributes[3]:

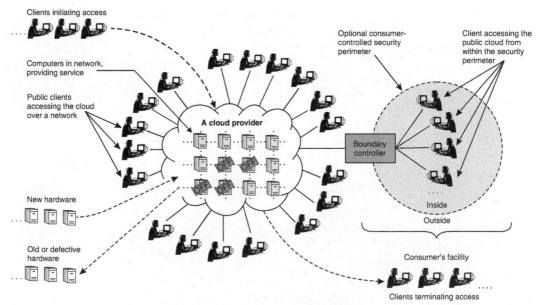

Figure 2: Cloud Consumption Model

---

2   Cloud Computing Synopsis - https://nvlpubs.nist.gov/nistpubs/Legacy/SP/nistspecialpublication800-146.pdf
3   Fog Compute - https://nvlpubs.nist.gov/nistpubs/SpecialPublications/NIST.SP.500-325.pdf

Contextual location awareness, and low latency: Fog/Edge computing offers the lowest-possible latency due to the fog nodes' awareness of their logical location in the context of the entire systems and of the latency costs for communicating with other nodes. The origins of fog computing can be traced back to early proposals supporting endpoints with rich services at the edge of the network, including applications with low latency requirements [3]. Because fog nodes are often co-located with the smart end-devices, analysis and response to data generated by these devices is much quicker than from a centralized cloud service or data center. For IoT Data Analytics platform, the Cloud platform should match the compute and response standards as required by Edge compute (please see Figure 3).

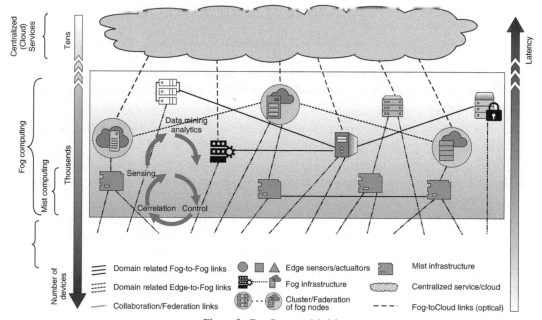

Figure 3: Fog Compute Model

Geographical distribution: In sharp contrast to the more centralized cloud, the services and applications targeted by the fog computing demand widely, but geographically-identifiable, distributed deployments. For instance, the fog computing will play an active role in delivering high quality streaming services to moving vehicles, through proxies and access points geographically positioned along highways and tracks.

Heterogeneity: Fog computing supports the collection and processing of data of different form factors, acquired through multiple types of network communication capabilities.

Interoperability and federation: Seamless support of certain services (real-time streaming services are a good example) requires the cooperation of different providers. Hence, fog computing components must be able to interoperate, and services must be federated across domains.

Real-time interactions. Fog computing applications involve real-time interactions rather than batch processing

The following IoT Capabilities[4] management is required for connecting IoT with the Cloud:

Device to Cloud Message: Device to Cloud message involves sending constant messages from IoT devices to the Cloud. For example, Dairy IoT Sensor posting messages every hour to the Cloud platform. Telemetry time series and alerts. For example, 256-KB sensor data batches sent every 5 minutes. Devices to Cloud message feature expects the cloud messaging to be reliable and available 24×7, 365 days.

---

4   IoT Capabilities - https://docs.microsoft.com/en-us/azure/iot-hub/iot-hub-compare-event-hubs

Protocols: HTTPS, AMQP, AMQP over web Sockets: Cloud Service management has to handle HTTPS, AMQP and Web Sockets seamlessly in order to support various devices and device form factors with constrained or non-constrained operating environments.

- **HTTPS:** Hypertext transfer Protocol Secure—this is for securely transferring data between Web Servers and Web Clients on a Secure channel over port number 443 and firewall friendly.

- **AMQP:** AMQP, which stands for Advanced Message Queuing Protocol, was designed as an open replacement for existing proprietary messaging middleware. Two of the most important reasons to use AMQP are reliability and interoperability. As the name implies, it provides a wide range of features related to messaging, including reliable queuing, topic-based publish-and-subscribe messaging, flexible routing, transactions, and security. AMQP exchanges route messages directly, in fanout form, by topic, and based on headers.

- **Web Sockets:** WebSocket makes real-time communication much more efficient. You can always use polling (and sometimes even streaming) over HTTP in order to receive notifications over HTTP. However, WebSocket saves bandwidth, CPU power, and latency. WebSocket is an innovation in performance. WebSocket makes communication between a client and server over the Web much simpler.

Protocols MQTT and MQTT Web Sockets: Message Queue Telemetry protocol is for the IoT device communication purposes with limited footprint and processing capacities needed.

Per-device identity: IoT Cloud and Data Analytics platform should enable support for per device identity. This includes device catalogue, master key management, authentication, authorization, PKI and other security processes.

File Upload from Devices: This is particularly important for device operation, diagnostics and remote management purposes. Many embedded devices maintain state machine operation semantics and having per device upload helps in solving device provisioning and operational challenges.

Device Twin and device management: Device twin is a sensor connected physical asset that is studied and processed for improving design. Device Twin requires huge data and analytics support in order to be successful.

IoT Edge: IoT Edge performs data collection and analytics on data in motion. IoT Edge needs to be fast and must support low latency operations.

IoT Data Analytics Platform

The following is a blue print of IoT Data Analytics platform (please see Figure 4) [4]:
   The major components on the Cloud include:
- Data Collection
  - o Batch
  - o Streams
- Data Ingestion (Streams & Sparks)

- Data Storage
  - In-memory
  - Traditional
- Advanced Analytics
  - Predictive
  - Prognostic
- Visualization

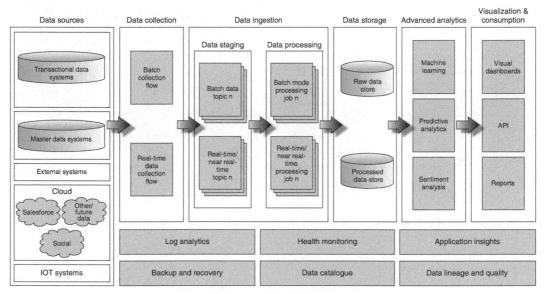

Figure 4: Data Platform

Data Processing Architecture

Like that of traditional data processing architectures, the IoT needs the same stack of processing layers: Data Processing, Model Engineering, Execution, and Deployment [5].

Data Acquisition Systems

Data Acquisitions (please see Figure 5) include ERP Databases, Legacy systems, and newly introduced IoT Sensor systems. The Data are ingested into either Batch databases for reporting and regulatory or compliance purposes or Stream processing servers immediate "Now" analytics.

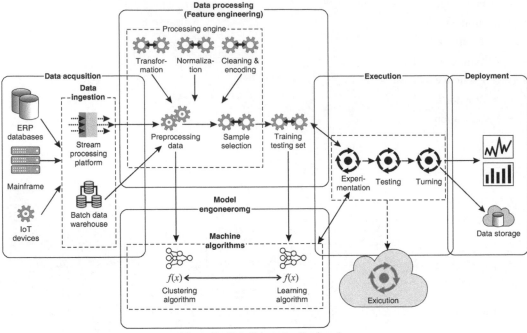

Figure 5: Data Processing Engineering

Insight Value Chain

Data in its raw and most basic form is virtually worthless until we give it a voice by gleaning valuable insights from it.[5] The multiplicate value chain framework connects data lineage from end to end (please see Figure 6); the value derivation is through the successful connection and synergies of all layers working together [6].

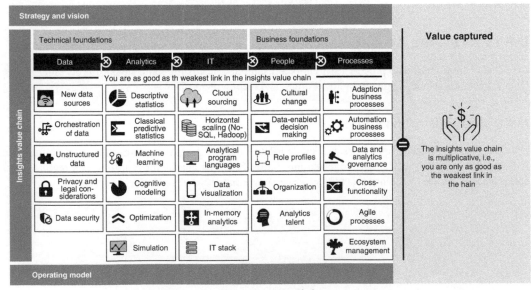

Figure 6: Value Chain

---

5   Achieving business impact with the data - https://www.mckinsey.com/~/media/mckinsey/business%20functions/ mckinsey%20analytics/our%20insights/achieving%20business%20impact%20with%20data/achieving-business-impact-with-data_final.ashx

Data must be thought of as the entire process of collecting, linking, cleaning, enriching, and augmenting internal information (potentially with additional external data sources). In addition, the security and privacy (e.g., GDPR) of the data throughout the process are fundamental.[6]

Analytics describes the set of digital methodologies (e.g., software) deployed to extract insights from data as well as the talent (e.g., data engineers and data scientists) capable of developing and applying these methods.

IT is the technical layer enabling the storing and processing of data, e.g., data lakes and two-speed IT architecture.

People from the front lines of sales to deep within the business are needed in order to run an analytics operation that turns data into insights and successfully implements those insights in the business. The crucial capability in today's Big Data world is being able to "translate" analytics- and data-driven insights into business implications and actions.

Processes must be assessed for their ability to deliver at scale. Some old processes might need to be adapted, some might need to be fully automated, and others might need to be made more agile.

> Capturing value from data requires excellence in each element of each link of the "insights value chain".

After reading this chapter, you should able to include and integrate:
- Cloud and IoT Systems
- Design Fault tolerant IoT

References

1. Fang Liu, Jin Tong, Jian Mao, Robert Bohn, John Messina, Lee Badger and Dawn Leaf, NIST Cloud Computing Reference Architecture, Sep 2011, https://ws680.nist.gov/publication/get_pdf.cfm?pub_id=909505, Access Date: 18 Sep 2018

2. Lee Badger, Tim Grance, Robert Patt-Corner, and Jeff Voas, Cloud Computing Synopsis and Recommendations, May 2012, https://nvlpubs.nist.gov/nistpubs/Legacy/SP/nistspecialpublication800-146.pdf, Access Date: April 2018

3. Michaela Iorga, Larry Feldman, Robert Barton, Michael J. Martin, Nedim Goren and Charif Mahmoudi, Fog Computing Conceptual Model, March 2018, https://nvlpubs.nist.gov/nistpubs/SpecialPublications/NIST.SP.500-325.pdf, Access Date: Sep 2018

4. Kelly Gremban, Reza Sherafat and et al., Connecting IoT Devices to Azure: IoT Hub and Event Hubs, March 2018, https://docs.microsoft.com/en-us/azure/iot-hub/iot-hub-compare-event-hubs, Access Date: Sep 2018

5. Carlton E. Sapp, Preparing and Architecting for Machine Learning, 17 January 2017, https://www.gartner.com/binaries/content/assets/events/keywords/catalyst/catus8/preparing_and_architecting_for_machine_learning.pdf, Access Date: Feb 2018

6. Niko Mohr and Holger Hürtgen, Achieving business impact with data, April 2018, https://www.mckinsey.com/~/media/mckinsey/business%20functions/mckinsey%20analytics/our%20insights/achieving%20business%20impact%20with%20data/achieving-business-impact-with-data_final.ashx, Access Date: Sep 2018

---

[6] Business Impact with the data

## CHAPTER 12

Future

IoT the Best Deterrent to Prevent the Effects of Climate Change

The agriculture industry has employed the first food harvesting citizen data scientists from the beginning of human civilization. Yet, the conundrum is today's agriculture industry lacks the data that is globally representative to address some of the most important challenges of our time: the global warming, infest of pesticides on humans, food security for 7+ billion population, food vulnerabilities in the face of rapid decrease of natural resources, agroeconomics issues, suicide of farmers and rapid depletion of natural species. One of the chief reasons for the lack of globally well represented data in agriculture is due to unavailability of compute technologies that's affordable, contextual, localized, connected and easy to plug-in & consume by small-scaled farmers.

With the emergence of Cloud Technologies and with the development of Internet of Things (IoT), the call for action is to express connect small-scale farmers to Cloud and IoT technologies and to globally collect agriculture datasets for solving humanity issues. That's our future!

Appendix

Appendix A: bench_time.c

Reading a hardware clock has a cost. The following table compares the performance of different hardware clocks on Linux 3.3 with Intel Core i7-2600 at 3.40GHz (8 cores). The bench_time.c program was used to fill these tables[1] [PEP0418].

```c
/*
 * Benchmark program written for the PEP 418.
 *
 * gcc bench_time.c -O3 -lrt -o bench_time && ./bench_time
 */
#include <time.h>
#include <stdio.h>
#include <sys/time.h>
#ifdef CLOCK_REALTIME
# define HAVE_CLOCK_GETTIME
#else
typedef int clockid_t;
#endif
#define NRUN 5
#define NLOOP 100000
#define UNROLL(expr) \
    expr; expr; expr; expr; expr; expr; expr; expr; expr; expr
#define NUNROLL 10
#ifdef HAVE_CLOCK_GETTIME
typedef struct {
    const char *name;
    clockid_t identifier;
} CLOCK;
CLOCK clocks[] = {
#ifdef CLOCK_REALTIME_COARSE
    {"CLOCK_REALTIME_COARSE", CLOCK_REALTIME_COARSE},
#endif
#ifdef CLOCK_MONOTONIC_COARSE
    {"CLOCK_MONOTONIC_COARSE", CLOCK_MONOTONIC_COARSE},
#endif
#ifdef CLOCK_THREAD_CPUTIME_ID
    {"CLOCK_THREAD_CPUTIME_ID", CLOCK_THREAD_CPUTIME_ID},
#endif
```

---

[1] PEP0418 - https://www.python.org/dev/peps/pep-0418/#rationale

```
#ifdef CLOCK_PROCESS_CPUTIME_ID
    {"CLOCK_PROCESS_CPUTIME_ID", CLOCK_PROCESS_CPUTIME_ID},
#endif
#ifdef CLOCK_MONOTONIC_RAW
    {"CLOCK_MONOTONIC_RAW", CLOCK_MONOTONIC_RAW},
#endif
#ifdef CLOCK_VIRTUAL
    {"CLOCK_VIRTUAL", CLOCK_VIRTUAL},
#endif
#ifdef CLOCK_UPTIME_FAST
    {"CLOCK_UPTIME_FAST", CLOCK_UPTIME_FAST},
#endif
#ifdef CLOCK_UPTIME_PRECISE
    {"CLOCK_UPTIME_PRECISE", CLOCK_UPTIME_PRECISE},
#endif
#ifdef CLOCK_UPTIME
    {"CLOCK_UPTIME", CLOCK_UPTIME},
#endif
#ifdef CLOCK_MONOTONIC_FAST
    {"CLOCK_MONOTONIC_FAST", CLOCK_MONOTONIC_FAST},
#endif
#ifdef CLOCK_MONOTONIC_PRECISE
    {"CLOCK_MONOTONIC_PRECISE", CLOCK_MONOTONIC_PRECISE},
#endif
#ifdef CLOCK_REALTIME_FAST
    {"CLOCK_REALTIME_FAST", CLOCK_REALTIME_FAST},
#endif
#ifdef CLOCK_REALTIME_PRECISE
    {"CLOCK_REALTIME_PRECISE", CLOCK_REALTIME_PRECISE},
#endif
#ifdef CLOCK_SECOND
    {"CLOCK_SECOND", CLOCK_SECOND},
#endif
#ifdef CLOCK_PROF
    {"CLOCK_PROF", CLOCK_PROF},
#endif
    {"CLOCK_MONOTONIC", CLOCK_MONOTONIC},
    {"CLOCK_REALTIME", CLOCK_REALTIME}
};
#define NCLOCKS (sizeof(clocks) / sizeof(clocks[0]))
void bench_clock_gettime(clockid_t clkid)
{
   unsigned long loop;
   struct timespec tmpspec;
   for (loop=0; loop<NLOOP; loop++) {
     UNROLL( (void)clock_gettime(clkid, &tmpspec) );
   }
}
#endif /* HAVE_CLOCK_GETTIME */
void bench_time(clockid_t clkid)
```

```
{
    unsigned long loop;
    for (loop=0; loop<NLOOP; loop++) {
        UNROLL( (void)time(NULL) );
    }
}
void bench_usleep(clockid_t clkid)
{
    unsigned long loop;
    for (loop=0; loop<NLOOP; loop++) {
        UNROLL( (void)usleep(1000) );
    }
}
void bench_gettimeofday(clockid_t clkid)
{
    unsigned long loop;
    struct timeval tmpval;
    for (loop=0; loop<NLOOP; loop++) {
        UNROLL( (void)gettimeofday(&tmpval, NULL) );
    }
}
void bench_clock(clockid_t clkid)
{
    unsigned long loop;
    for (loop=0; loop<NLOOP; loop++) {
        UNROLL( (void)clock() );
    }
}
void benchmark(const char *name, void (*func) (clockid_t clkid), clockid_t
clkid)
{
    unsigned int run;
    double dt, best;
#ifdef HAVE_CLOCK_GETTIME
    struct timespec before, after;
#else
    struct timeval before, after;
#endif
    struct timeval tmpval;
    best = -1.0;
    for (run=0; run<NRUN; run++) {
#ifdef HAVE_CLOCK_GETTIME
        clock_gettime(CLOCK_MONOTONIC, &before);
        (*func) (clkid);
        clock_gettime(CLOCK_MONOTONIC, &after);
            dt = (after.tv_sec - before.tv_sec) * 1e9;
        if (after.tv_nsec >= before.tv_nsec)
            dt += (after.tv_nsec - before.tv_nsec);
        else
            dt -= (before.tv_nsec - after.tv_nsec);
```

```
#else
     gettimeofday(&before, NULL);
     (*func) (clkid);
      gettimeofday(&after, NULL);

      dt = (after.tv_sec - before.tv_sec) * 1e9;
      if (after.tv_usec >= before.tv_usec)
          dt += (after.tv_usec - before.tv_usec) * 1e3;
      else
          dt -= (before.tv_usec - after.tv_usec) * 1e3;
#endif
      dt /= NLOOP;
      dt /= NUNROLL;

      if (best != -1.0) {
         if (dt < best)
             best = dt;
      }
      else
          best = dt;
   }
   printf("%s: %.0f ns\n", name, best, NLOOP);
}
int main()
{
#ifdef HAVE_CLOCK_GETTIME
   clockid_t clkid;
   int i;

   for (i=0; i<NCLOCKS; i++) {
       benchmark(clocks[i].name, bench_clock_gettime, clocks[i].identifier);
   }
#endif
   benchmark("clock()", bench_clock, 0);
   benchmark("gettimeofday()", bench_gettimeofday, 0);
   benchmark("time()", bench_time, 0);
   return 0;
+
-}
```

Appendix: California Reservoir Sensor Details

Source: http://cdec.water.ca.gov/dynamicapp/getAll

Sensor Num	Name	Short Name	Units
211	% GATE OPENED 01	% UP 01	%
212	% GATE OPENED 02	% UP 02	%
213	% GATE OPENED 03	% UP 03	%
214	% GATE OPENED 04	% UP 04	%
215	% GATE OPENED 05	% UP 05	%
216	% GATE OPENED 06	% UP 06	%
217	% GATE OPENED 07	% UP 07	%
218	% GATE OPENED 08	% UP 08	%
219	% GATE OPENED 09	% UP 09	%
220	% GATE OPENED 10	% UP 10	%
245	ALBEDO	ALBEDO	%
17	ATMOSPHERIC PRESSURE	BAR PRE	INCHES
14	BATTERY VOLTAGE	BAT VOL	VOLTS
115	BATTERY VOLTAGE AUX	BAT VOLA	VOLTS
28	CHLOROPHYLL	CHLORPH	ug/L
164	CONTROL SWITCH, 0-OFF, 1-ON	ON/OFF	ON/OFF
95	CREEK BED ELEV FROM MEAN SEA L	CB ELEV	FEET
170	DEPTH OF READING BLW SURFACE	DEPTH B	FEET
48	DISCHARGE, POWER GENERATION	DIS PWR	CFS
71	DISCHARGE, SPILLWAY	SPILL	CFS
85	DISCHARGE,CONTROL REGULATING	RIV REL	CFS
179	DISSLVD ORG. CARBON, OXIDATION	D ORGCO	mg/L
109	DISSOLVED ORG. CARBON, COMBUST	D ORGCZ	mg/L
102	ELECTRICAL COND BOTTOM MICRO S	EL CONDB	uS/cm
92	ELECTRICAL COND BOTTOM MILLI S	EL CONDB	mS/cm
100	ELECTRICAL CONDUCTIVTY MICRO S	EL COND	uS/cm
5	ELECTRICAL CONDUCTIVTY MILLI S	EL CND	mS/cm
176	ELEVATION PROJ TAIL WATER STG	ELEV TW	FEET
175	EVAPORATION RATE	EVAP RAT	IN/DAY
64	EVAPORATION, PAN INCREMENT	EVP PAN	INCHES

Appendix: Sacramento River At Knights Landing

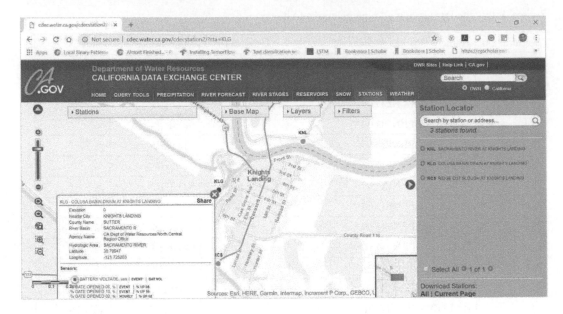

Sensors

⊖	**KNL**	SACRAMENTO RIVER AT KNIGHTS LANDING
Latitude: 38.8033485412598		Longitude: -121.71639251709
Elevation 30 Nearby City KNIGHT'S LANDING County Name SUTTER River Basin SACRAMENTO R Agency Name CA Dept of Water Resources/North Region Office Hydrologic Area SACRAMENTO RIVER		
Sensors: • BATTERY VOLTAGE **(Hourly)** • RIVER STAGE **(Event)** • RIVER STAGE **(Hourly)**		

⊖	**KLG**	COLUSA BASIN DRAIN AT KNIGHTS LANDING

Latitude: 38.79947   Longitude: -121.725203

Elevation	30
Nearby City	KNIGHTS LANDING
County Name	SUTTER
River Basin	SACRAMENTO R
Agency Name	CA Dept of Water Resources/North Central Region Office
Hydrologic Area	SACRAMENTO RIVER

Sensors:
- BATTERY VOLTAGE **(Event)**
- % GATE OPENED 08 **(Event)**
- % GATE OPENED 10 **(Event)**
- % GATE OPENED 02 **(Hourly)**
- RIVER STAGE **(Event)**
- % GATE OPENED 02 **(Event)**
- % GATE OPENED 02 **(Event)**
- RIVER STAGE **(Hourly)**
- % GATE OPENED 01 **(Event)**
- % GATE OPENED 09 **(Event)**
- % GATE OPENED 04 **(Hourly)**
- % GATE OPENED 08 **(Hourly)**
- % GATE OPENED 09 **(Hourly)**
- % GATE OPENED 10 **(Hourly)**
- % GATE OPENED 03 **(Event)**
- % GATE OPENED 04 **(Event)**
- % GATE OPENED 05 **(Event)**
- % GATE OPENED 01 **(Hourly)**
- % GATE OPENED 05 **(Hourly)**
- % GATE OPENED 06 **(Hourly)**
- % GATE OPENED 07 **(Event)**
- % GATE OPENED 03 **(Hourly)**
- % GATE OPENED 07 **(Hourly)**

⊖	**RCS**	RIDGE CUT SLOUGH AT KNIGHTS LANDING

Latitude: 38.7935562133789   Longitude: -121.72534942627

Elevation	0
Nearby City	KNIGHT'S LANDING
County Name	YOLO
River Basin	SACRAMENTO R
Agency Name	CA Dept of Water Resources/North Region Office
Hydrologic Area	SACRAMENTO RIVER

Sensors:
- WATER VOLTAGE **(Event)**
- FLOW, RIVER DISCHARGE **(Event)**
- RIVER STAGE **(Event)**
- TEMPERATURE, WATER **(Event)**

Query Interface

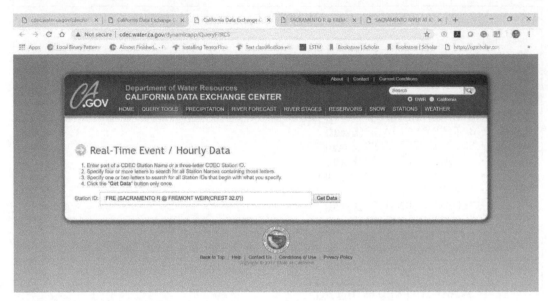

http://cdec.water.ca.gov/dynamicapp/QueryF?RCS

Scaremento Stations

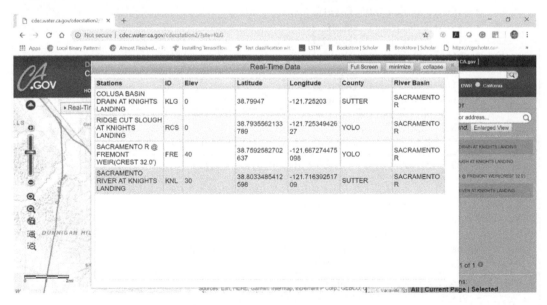

Source: http://cdec.water.ca.gov/cdecstation2/?sta=KLG

SACRAMENTO R @ FREMONT WEIR(CREST 32.0') (FRE)

Elevation: 40.0' · SACRAMENTO R basin · Operator: CA Dept of Water Resources/North Central Region Office
Datum 0 = 0.0' NAVD
River Stage Definitions: Monitor stage **32.0'** Flood stage **39.5'** Danger stage **40.5'**
 Peak Stage of Record 41.02' on 01/02/1997 00:00

Query executed Friday at 21:11:42

Provisional data, subject to change.
Select a sensor type for a plot of data.

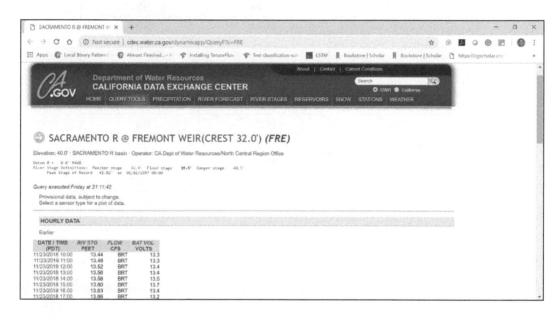

Real Time Sensor Data – Sacramento R @ Fremont WEIR

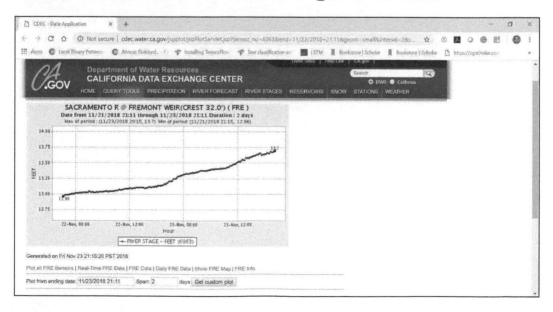

Source: http://cdec.water.ca.gov/jspplot/jspPlotServlet.jsp?sensor_no=6063&end=11/23/2018+21:11&geom=small&interval=2&cookies=cdec01

SACRAMENTO R @ FREMONT WEIR(CREST 32.0') (FRE)		
DATE TIME (PDT)	**RIV STG FEET**	**FLOW CFS**
11/23/2018 09:15	13.42	BRT
11/23/2018 09:30	13.43	BRT
11/23/2018 09:45	13.44	BRT
11/23/2018 10:00	13.44	BRT
11/23/2018 10:15	13.46	BRT
11/23/2018 10:30	13.46	BRT
11/23/2018 10:45	13.47	BRT
11/23/2018 11:00	13.48	BRT
11/23/2018 11:15	13.49	BRT
11/23/2018 11:30	13.49	BRT
11/23/2018 11:45	13.51	BRT
11/23/2018 12:00	13.52	BRT
11/23/2018 12:15	13.53	BRT
11/23/2018 12:30	13.52	BRT
11/23/2018 12:45	13.54	BRT
11/23/2018 13:00	13.56	BRT
11/23/2018 13:15	13.55	BRT

SNOW Sensors

http://cdec.water.ca.gov/jspplot/jspPlotServlet.jsp?sensor_no=5239&end=&geom=small&interval=730&cookies=cdec01

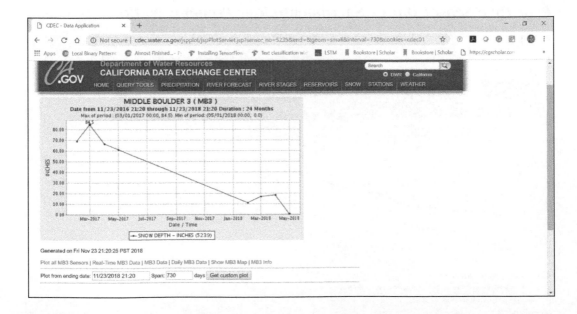

Appendix - SQLite Build - Amalgamation

To build the SQLite for embedded projects, one can create an amalgamation project that can create a lite weight SQLite database.

Developers sometimes experience trouble debugging the 185,000-line-long amalgamation source file because some debuggers are only able to handle source code line numbers less than 32,768. The amalgamation source code runs fine. One just cannot single-step through it in a debugger.[2]

To compile:[3]

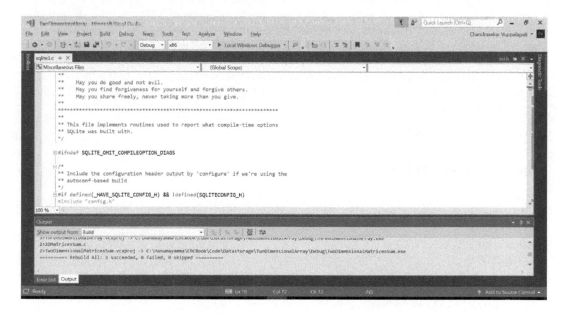

---

[2]   SQLite Amalgamation - https://www.sqlite.org/amalgamation.html

[3]   153 To Compile SQLite - https://www.sqlite.org/howtocompile.html

Appendix – DO-178B (Software Considerations in Airborne Systems and Equipment Certification)

DO-178B, Software Considerations in Airborne Systems and Equipment Certification is a guideline dealing with the safety of safety-critical software used in certain airborne systems. Although technically a guideline, it was a de facto standard for developing avionics software systems until it was replaced in 2012 by DO-178C.

The FAA applies DO-178B as the document it uses for guidance to determine if the software will perform reliably in an airborne environment, [1] when specified by the Technical Standard Order (TSO) for which certification is sought. In the United States, the introduction of TSOs into the airworthiness certification process, and by extension DO-178B, is explicitly established in Title 14: Aeronautics and Space of the Code of Federal Regulations (CFR), also known as the Federal Aviation Regulations, Part 21, Subpart O.

It was jointly developed by the safety-critical working group RTCA SC-167 of RTCA and WG-12 of EUROCAE. RTCA published the document as RTCA/DO-178B, while EUROCAE published the document as ED-12B.

Source: https://en.wikipedia.org/wiki/DO-178B

Appendix – STDIO.H File[4]

NAME
stdio.h - standard buffered input/output
SYNOPSIS
#include <stdio.h>
DESCRIPTION
The *<stdio.h>* header defines the following macro names as positive integral constant expressions:
BUFSIZ
      Size of *<stdio.h>* buffers.
FILENAME_MAX
      Maximum size in bytes of the longest filename string that the implementation guarantees can be opened.
FOPEN_MAX
      Number of streams which the implementation guarantees can be open simultaneously. The value will be at least eight.
_IOFBF
      Input/output fully buffered.
_IOLBF
      Input/output line buffered.
_IONBF
      Input/output unbuffered.
L_ctermid
      Maximum size of character array to hold *ctermid*() output.
L_cuserid
      Maximum size of character array to hold *cuserid*() output. **(LEGACY)**
L_tmpnam
      Maximum size of character array to hold *tmpnam*() output.
SEEK_CUR
      Seek relative to current position.
SEEK_END
      Seek relative to end-of-file.

---

[4]   Stdio.h - http://pubs.opengroup.org/onlinepubs/7908799/xsh/stdio.h.html

SEEK_SET

>Seek relative to start-of-file.

TMP_MAX

>Minimum number of unique filenames generated by *tmpnam*(). Maximum number of times an application can call *tmpnam*() reliably. The value of TMP_MAX will be at least 10,000.

The following macro name is defined as a negative integral constant expression:

EOF

>End-of-file return value.

The following macro name is defined as a null pointer constant:

NULL

>Null pointer.

The following macro name is defined as a string constant:

P_tmpdir

>default directory prefix for *tempnam*().

The following macro names are defined as expressions of type pointer to FILE:

stderr

>Standard error output stream.

stdin

>Standard input stream.

stdout

>Standard output stream.

The following data types are defined through **typedef**:

FILE

>A structure containing information about a file.

fpos_t

>Type containing all information needed to specify uniquely every position within a file.

va_list

>As described in *<stdarg.h>*.

size_t

>As described in *<stddef.h>*.

The following are declared as functions and may also be defined as macros. Function prototypes must be provided for use with an ISO C compiler.

```
void      clearerr(FILE *);
char      *ctermid(char *);
char      *cuserid(char *);(LEGACY)
int       fclose(FILE *);
FILE      *fdopen(int, const char *);
int       feof(FILE *);
int       ferror(FILE *);
int       fflush(FILE *);
int       fgetc(FILE *);
int       fgetpos(FILE *, fpos_t *);
char      *fgets(char *, int, FILE *);
int       fileno(FILE *);
void      flockfile(FILE *);
FILE      *fopen(const char *, const char *);
int       fprintf(FILE *, const char *, ...);
int       fputc(int, FILE *);
int       fputs(const char *, FILE *);
size_t    fread(void *, size_t, size_t, FILE *);
FILE      *freopen(const char *, const char *, FILE *);
int       fscanf(FILE *, const char *, ...);
```

```
int      fseek(FILE *, long int, int);
int       fseeko(FILE *, off_t, int);
int      fsetpos(FILE *, const fpos_t *);
long     int ftell(FILE *);
off_t     ftello(FILE *);
int      ftrylockfile(FILE *);
void     funlockfile(FILE *);
size_t    fwrite(const void *, size_t, size_t, FILE *);
int      getc(FILE *);
int      getchar(void);
int      getc_unlocked(FILE *);
int      getchar_unlocked(void);
int      getopt(int, char * const[], const char); (LEGACY)
char     *gets(char *);
int      getw(FILE *);
int      pclose(FILE *);
void     perror(const char *);
FILE     *popen(const char *, const char *);
int      printf(const char *, ...);
int      putc(int, FILE *);
int      putchar(int);
int      putc_unlocked(int, FILE *);
int       putchar_unlocked(int);
int      puts(const char *);
int      putw(int, FILE *);
int      remove(const char *);
int      rename(const char *, const char *);
void      rewind(FILE *);
int      scanf(const char *, ...);
void     setbuf(FILE *, char *);
int      setvbuf(FILE *, char *, int, size_t);
int      snprintf(char *, size_t, const char *, ...);
int      sprintf(char *, const char *, ...);
int      sscanf(const char *, const char *, int ...);
char     *tempnam(const char *, const char *);
FILE     *tmpfile(void);
char     *tmpnam(char *);
int       ungetc(int, FILE *);
int      vfprintf(FILE *, const char *, va_list);
int      vprintf(const char *, va_list);
int      vsnprintf(char *, size_t, const char *, va_list);
int      vsprintf(char *, const char *, va_list);
```

The following external variables are defined:

extern char *optarg;)

extern int opterr;)

extern int optind;) **(LEGACY)**

extern int optopt;)

Inclusion of the *<stdio.h>* header may also make visible all symbols from *<stddef.h>*.

APPLICATION USAGE

None.

FUTURE DIRECTIONS

None.

Appendix: Milk Producing Data Centers

Dairy Cows & Peak of Health

Benefits

Connected Dairy Analytics (CDA) monitor Dairy cattle electronically to make sure the cattle are always in the peak of health.

CDA will help to identify Dairy cattle related health issues such as Heat Stress and Bovine Respiratory Disease (BRD). The Heat stress (HS) causes cattle to produce less milk with the same nutritional input, which effectively increases farmers' production costs. The economic toll due to higher-temperature, heat stress is a $1 billion annual problem. Not only in the United States, but also around the globe heat stress causes an adverse impact on dairy productivity.

Connected Dairy will yield huge operational efficiencies, cost savings, and actionable insights to address Dairy cattle related critical issues. Connected dairy, importantly, is a data enabled insightful tool that facilitates better management of Dairy activities. Finally, connected dairy provides forecasting insights that provides window of time opportunity to dairy operational management so that they can better plan to handle any un-expected weather related abnormalities, dairy cattle health and emergencies

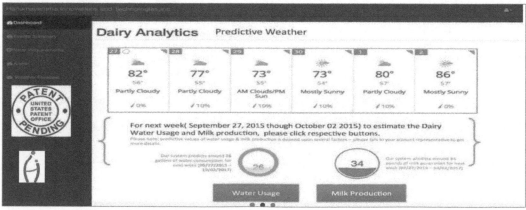

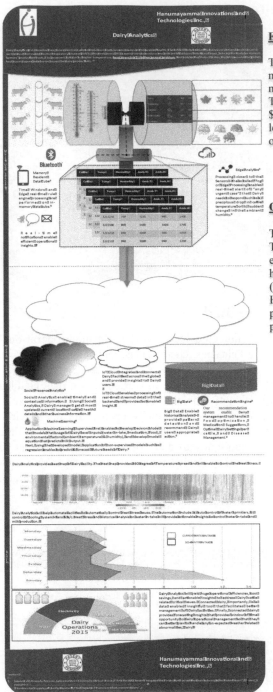

Electronic Monitoring

The opportunity for the dairy industry is to electronically monitor cattle temperature and implement appropriate measures so that the impact of HS can be minimized. The U.S. Department of Agriculture estimates nearly $2.4 billion a year in losses from animal illnesses that lead to death can be prevented by electronically checking on cattle's' vital signs.

Our Approach

The 'Smart Connected Objects", aka, 'the Internet of Things (IoT)', that enables dairies to minimize the economic impact of HS and, at the same, capture the higher Return on Assets (ROA) & Return on Investment (ROI) by improving operational efficiencies.

Happy Cow, more importantly, means happier, more profitable, dairy industry and richer and creamer dairy products.

Reduce medication costs by about 15% per animal and save more sick cattle from death. WSJ: High-Tech Tagging Comes to the Ranch

Contact details:
Hanumayamma Innovations and Technologies Inc.,
628 Crescent Terrace,
Fremont, CA 94536
Phone: (510) 791 – 5759
Email: cvuppalapati@hanuinnotech.com
Fax: (510)-857-5794
Website: http://www.hanuinnotech.com

Pthread Support in Microsoft Windows Services for UNIX Version 3.5[5]

The Microsoft® Windows® Services for UNIX (SFU) 3.5 product is a collection of software packages for UNIX users and administrators who need to work with or on Windows platforms. It includes cross-platform network services that allow you to integrate your Windows® and UNIX-based environments together. It also includes a complete UNIX system environment called Interix that installs and runs on Windows, co-existing with the Windows subsystem. This environment comes with hundreds of UNIX utilities, such as ksh, csh, awk and telnet, and a complete C and C++ programming development environment for UNIX applications. With the release of SFU 3.5, this development environment now includes support for POSIX threads (Pthreads) and the POSIX semaphore functions.

To download pThread support, please enable Windows Subsystem for Linux.

If you encounter the following error: please follow steps 1 through 5.

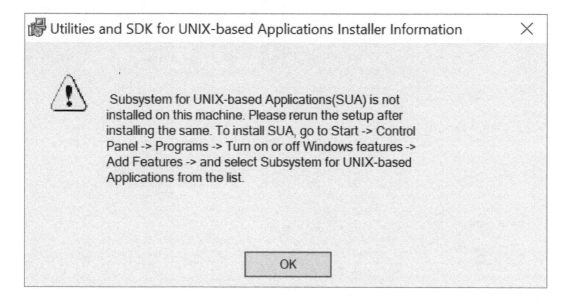

5   PThread Support for Windows - https://docs.microsoft.com/en-us/previous-versions/tn-archive/bb463209(v=technet.10)

Step 1: Go to Windows Control Panel

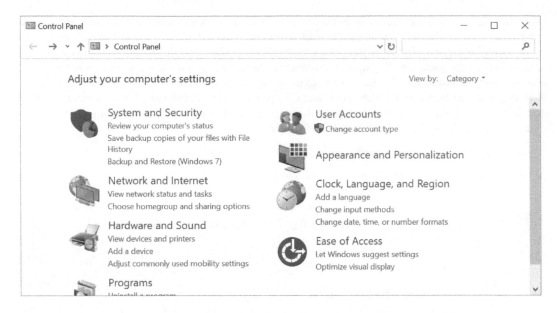

Step 2: Click Programs

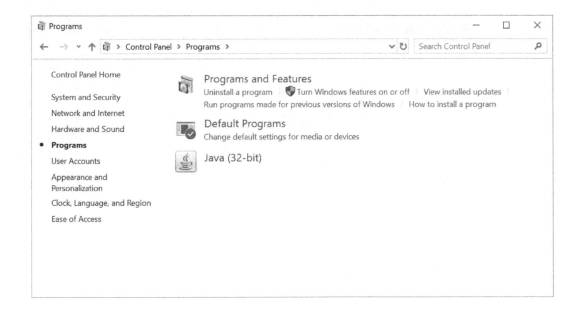

Step 3: Click Turn Windows on or Off Features

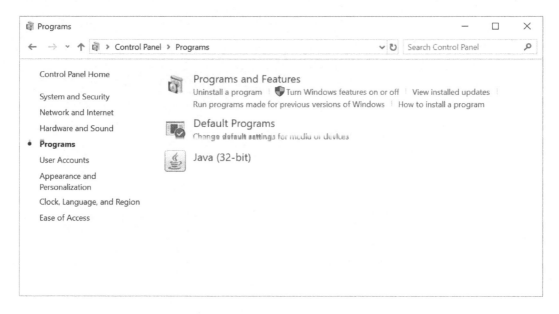

Step 4: Select Windows Subsystem for Linux

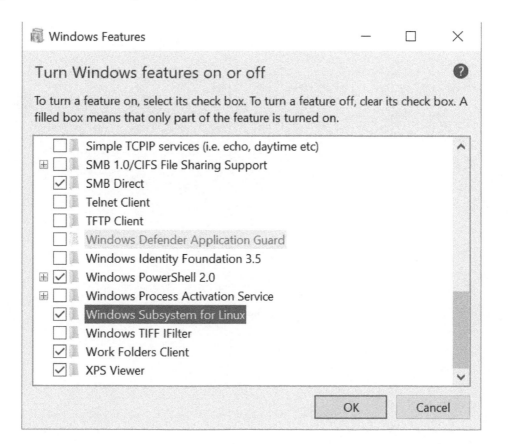

Step 5: Reboot the System

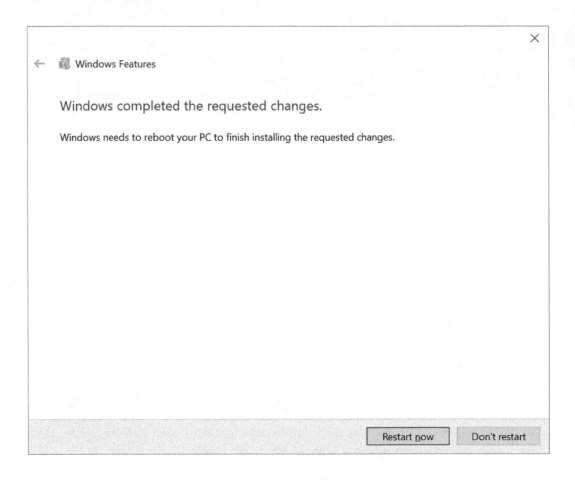

Index